State-space methods form the basis of modern control theory. This textbook is devoted to a description of these methods in the analysis of linear multiple-input, multiple-output dynamic systems. Throughout, continuous-time and discrete-time systems are treated in parallel, as are time-varying and classical time-invariant systems, thereby highlighting their similarities and differences.

Following a chapter, which sets out the basic concepts and definitions, state equations of finite dimensional systems, and their solution, are discussed in detail. The principles of time-domain and frequency-domain analysis are then presented, as are the properties and applications of the $\mathcal{Z}$-transformation. One chapter deals with the controllability and observability of linear systems and a separate chapter discusses stability. An extensive tutorial review of the key results of matrix theory and linear algebra is given in the Appendix.

In addition to illustrative examples, which are worked out in detail, numerous problems are available at the end of each chapter. This text should be of great value to graduate students of aeronautical, electrical, and mechanical engineering or anyone who wants to acquire a solid background in the fundamentals of linear systems and modern control systems.

Principles of linear systems

Principles of linear systems

PHILIP E. SARACHIK
Polytechnic University, Brooklyn, New York

CAMBRIDGE
UNIVERSITY PRESS

CAMBRIDGE UNIVERSITY PRESS
Cambridge, New York, Melbourne, Madrid, Cape Town, Singapore, São Paulo

Cambridge University Press
The Edinburgh Building, Cambridge CB2 8RU, UK

Published in the United States of America by Cambridge University Press, New York

www.cambridge.org
Information on this title: www.cambridge.org/9780521570572

First published 1997

A catalogue record for this publication is available from the British Library

Library of Congress Cataloguing in Publication data
Sarachik, Philip E.
Principles of linear systems / Philip E. Sarachik.
p. cm.
Includes bibliographical references and index.
ISBN 0-521-57057-3 (hardcover). ISBN 0-521-57606-7 (pbk.)
1. System analysis. 2. State-space methods. 3. Linear systems.
I. Title.
QA402.S35155 1996
629.8′32 – dc20 96-12843
CIP

ISBN 978-0-521-57057-2 hardback
ISBN 978-0-521-57606-2 paperback

Transferred to digital printing 2008

CONTENTS

1	**Basic concepts, terms, and definitions**	*page* 1
1.1	Introduction	1
1.2	Input-output relations	2
1.3	Classification of systems	3
	1.3.1 Physical realizability	3
	1.3.2 Dynamic	4
	1.3.3 Time invariance	6
	1.3.4 Linearity	7
	1.3.5 Type of signals	9
1.4	The techniques of linear system analysis	10
	Problems	12
2	**State equations of finite dimensional linear systems**	15
2.1	Introduction	15
2.2	State differential equations of circuits	16
2.3	State differential equations of mechanical systems	22
2.4	Choice of state variables	25
2.5	Simulation diagrams	26
2.6	Systems governed by a single dynamic equation	28
2.7	Systems of multiple dynamic equations	34
	Problems	37
3	**Principles of time-domain analysis**	43
3.1	Introduction	43
3.2	Elementary analog signals	43
3.3	Impulses and the impulse response	44
3.4	Step function and step response	44
3.5	Relation between the step and impulse response	45
3.6	Graphical interpretation of the convolution integral	47

3.7 The discrete delta and delta response 51
3.8 Systems with multiple inputs and outputs 53
Problems 54

4 Solution of the dynamic state equations 60
4.1 Introduction 60
4.2 Analog systems 60
4.3 The transition matrix 63
4.3.1 The physical meaning of $\phi(t, t_0)$ 63
4.3.2 $\phi(t)$ for constant system matrix A 66
4.3.3 Calculating e^{At} 67
4.3.4 $\phi(t, t_0)$ for time-varying system matrix $A(t)$ 80
4.4 Discrete-time systems 85
4.5 Analog systems with sampled inputs 90
Problems 92

5 Frequency-domain analysis for fixed analog systems: application of the Laplace transform 98
5.1 Basic principles 98
5.2 Application 99
5.3 Solution of the state differential equations using $\mathscr{L}$-transforms 101
5.4 Solutions of systems of differential equations 103
5.5 Leverrier's method 106
Problems 108

6 The $\mathcal{Z}$-transformation of discrete-time signals 111
6.1 Introduction 111
6.2 Definition of the $\mathcal{Z}$-transform 111
6.3 The inverse transformation 114
6.4 Properties of $\mathcal{Z}$-transforms 118
Problems 122

7 Frequency-domain analysis of discrete systems and application of $\mathcal{Z}$-transforms 125
7.1 Introduction 125
7.2 Application to the indirect analysis approach 126
7.3 Solution of the state equations using $\mathcal{Z}$-transforms 126
7.4 Solution of systems of difference equations 129
7.5 Sampled signals and sampled data systems 131
7.6 Method for obtaining the $\mathcal{Z}$-transform from the $\mathscr{L}$-transform 138
7.7 Analysis of systems containing impulse-modulated signals 139
7.8 $\mathcal{Z}$-transforms of shifted time functions 144
Problems 146

8 Controllability and observability of linear systems 151
8.1 Introduction and definition of controllability 151
8.2 Controllability conditions for analog systems 152
8.3 Controllability conditions for discrete systems 155
8.4 Controllability test for fixed discrete systems 156
8.5 Controllability test for fixed analog systems 157
8.6 Observability: definition and conditions 158
8.7 Duality between observability and controllability 160
8.8 Sufficient conditions for time-varying analog systems 162
8.9 The structure of noncontrollable and nonobservable systems 164
8.9.1 The controllable subspace 165
8.9.2 The observable subspace 169
8.9.3 Kalman decomposition 171
Problems 180

9 Stability of linear systems 184
9.1 Introduction and general definitions 184
9.2 Zero-input stability of linear systems 185
9.3 Zero-state stability of linear systems 188
9.4 Conditions for BIBO stability of fixed linear systems 190
9.5 Lyapunov's second method 192
9.6 Lyapunov functions for fixed linear systems 196
9.7 Stability tests for fixed analog systems 198
9.7.1 The continued fraction test 199
9.7.2 The Routh test 202
9.7.3 Agashe's modification of the Routh test 207
9.7.4 The Hurwitz test 213
9.8 Stability tests for fixed discrete systems 214
9.8.1 Mapped Routh-Hurwitz test for discrete systems 214
9.8.2 Modified continued fraction test for discrete systems 215
9.8.3 A tabular test for discrete systems 217
Problems 218

Appendix Review of matrix theory 221
A.1 Terminology 221
A.2 Equality 222
A.3 Addition 222
A.4 Multiplication by a scalar 222
A.4.1 Subtraction 223
A.4.2 Commutative and associative laws for addition and scalar multiplication 223
A.5 Matrix multiplication 223
A.5.1 Properties of multiplication 224

A.6 Transpose 225
A.7 Special matrices 226
A.8 Partitioning 227
A.9 Matrix inversion 229
A.9.1 The adjoint matrix 230
A.10 Linear spaces 232
A.11 Linear independence 232
A.12 Rank of a matrix 236
A.13 Simultaneous linear equations 239
A.13.1 Motivation and definitions 239
A.13.2 Equivalent systems of equations and elementary operations 240
A.13.3 Gaussian elimination 240
A.13.4 Fundamental theorem for linear equations 243
A.13.5 Properties of the general solution 243
A.13.6 Rank using Gaussian elimination 246
A.14 Eigenvalues and eigenvectors 247
A.14.1 The eigenvector problem 247
A.14.2 Diagonalization of matrices 251
A.15 Generalized eigenvectors 253
A.16 Jordan canonical form 255
A.17 The Cayley-Hamilton theorem 259
A.18 Leverrier's algorithm 261
A.19 The minimum polynomial 264
A.20 Functions of a matrix 265
A.21 Quadratic forms 270
A.21.1 Definite quadratic forms 270

References 276

Index 279

1

Basic concepts, terms, and definitions

1.1 Introduction

Since this book is concerned with the Principles of Linear Systems, it is therefore natural to first give a definition of a system.

Definition 1.1 *A* **physical system** *is an interconnection of physical components that perform a specific function. These components may be electrical, mechanical, hydraulic, thermal and so forth.*

Associated with every system is a variety of physical quantities such as electrical voltages and currents, mechanical forces and displacements, flow rates, and temperatures, which generally change with time. We will call them *signals* although some of them may be nonelectrical in nature. Some of the signals can be directly changed with time in order to effect indirectly desired changes in some other signals of the system that happen to be of particular interest. The former set of signals are called *inputs* or *excitations;* the latter are called *outputs* or *responses.*

Clearly, the set of inputs and the set of outputs of a system are not necessarily uniquely defined, and some of them may have to be chosen. After such a choice has been made, a definite situation exists where the system receives inputs and transforms them into outputs. The idea of this transformation can be conveyed pictorially through the use of a generalized "black box" representation as shown in Figure 1.1.

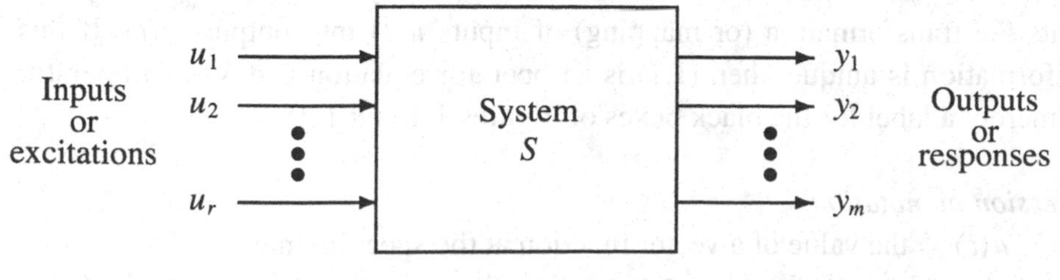

Fig. 1.1.

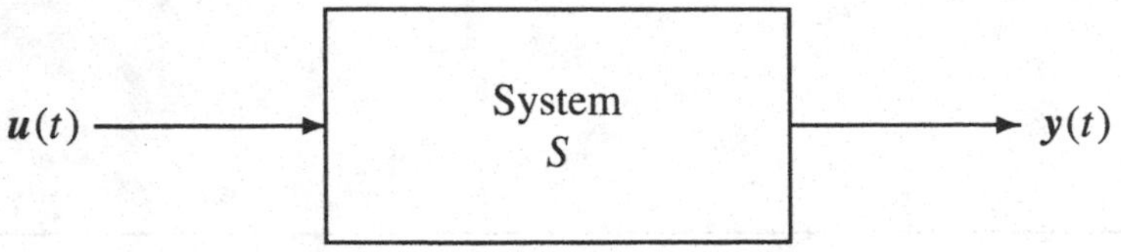

Fig. 1.2.

In Figure 1.1 the inputs $u_1, u_2, \ldots, u_r$ and the outputs $y_1, y_2, \ldots, y_m$ are in general time functions, and if we use the vector notation

$$\boldsymbol{u}(t) = \begin{bmatrix} u_1(t) \\ u_2(t) \\ u_3(t) \\ \vdots \\ u_r(t) \end{bmatrix} \qquad \boldsymbol{y}(t) = \begin{bmatrix} y_1(t) \\ y_2(t) \\ y_3(t) \\ \vdots \\ y_m(t) \end{bmatrix}$$

we can represent the system as in Figure 1.2.

1.2 Input-output relations

It is an essential part of the system concept that changes of the input cause changes to the output. An important problem in the study of systems is to determine what is the output response of a system due to a given input. One way to solve this problem is to experiment with the given system by applying the given input to the system and recording the resulting output. However, experimentation of this kind is usually not a satisfactory approach to the problem. An alternate approach is to express the signals as time functions and use the mathematical relations describing the components of the system combined in accordance with their interconnections. Thus a mathematical relation between the input and the output can be obtained, which may be written as

$$\boldsymbol{y}(t) = S\{\boldsymbol{u}(t)\}. \tag{1.1}$$

This is called the *input-output relation* for the given system. Equation (1.1) represents the transformation (or mapping) of inputs $u(t)$ into outputs $y(t)$. If this transformation is unique, then (1.1) is an operator equation and S is an operator (not merely a label for the black boxes of Figures 1.1 and 1.2).

Digression on notation

$\boldsymbol{u}(t) \triangleq$ the value of a vector function at the specific time t.

$\underset{\sim}{\boldsymbol{u}}[t_0, t_1] \triangleq \boldsymbol{u}(t)$ for all $t_0 \le t \le t_1$ (i.e., the entire time function defined over $[t_0, t_1]$).

$\underset{\sim}{u} \triangleq$ the entire time function without specifying the interval of definition (i.e., $\boldsymbol{u}(t)$ for all t on an unspecified interval of time). This is also written as $\{\boldsymbol{u}(t)\}$. Now, using the above notation, (1.1) may be written as

$$\underset{\sim}{y} = S\{\underset{\sim}{u}\} \tag{1.2}$$

or simply

$$\underset{\sim}{y} = S\underset{\sim}{u}. \tag{1.3}$$

The three expressions will be used interchangeably and according to notational needs.

1.3 Classification of systems

1.3.1 Physical realizability

Definition 1.2 *A system S with an input-output relation $\underset{\sim}{y} = S\{\underset{\sim}{u}\}$ is called* **physically realizable** *if a physical system can be built whose inputs and outputs are related via $\underset{\sim}{y} = S\{\underset{\sim}{u}\}$.*

Definition 1.3 *A system is called* **real** *if its response to any real input is real.*

An example of a nonreal system is

$$y(t) = ju(t)$$

where j is the imaginary unit.

Definition 1.4 *A system is called* **causal** *or* **nonanticipative** *if for any t, $\mathbf{y}(t)$ does not depend on any $\mathbf{u}(t')$ for $t' > t$ (i.e., if $\mathbf{y}(t)$ does not depend on future values of $\underset{\sim}{u}$). Otherwise, it is called* **noncausal** *or* **anticipative**.

In other words, a system is causal if the output at any time t may depend only on input values occurring for $t' \leq t$. Obviously a noncausal or nonreal system is not realizable.

Examples of an anticipative or noncausal system:

1 $y(t) = S\{u(t)\} = u(t+1)$ because the value of y at any time t depends on the value of u at time $t + 1 > t$.
2 $y(t) = S\{u(t)\} = u(t^2)$ because for times $t < 0$ or $t > 1$, $y(t)$ depends on future values of $\{u(t)\}$.
3 $y(t) = \int_{t-2}^{t+4} u(\tau)\, d\tau$ because the value of y at any time t depends on the values of u for times up to $t + 4$.

A system described by $y(t) = S\{u(t)\} = u^2(t)$ is causal.

1.3.2 Dynamic

Definition 1.5 *A system is called* **dynamic** *(or is said to have memory) if* $\mathbf{y}(t_0)$ *depends on some values of* $\mathbf{u}(t)$ *for* $t \neq t_0$. *A system for which* $\mathbf{y}(t_0)$ *does not depend on* $\mathbf{u}(t)$ *for* $t \neq t_0$ *is called* **instantaneous** *(or is said to have zero memory).*

An example of an instantaneous system:

$$y(t) = g[u(t), t] \quad \text{where } g[.,.] \text{ is a function.}$$

Definition 1.6 *A causal system whose output at time t is completely determined by the input in the interval* $[t - T, t]$ *for* $0 \leq T$, *is said to have a memory of length T (Figs. 1.3–1.4).*

Example 1.1
Instantaneous system (zero memory):

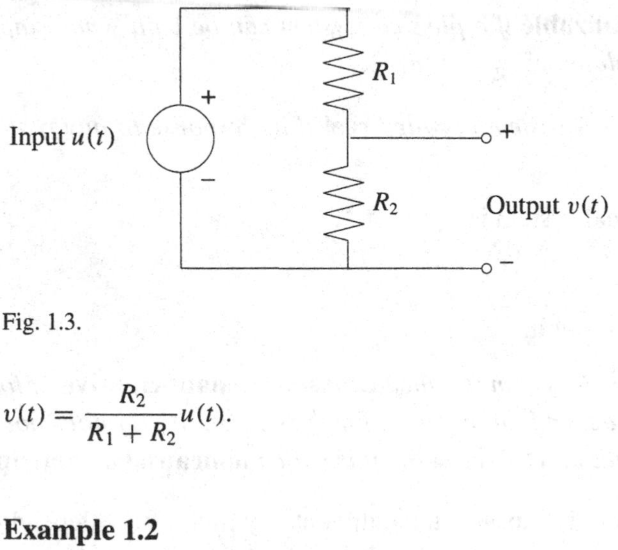

Fig. 1.3.

$$v(t) = \frac{R_2}{R_1 + R_2} u(t).$$

Example 1.2
Dynamic system (infinite memory):

$$v(t) = \frac{1}{RC} \int_{-\infty}^{t} e^{-\frac{(t-\tau)}{RC}} u(\tau)\, d\tau \quad \text{for all } t,$$

assuming there is no charge on the capacitor at $t = -\infty$. If there is an initial charge $q(t_0)$ at time t_0, then

$$v(t) = \frac{q(t_0)}{C} e^{-\frac{(t-t_0)}{RC}} + \frac{1}{RC} \int_{t_0}^{t} e^{-\frac{(t-\tau)}{RC}} u(\tau)\, d\tau, \quad t \geq t_0.$$

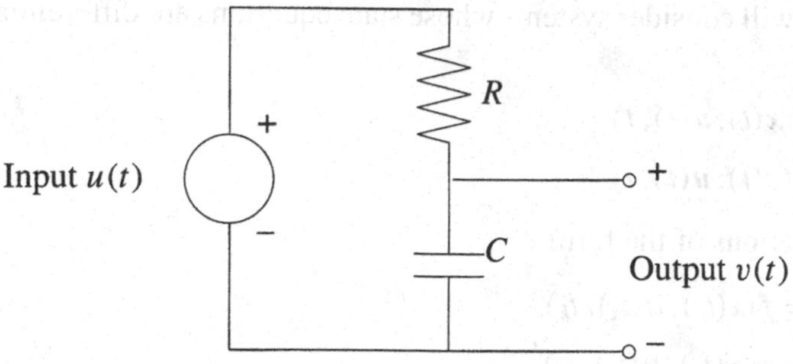

Fig. 1.4.

Note: In order to determine $v(t)$ for $t \geq t_0$, only $\underset{\sim}{u}[t_0, t]$ must be known since the effect of the input for $t < t_0$ is summarized by the value $v(t_0) = q(t_0)/C$ of the voltage across the capacitor at time t_0. Because of this property, the voltage $v(t)$ or the charge $q(t)$ is called the state of the system.

Definition 1.7 *The* **state** *of a nonanticipative dynamic system at time t_0 is the smallest set of quantities $\boldsymbol{x}(t_0)$, which summarize all information about $\underset{\sim}{u}(-\infty, t_0)$ needed to determine $\underset{\sim}{y}[t_0, t_1]$ when $\underset{\sim}{u}[t_0, t_1]$ is known.*

We can now modify the input-output relation of a nonanticipative dynamic system to include the state as follows:

$$y(t) = A\{\boldsymbol{x}(t_0); \underset{\sim}{u}[t_0, t]\} \quad \text{for } t \geq t_0 \tag{1.4}$$

or

$$\underset{\sim}{y}[t_0, t] = A\{\boldsymbol{x}(t_0); \underset{\sim}{u}[t_0, t]\}. \tag{1.5}$$

We call (1.4) or (1.5) the *input-output-state relation.*

Now, inherent in the concept of state is the requirement that for any $t_1 \in [t_0, t]$ it must be possible to define $\boldsymbol{x}(t_1)$ such that for

$$\underset{\sim}{y}'[t_1, t] = A\{\boldsymbol{x}(t_1); \underset{\sim}{u}'[t_1, t]\}$$

then

$$\underset{\sim}{y}'[t_1, t] \equiv \underset{\sim}{y}[t_1, t] \quad \text{when} \quad \underset{\sim}{u}'[t_1, t] \equiv \underset{\sim}{u}[t_1, t].$$

In other words, the state itself at t_1 must be uniquely determined by an earlier state at t_0 and the input $\underset{\sim}{u}[t_0, t_1]$. It implies that for consistency, a condition of the form

$$\boldsymbol{x}(t_1) = R\{\boldsymbol{x}(t_0); \underset{\sim}{u}[t_0, t_1]\} \tag{1.6}$$

must be satisfied. We call (1.6) the *state transition equation* of the system.

In general, we will consider systems whose state equations are differential equations of the form

$$\begin{aligned} \dot{\boldsymbol{x}}(t) &= \boldsymbol{f}(\boldsymbol{x}(t), \boldsymbol{u}(t), t) \\ \boldsymbol{y}(t) &= \boldsymbol{g}(\boldsymbol{x}(t), \boldsymbol{u}(t), t) \end{aligned} \tag{1.7}$$

or difference equations of the form

$$\begin{aligned} \boldsymbol{x}(t_{k+1}) &= \boldsymbol{f}(\boldsymbol{x}(t_k), \boldsymbol{u}(t_k), t_k) \\ \boldsymbol{y}(t_k) &= \boldsymbol{g}(\boldsymbol{x}(t_k), \boldsymbol{u}(t_k), t_k). \end{aligned} \tag{1.8}$$

These are called the *state (differential or difference) equations*.

Definition 1.8 *The* **zero state** θ *of a dynamic system is the state for which* $\underset{\sim}{y}[t_0, \infty] = \underset{\sim}{0}$ *when* $\boldsymbol{x}(t_0) = \theta$ *and the input* $\underset{\sim}{u}[t_0, \infty] = \underset{\sim}{0}$ *(i.e.,* $A\{\theta, \underset{\sim}{0}\} = \underset{\sim}{0}$*).*

Definition 1.9 *An* **equilibrium state** $\boldsymbol{x}^*$ *is a state for which* $\boldsymbol{x}(t) = \boldsymbol{x}^*$ *for all* $t \geq t_0$ *when* $\boldsymbol{x}(t_0) = \boldsymbol{x}^*$ *and* $\boldsymbol{u}[t_0, \infty] = \underset{\sim}{\boldsymbol{0}}$ *(i.e.,* $R\{\boldsymbol{x}^*; \underset{\sim}{\boldsymbol{0}}\} = \boldsymbol{x}^*$*).*

The *zero-state response* is given by

$$\boldsymbol{y}_{zs}(t) = A\{\theta; \underset{\sim}{\boldsymbol{u}}\}. \tag{1.9}$$

The *zero-input response* is given by

$$\boldsymbol{y}_{zi}(t) = A\{\boldsymbol{x}(t_0); \underset{\sim}{\boldsymbol{0}}\}. \tag{1.10}$$

Note: Unless otherwise indicated, the input-output relation $S\{\underset{\sim}{\boldsymbol{u}}\}$ of a dynamic system refers to the zero-state response so $\underset{\sim}{\boldsymbol{y}}_{zs} = S\{\underset{\sim}{\boldsymbol{u}}\}$.

1.3.3 Time invariance

Definition 1.10 *A system is called* **time-invariant** *or* **fixed** *if*

$$\boldsymbol{y}(t) \triangleq A\{\boldsymbol{x}(t_0) = \alpha; \underset{\sim}{\boldsymbol{u}}[t_0, t]\}$$

and

$$\boldsymbol{y}'(t) \triangleq A\{\boldsymbol{x}(t_0 + T) = \alpha; \underset{\sim}{\boldsymbol{u}}[t_0 + T, t + T]\}$$

implies that $\boldsymbol{y}'(t) = \boldsymbol{y}(t + T)$ *for all* $T, \alpha, \underset{\sim}{\boldsymbol{u}}[t_0, t], t_0,$ *and* t. *If this is not true, the system is called* **time-varying**.

Example 1.3
For the system

$$y(t) = S\{u(t)\} \triangleq \int_0^1 tu(\tau)\,d\tau \qquad \text{for all t,}$$

we see that the time-shifted output is

$$y(t+T) = \int_0^1 (t+T)u(\tau)\,d\tau,$$

but the response to the time-shifted input is

$$y'(t) = S\{u(t+T)\} = \int_0^1 tu(\tau+T)\,d\tau = \int_T^{T+1} tu(\lambda)\,d\lambda.$$

Since $y'(t) \neq y(t+T)$, the system is time-varying.

1.3.4 Linearity

Definition 1.11
a *A zero-memory system is* **homogeneous** *if*

$$S\{k\underset{\sim}{u}\} = kS\{\underset{\sim}{u}\} \quad \textit{for all k and } \underset{\sim}{u}. \tag{1.11}$$

b *A zero-memory system is* **additive** *if*

$$S\{\underset{\sim}{u}_1 + \underset{\sim}{u}_2\} = S\{\underset{\sim}{u}_1\} + S\{\underset{\sim}{u}_2\} \quad \textit{for any } \underset{\sim}{u}_1 \textit{ and } \underset{\sim}{u}_2. \tag{1.12}$$

Note:

1 When homogeneous or additive (without any qualifying adjective) is used in referring to dynamic systems, it is implied that the term refers to the zero-state response.
2 (b) implies (a) for k rational (see Problem 1.11).
3 $y = \sqrt[3]{u^3 + \dot{u}^3}$ satisfies (a) but not (b).

Definition 1.12 *A zero-memory system is* **linear** *if it is homogeneous and additive.*

This is the entire definition of linearity for zero-memory systems. However, if a system is dynamic, the concept of linearity is more complex. The following example was first proposed by D. Cargille, a student at the University of California at Berkeley (Zadeh and Desoer 1963:140).

Example 1.4

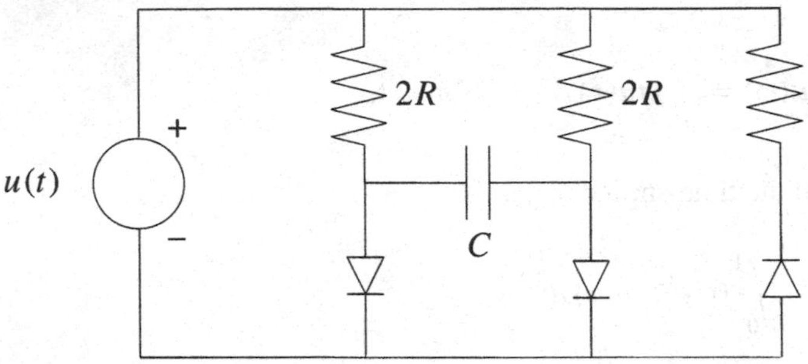

Fig. 1.5.

Consider the circuit shown in Figure 1.5.

For zero initial charge on the capacitor (i.e., zero initial voltage across its terminals), complete symmetry exists in the two branches bridged through the capacitor. Hence the system is equivalent to the circuit of Figure 1.6 which is linear.

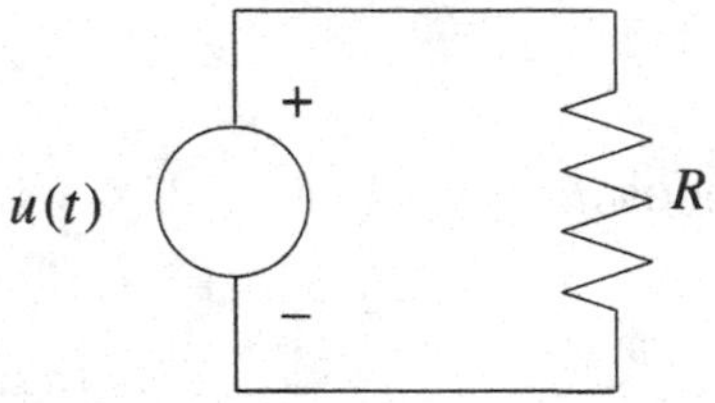

Fig. 1.6.

However, if an initial charge exists, the symmetry is destroyed, and the system is not linear. Therefore, the above definition is not enough for dynamic systems.

In order to define linearity for dynamic systems, we must extend our definition. We begin by utilizing $A\{\theta; \underset{\sim}{u}\} = S\{\underset{\sim}{u}\}$ for dynamic systems, so the previous definition gives:

Definition 1.13 *A dynamic system is called* **zero-state linear** *if* $A\{\theta; \alpha \underset{\sim}{u}_1 + \beta \underset{\sim}{u}_2\} = \alpha A\{\theta; \underset{\sim}{u}_1\} + \beta A\{\theta; \underset{\sim}{u}_2\}$ *for all* α *and* β *and any* $\underset{\sim}{u}_1$ *and* $\underset{\sim}{u}_2$ *(i.e., if it is zero-state additive and zero-state homogeneous).*

Definition 1.14 *A dynamic system is said to have the* **decomposition property** *if*

$$A\{\boldsymbol{x}(t_0); \underset{\sim}{u}\} = A\{\boldsymbol{x}(t_0); \underset{\sim}{0}\} + A\{\theta; \underset{\sim}{u}\} \tag{1.13}$$

for all $\boldsymbol{x}(t_0)$ *and all* $\underset{\sim}{u}$.

Definition 1.15 *A dynamic system is* **zero-input linear** *if it is zero-input homogeneous and additive. That is,*

$$\begin{aligned} A\{k\boldsymbol{x}(t_0); \underset{\sim}{0}\} &= kA\{\boldsymbol{x}(t_0); \underset{\sim}{0}\} \qquad \text{and} \\ A\{\boldsymbol{x}_1(t_0) + \boldsymbol{x}_2(t_0); \underset{\sim}{0}\} &= A\{\boldsymbol{x}_1(t_0); \underset{\sim}{0}\} + A\{\boldsymbol{x}_2(t_0); \underset{\sim}{0}\} \end{aligned} \tag{1.14}$$

for all $k, \boldsymbol{x}, \boldsymbol{x}_1$, and $\boldsymbol{x}_2$.

Example 1.5
The system

$$y(t) = \left(e^{-(t-t_0)}x(t_0)\right)^2 + \int_{t_0}^{t} e^{-(t-\tau)}u(\tau)\,d\tau, \qquad \text{for } t \geq t_0$$

is zero-state linear and has the decomposition property, but is not zero-input linear. Although this example is contrived, it illustrates that each part of the response must have the linearity property.

Definition 1.16 *A dynamic system is called linear if*
1 It is zero-state linear.
2 It has the decomposition property.
3 It is zero-input linear.

A system that is not linear is called nonlinear.

We noted that additivity by itself implies homogeneity for rational numbers k. It can be proven that additivity with an additional property of either continuity or boundedness implies the homogeneity property for any real k. In much of the Russian literature, linearity is defined in terms of additivity and continuity (see Liusternik and Sobolev 1961:59 or Kantorovich and Akilov 1964:100). Linearity as defined here is the usual one used in western literature (Schwarz and Friedland 1965, Zadeh and Desoer 1963). Although systems that are linear by this definition are not guaranteed to be continuous (and thus not bounded), physical systems usually have these additional properties.

1.3.5 Type of signals

Definition 1.17 *A* **continuous-time** *or* **analog** *system is one whose signals (i.e., input, output, and state variables) are continuous-time functions. A* **discrete** *system is one whose signals (i.e., input, output, and state variables) are discrete-time functions.*[1] *A sampled-data system is one in which both*

[1] A function f defined for all values of t on an interval of the real line is called a continuous-time function or analog function. A function f defined on a countable set of values of the real line is called a discrete-time function or just a discrete function.

continuous-time and discrete-time signals exist or, frequently, one in which the analog signals can be uniquely specified via discrete signals.

1.4 The techniques of linear system analysis

In the following, for the most part, we will be concerned with physical, dynamic, both fixed and time-varying, linear, analog, and discrete systems. Such systems are described by linear differential or difference equations. We are interested in solving these equations in order to obtain the input-output-state relations for the corresponding systems. There are two approaches to this problem: direct and indirect.

The *direct approach* is to obtain the solution as the sum of the homogeneous solution and the particular integral. This approach may yield quick answers for certain forms of inputs, and it has advantages whenever one is interested in the output due to a specific input for which the particular integral can be determined easily. The answer is in the form of an explicit time function, but its relation to the input is lost completely. Consequently, one cannot readily determine the class of outputs corresponding to a given class of inputs by using the direct approach.

It is the *indirect approach* that yields the input-output relation for a given system; this relation is of great importance in the theory of linear systems. The indirect approach consists of the following steps:

1 Resolution of an arbitrary input into a linear combination of elementary functions.
2 Determination of the response to a typical elementary function.
3 Composition or recombination of a linear combination of elementary responses to obtain the response to the arbitrary input.

Before outlining two important applications of the indirect approach, we will make two assertions. The proof of these assertions is not presented here. First, we can easily prove by induction that for a linear system

$$S\left\{\sum_{i=1}^{n} \alpha_i\{u_i(t)\}\right\} = \sum_{i=1}^{n} \alpha_i S\{u_i(t)\}$$

for finite n.

Assertion 1

It is true that for a bounded or continuous linear system

$$S\left\{\sum_{i=1}^{\infty} \alpha_i\{u_i(t)\}\right\} = \sum_{i=1}^{\infty} \alpha_i S\{u_i(t)\} \tag{1.15}$$

provided $\lim_{n\to\infty} \sum_{i=1}^{n} \alpha_i u_i(t)$ exists uniformly in t.

Assertion 2
It is true that for a bounded or continuous linear system

$$S\left\{\int_a^b \alpha(\tau)\{u(t,\tau)\}\,d\tau\right\} = \int_a^b \alpha(\tau)S\{u(t,\tau)\}\,d\tau \tag{1.16}$$

provided the Riemann integral $\int_a^b \alpha(\tau)u(t,\tau)\,d\tau$ exists uniformly in t as the limit of a sum.

Application 1: *Time-domain analysis*
If the elementary functions are unit impulses, then $u(t)$ can be expressed by the well-known formula

$$u(t) = \int_{-\infty}^{\infty} u(\tau)\delta(t-\tau)\,d\tau.$$

Then, using linearity and Assertion 2, we obtain

$$y(t) = S\{u(t)\} = S\left\{\int_{-\infty}^{\infty} u(\tau)\{\delta(t-\tau)\}\,d\tau\right\} = \int_{-\infty}^{\infty} u(\tau)S\{\delta(t-\tau)\}\,d\tau.$$

Defining the impulse response by $h(t,\tau) \triangleq S\{\delta(t-\tau)\}$ we obtain

$$y(t) = \int_{-\infty}^{\infty} h(t,\tau)u(\tau)\,d\tau$$

which is the input-output relation for a system with impulse response $h(t,\tau)$.

Application 2: *Frequency-domain analysis*
Here the input is expressed in terms of exponential time functions. Such an expression is given by the well-known inverse Laplace transform formula

$$u(t) = \frac{1}{2\pi j}\int_{c-j\infty}^{c+j\infty} U(s)e^{st}\,ds.$$

Then, using linearity and Assertion 2, we obtain

$$\begin{aligned} y(t) = S\{u(t)\} &= S\left\{\frac{1}{2\pi j}\int_{c-j\infty}^{c+j\infty} U(s)e^{st}\,ds\right\} \\ &= \frac{1}{2\pi j}\int_{c-j\infty}^{c+j\infty} U(s)S\{e^{st}\}\,ds. \end{aligned}$$

We will prove later that the response of a linear and fixed analog system to an exponential input e^{st} is of the form $S\{e^{st}\} = H(s)e^{st}$, where $H(s)$ is the system transfer function. Hence

$$y(t) = \frac{1}{2\pi j} \int_{c-j\infty}^{c+j\infty} U(s)H(s)e^{st}\, ds$$

where the product $U(s)H(s)$ must be identified as the Laplace transform $Y(s)$ of the time function $y(t)$. Obviously $Y(s) = H(s)U(s)$ is the well-known transfer function equation. Knowledge of $H(s)$ suffices to yield $Y(s)$ and then $y(t)$ for any $U(s)$. This is the familiar Laplace transform approach, which concentrates on treating the functions of s instead of the time functions themselves.

Problems

1.1 An analog system has the zero-state response

$$y(t) = S\{u(t)\} = \int_{-1}^{1} \tau^2 u(t+\tau)\, d\tau.$$

Determine whether the system *is* or *is not* (a) causal, (b) fixed, (c) zero-state linear. Give an answer and a reason for each of (a), (b), and (c).

1.2 Determine whether each of the following systems is linear, fixed, or zero-memory. Give an answer for each category and state the reason for your answer in each case.

a) A system characterized by $y(t) = u(t) - u(t-1)$.
b) A system in which $y(t) = 2(t+1)$ for all t (independent of the input).
c) The network

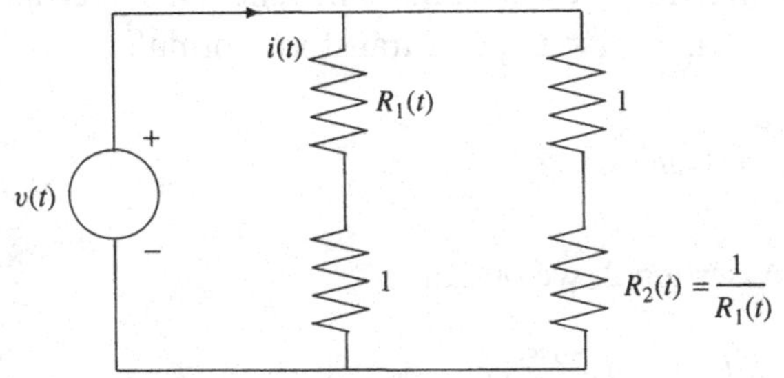

with $u(t) = v(t)$ and $y(t) = i(t)$.

1.3 Specify whether the system governed by the difference equation

$$y(k+1) + 2y(k) = u(k+2) + k^2 u(k)$$

is or *is not* (a) causal, (b) fixed (i.e., time-invariant), (c) zero-state linear. Give an answer for each of (a), (b), and (c) and justify it.

1.4 Which of the following systems are linear?

a) $y(t) = \max[u_1(t), u_2(t)]$

b) $y(t) = \begin{cases} 0 & \text{for } t < T \\ t^2 u(t) & \text{for } t \geq T \end{cases}$

c) $y(t) = t \sin[t u(t)]$

d) $y(t) = t[1 + u(t)]$

1.5 A system has the zero-state response

$$y(t) = S\{u(t)\} = \int_{-\infty}^{\infty} (\tau - t)^3 \, u(\tau) 1(t - \tau - 1) \, d\tau$$

where the unit-step function $1(\lambda)$ is defined by

$$1(\lambda - a) \triangleq \begin{cases} 0 & \text{for } \lambda < a \\ 1 & \text{for } \lambda \geq a \end{cases}$$

Determine whether this system *is* or *is not* (a) causal, (b) time-varying, (c) zero-memory, (d) zero-state linear. Give an answer for each category and justify your answer.

1.6 Specify for each of the following systems whether it is linear, causal, fixed, or zero-memory or whether insufficient information is given. Give an answer for each category and state the reason for your answer in each case.

a) $y(t) = 3u(t) + \dfrac{du(t)}{dt} + u(t-2) \quad \text{for } t > 0$

b) $e^{-k} y(k+1) + y(k) = u(k+2) + k^2 u(k)$

1.7 Show that the system with input-output relation

a) $y(t) = S\{u(t)\} = \int_{t-T}^{t} (t - \tau) u(\tau) \, d\tau$ is time-invariant.

b) $y(t) = S\{u(t)\} = \int_{t-T}^{t} \tau u(\tau) \, d\tau$ is time-varying.

1.8 A system is found to have the zero-state response $\{y(t)\}$ when the input $\{u(t)\}$ is applied (both are shown below).

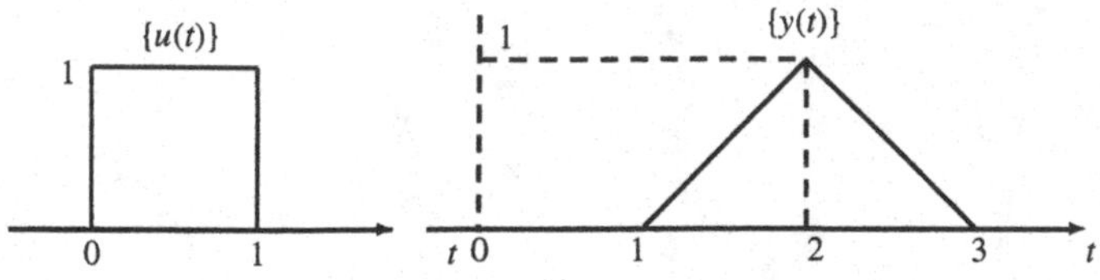

Is this system (a) causal, (b) time-varying, (c) zero-memory, and/or (d) zero-state linear? Explain your answers.

1.9 In the circuit below $u(t)$ is the input voltage, $y(t)$ is the output voltage, and the switch S opens and closes periodically every T seconds.

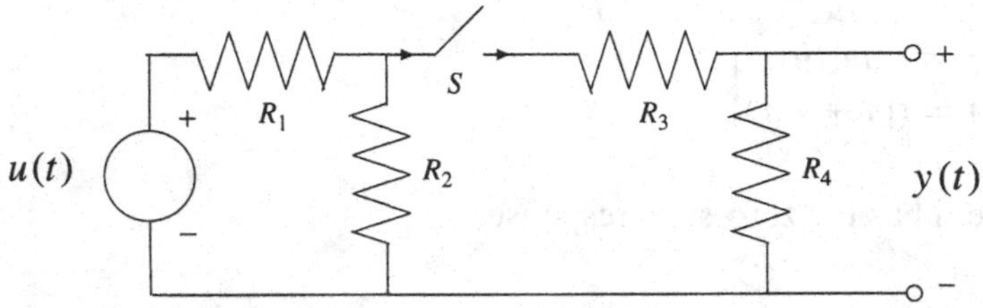

Is this circuit (a) causal, (b) time-varying, (c) zero-memory, and/or (d) zero-state linear? Explain your answers.

1.10 Specify whether each of the systems below is causal, zero-state linear, and/or time-invariant.
a) $y(t) = S\{u(t)\} = \int_t^{t+2} t^2 u(\tau)\, d\tau$
b) $y(t) = S\{u(t)\} = \int_{t-2}^{t} \tau u^2(\tau)\, d\tau$
c) $y(t) = S\{u(t)\} = \int_{t+2}^{t} (t-\tau)^2 u(\tau)\, d\tau$

1.11 Show that for α, a rational number (i.e., $\alpha = n/m$ where n and m are integers), the additivity property $S\{\underset{\sim}{u}_1 + \underset{\sim}{u}_2\} = S\{\underset{\sim}{u}_1\} + S\{\underset{\sim}{u}_2\}$ implies the homogeneity property $S\{\alpha \underset{\sim}{u}\} = \alpha S\{\underset{\sim}{u}\}$.

2

State equations of finite dimensional linear systems

2.1 Introduction

As we mentioned earlier in our discussion of dynamic systems, we will be primarily concerned with systems whose state $\boldsymbol{x}$ can be represented by a finite set of numbers (or equivalently an n-vector) and whose dynamic behavior (i.e., the changes in the state) are governed by either the state differential equations

$$\dot{\boldsymbol{x}}(t) = \boldsymbol{f}(\boldsymbol{x}(t), \boldsymbol{u}(t), t) \quad \text{with} \quad \boldsymbol{y}(t) = \boldsymbol{g}(\boldsymbol{x}(t), \boldsymbol{u}(t), t) \tag{2.1}$$

for analog systems, or by the state difference equations

$$\boldsymbol{x}(t_{k+1}) = \boldsymbol{f}(\boldsymbol{x}(t_k), \boldsymbol{u}(t_k), t_k) \quad \text{with} \quad \boldsymbol{y}(t_k) = \boldsymbol{g}(\boldsymbol{x}(t_k), \boldsymbol{u}(t_k), t_k) \tag{2.2}$$

for discrete systems. Such systems are sometimes called *lumped* systems to distinguish them from *distributed* systems which are describable by partial differential equations or differential-difference equations and whose states can only be expressed as functions rather than by finite dimensional vectors. The number of components of the state vector is called the order of the system, and it is designated by n. The input $\boldsymbol{u}(t)$ and output $\boldsymbol{y}(t)$ will have r and m components respectively. If the vector functions $\boldsymbol{f}$ and $\boldsymbol{g}$ are linear in $\boldsymbol{x}$ and $\boldsymbol{u}$ then the equations become

$$\begin{aligned} \dot{\boldsymbol{x}}(t) &= A(t)\boldsymbol{x}(t) + B(t)\boldsymbol{u}(t) \\ \boldsymbol{y}(t) &= C(t)\boldsymbol{x}(t) + D(t)\boldsymbol{u}(t) \end{aligned} \quad \text{or} \quad \begin{aligned} \boldsymbol{x}(t_{k+1}) &= A(t_k)\boldsymbol{x}(t_k) + B(t_k)\boldsymbol{u}(t_k) \\ \boldsymbol{y}(t_k) &= C(t_k)\boldsymbol{x}(t_k) + D(t_k)\boldsymbol{u}(t_k) \end{aligned} \tag{2.3}$$

where the matrices $A, B, C,$ and D are $n \times n, n \times r, m \times n,$ and $m \times r$ respectively, and are constant for time-invariant systems and functions of t for time-varying systems. It will be shown later that in this case the systems are linear by our definitions. The above equations are called linear dynamical (differential or difference) state equations in standard or normal form.

2.2 State differential equations of circuits

We start by showing how passive electrical circuits can be described by state differential equations in normal form. The basic elements of such systems are resistors, capacitors, and inductors, as shown in Figure 2.1, and for these systems there is a "natural" choice of state variables.

As is well known, the terminal relation describing a resistor is $v_R(t) = R(t)i_R(t)$ where $v_R(t)$ is the voltage across the resistor, $i_R(t)$ is the current through the resistor, and $R(t)$ is the resistance. Thus the resistor is a zero-memory element.

To get the basic terminal relation for a capacitor, we note that $i_C(t) = \dot{q}_C(t)$ where $i_C(t)$ is the current through the capacitor and $q_C(t)$ is the electrical charge in the capacitor. Since the charge $q_C(t)$ is related to the voltage $v_C(t)$ across the capacitor by $q_C(t) = C(t)v_C(t)$, where $C(t)$ is the capacitance, we find that

$$\frac{d[C(t)v_C(t)]}{dt} = \frac{dC(t)}{dt}v_C(t) + C(t)\frac{dv_C(t)}{dt} = i_C(t)$$

or

$$C(t)\frac{dv_C(t)}{dt} = i_C(t) - \frac{dC(t)}{dt}v_C(t) \tag{2.4}$$

can be used to describe the basic terminal relation for a capacitor. When the capacitor value does not change with time (i.e., C is a constant) then $\dot{C} = 0$, so the terminal relation becomes

$$C\frac{dv_C(t)}{dt} = i_C(t). \tag{2.5}$$

The terminal relation for an inductor is found in much the same way. We first observe that the voltage $v_L(t)$ across an inductor is equal to the rate of change of flux $\phi(t)$ stored in the inductor; that is, $v_L(t) = \dot{\phi}(t)$. Since $\phi(t) = L(t)i_L(t)$, where $L(t)$ is the inductance, we see that

$$\frac{d[L(t)i_L(t)]}{dt} = \frac{dL(t)}{dt}i_L(t) + L(t)\frac{di_L(t)}{dt} = v_L(t)$$

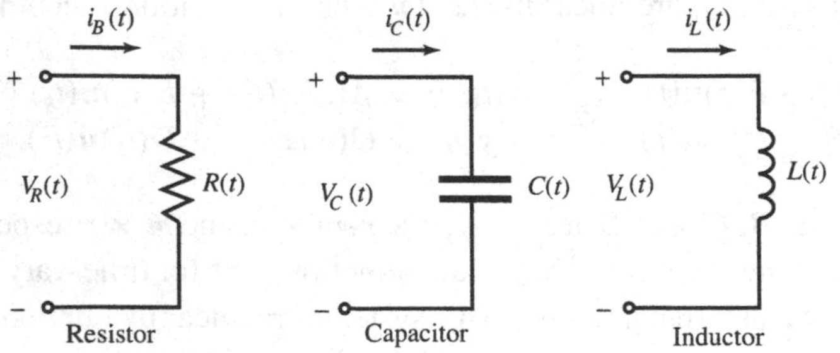

Fig. 2.1.

or

$$L(t)\frac{di_L(t)}{dt} = v_L(t) - \frac{dL(t)}{dt}i_L(t) \tag{2.6}$$

can be used to describe the basic terminal relation for an inductor. When the inductor value does not change with time (i.e., L is a constant) then $\dot{L} = 0$, so the terminal relation becomes

$$L\frac{di_L(t)}{dt} = v_L(t). \tag{2.7}$$

The Kirchhoff voltage law (KVL) and Kirchhoff current law (KCL) equations specify how these components combine. The KVL requires that the voltages around a closed loop sum to zero. The KCL requires that the algebraic sum of currents entering a node equals zero. From (2.4) or (2.5) we see that a "natural" choice for the state variable when a capacitor is present is the voltage across the capacitor since the derivative of $v_C(t)$ appears in the terminal relation and the derivative of the state variables is required for the normal form of the state equations. This comes about because the energy stored in the electric field of the capacitor (due to the charge) cannot change instantaneously, and thus the voltage across the capacitor cannot change instantaneously either. Similarly, a "natural" choice for the state variable when an inductor is present is the current through the inductor since (2.6) or (2.7) shows that the derivative of $i_L(t)$ appears in the terminal relation. Here the energy is stored in the magnetic field of the inductor (due to the flux), and it cannot change instantaneously. Thus the current through the inductor cannot change instantaneously either. In other words, the inductor currents and capacitor voltages are natural initial conditions that describe the state of the system.

The following example will illustrate how the state equations can be found for a simple circuit.

Example 2.1

Consider the circuit in Figure 2.2,

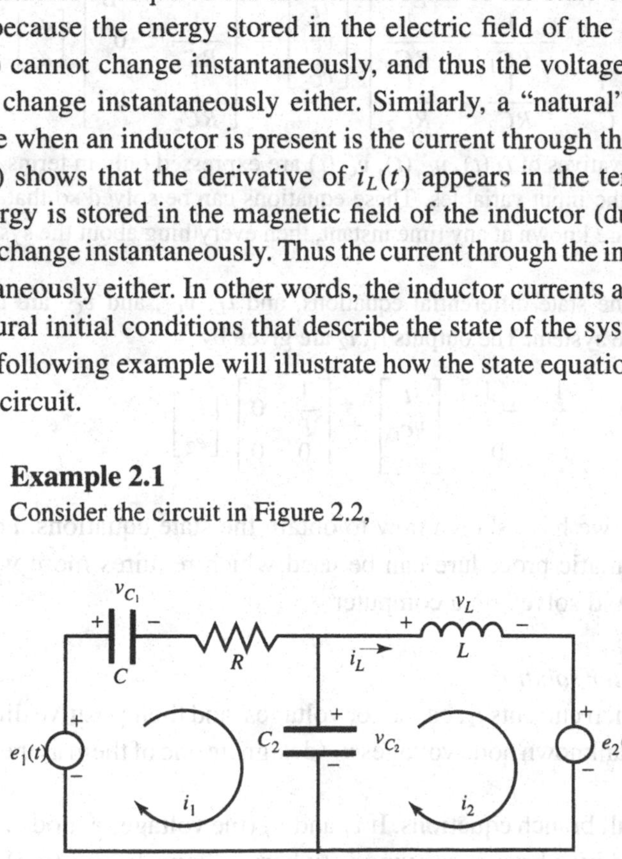

Fig. 2.2.

where $e_1(t)$ and $e_2(t)$ are the inputs, and the loop currents $i_1(t)$ and $i_2(t)$ are the outputs.

We could write the two loop equations and combine them to get a third-order differential equation in either $i_1(t)$ or $i_2(t)$. Three initial conditions are then required to obtain a solution.

Alternatively, we could also describe the system in terms of the variables $v_{C_1}(t)$, $v_{C_2}(t)$, and $i_L(t)$ whose initial values must be known. That is,

$$\dot{i}_L = \frac{v_L}{L} = \frac{v_{C_2} - e_2}{L}$$

$$\dot{v}_{C_1} = \frac{i_1}{C_1} = \frac{1}{C_1}\left[\frac{e_1 - v_{C_1} - v_{C_2}}{R}\right]$$

$$\dot{v}_{C_2} = \frac{i_1 - i_2}{C_2} = \frac{1}{C_2}\left[\frac{e_1 - v_{C_1} - v_{C_2}}{R} - i_L\right]$$

which can be written in vector matrix form as

$$\begin{bmatrix} \dot{i}_L \\ \dot{v}_{C_1} \\ \dot{v}_{C_2} \end{bmatrix} = \begin{bmatrix} 0 & 0 & \frac{1}{L} \\ 0 & -\frac{1}{RC_1} & -\frac{1}{RC_1} \\ -\frac{1}{C_2} & -\frac{1}{RC_2} & -\frac{1}{RC_2} \end{bmatrix} \begin{bmatrix} i_L \\ v_{C_1} \\ v_{C_2} \end{bmatrix} + \begin{bmatrix} 0 & -\frac{1}{L} \\ \frac{1}{RC_1} & 0 \\ \frac{1}{RC_2} & 0 \end{bmatrix} \begin{bmatrix} e_1 \\ e_2 \end{bmatrix}.$$

Note: The derivatives of $i_L(t)$, $v_{C_1}(t)$, $v_{C_2}(t)$ are expressed only in terms of these variables and the input variables. These equations can be solved so that if $i_L(t)$, $v_{C_1}(t)$, $v_{C_2}(t)$ are known at any time instant, then everything about the system can be determined.

These are the state differential equations, and i_L, v_{C_1}, and v_{C_2} are the state variables of the system. The outputs i_1, i_2 are given by

$$\begin{bmatrix} i_1(t) \\ i_2(t) \end{bmatrix} = \begin{bmatrix} 0 & -\frac{1}{R} & -\frac{1}{R} \\ 1 & 0 & 0 \end{bmatrix} \begin{bmatrix} i_L \\ v_{C_1} \\ v_{C_2} \end{bmatrix} + \begin{bmatrix} \frac{1}{R} & 0 \\ 0 & 0 \end{bmatrix} \begin{bmatrix} e_1 \\ e_2 \end{bmatrix}.$$

For simple circuits we have shown how to obtain the state equations. For large circuits a more systematic procedure can be used which requires more work but can be programmed and solved on a computer.

Procedure for computer solution

Step 1 Label all branch currents i_k, capacitor voltages, and their positive directions as well as all unknown node voltages v_j (designate one of them as the ground node).

Step 2 Write down all branch equations. If v_i and v_j (the voltages at nodes i and j) are connected by a branch carrying i_k (where i_k flows from i to j) then

a For resistive branches (R_k) the equation is $v_i - v_j = i_k R_k$.
b For inductive branches (L_k) the equation is $v_i - v_j = L_k(di_k/dt)$.
c For capacitative branches (C_k) $v_i - v_j = v_C$ (where v_i is at the + terminal).
d For voltage sources (e_k) $v_i - v_j = e_k$ (where v_i is at the + terminal).

Step 3 Write out all capacitor current equations $C_k(dv_{C_k}/dt) = \pm i_k$ (+ if the current flows into the positive side of v_C).

Step 4 Write out all node current equations except for those nodes with known voltages; that is, $\sum_k(\pm i_k) = 0$ (where + is assigned to currents leaving a node). Thus far the procedure leads to equal numbers N of unknowns and equations although many are redundant. Note that among these equations there will be exactly as many differential equations as energy storage elements (i.e., inductors and capacitors); call this number n. The remaining $l \triangleq N - n$ equations are algebraic equations. The n variables representing inductor currents or capacitor voltages are chosen to be the state variables. The input variables are the independent voltage or current sources. The output variables are those unknown variables specified as such. All the remaining unknowns are called auxiliary variables, and these eventually will be eliminated from the equations. The procedure thus leads to a set of equations having the form

$$\begin{aligned} \mathbf{0} &= F_{11}\mathbf{z} + F_{12}\mathbf{x} + G_1\mathbf{u} \\ \dot{\mathbf{x}} &= F_{21}\mathbf{z} + F_{22}\mathbf{x} + G_2\mathbf{u} \end{aligned} \tag{2.8}$$

where the first group of equations denotes l algebraic equations and the second group n differential equations. These can be written as

$$\begin{bmatrix} 0 \\ I \end{bmatrix} \dot{\mathbf{x}} = \begin{bmatrix} F_{11} & F_{12} \\ F_{21} & F_{22} \end{bmatrix} \begin{bmatrix} \mathbf{z} \\ \mathbf{x} \end{bmatrix} + \begin{bmatrix} G_1 \\ G_2 \end{bmatrix} \mathbf{u} \tag{2.9}$$

where the l-vector $\mathbf{z}$ denotes the auxiliary variables and I is the identity matrix.

Step 5 Write down the designated output variables in terms of the auxiliary, the state, and the input variables. This leads to a set of m output equations having the form

$$\mathbf{y} = H_1\mathbf{z} + H_2\mathbf{x} + J\mathbf{u}. \tag{2.10}$$

As these equations now stand they are not in normal form because of the presence of the auxiliary variables $\mathbf{z}$.

Step 6 Eliminate the auxiliary variables from these sets of equations. This can be done formally by finding the inverse of the matrix F_{11} and solving the algebraic equations for z. When F_{11} is singular (i.e., it has no inverse) then the equations are degenerate and cannot be put in normal form. (This occurs when the circuit contains a loop of capacitors or a node connecting only inductors.)

In practice the easiest method of eliminating the unwanted variables (and simultaneously determining whether the system is degenerate) is to use the Gaussian elimination method on the system of equations (2.9, 2.10)

$$\begin{bmatrix} 0 & 0 \\ I & 0 \\ 0 & I \end{bmatrix} \begin{bmatrix} \dot{x} \\ y \end{bmatrix} = \begin{bmatrix} F_{11} & F_{12} \\ F_{21} & F_{22} \\ H_1 & H_2 \end{bmatrix} \begin{bmatrix} z \\ x \end{bmatrix} + \begin{bmatrix} G_1 \\ G_2 \\ J \end{bmatrix} u \tag{2.11}$$

in order to eliminate the variables z_i.

Gaussian elimination procedure

Step 1 Identify the element f_{ij} of F_{11} which has the largest magnitude and move the jth variable (column) and ith equation (row) into the (1, 1) position.

Step 2 Divide the first row by f_{11} and then subtract f_{i1} times this row from the ith row for $i = 2, \ldots, N + m$. (Note this places all zeros in the column below f_{11}.)

Step 3 Delete the first row and column. Replace N by $N - 1$ and l by $l - 1$. If the new F_{11} has only zero elements, stop. If some elements are not zero, go to Step 1.

Example 2.2

The preceding procedure will be illustrated by our earlier example (see Figure 2.3).

There are 6 branches and 4 nodes (not counting the reference ground node).

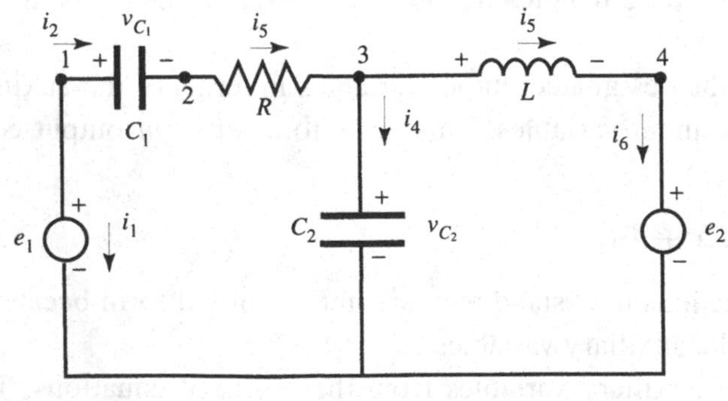

Fig. 2.3.

Branch equations

1 $e_1 = v_1$ 2 $v_{C_1} = v_1 - v_2 = e_1 - v_2$

3 $Ri_3 = v_2 - v_3$ 4 $v_{C_2} = v_3$

5 $L\dot{i}_5 = v_3 - v_4 = v_3 - e_2$

We have replaced the unknowns v_1 and v_4 by the knowns e_1 and e_2 in order to reduce the number of equations.

Capacitor current equations

$C_1\dot{v}_{C_1} = i_2$

$C_2\dot{v}_{C_2} = i_4$

Node equations

(Other than at 1 and 4)

node 2 $i_3 - i_2 = 0$

node 3 $i_4 + i_5 - i_3 = 0$

The unknowns are $v_2, v_3, v_{C_1}, v_{C_2}, i_2, i_3, i_4, i_5$ (8 unknowns), and there are 8 equations, so $N = 8$. Let $x_1 = i_5, x_2 = v_{C_1}, x_3 = v_{C_2}$ be the state variables and $z_1 = i_2, z_2 = i_3, z_3 = i_4, z_4 = v_2, z_5 = v_3$ be the auxiliary variables, and $u_1 = e_1$ and $u_2 = e_2$ the input variables.

In vector matrix form this gives

$$\begin{bmatrix} 0 & 0 & 0 \\ 0 & 0 & 0 \\ 0 & 0 & 0 \\ 0 & 0 & 0 \\ 0 & 0 & 0 \\ \cdots & \cdots & \cdots \\ 1 & 0 & 0 \\ 0 & 1 & 0 \\ 0 & 0 & 1 \end{bmatrix} \begin{bmatrix} \dot{x}_1 \\ \dot{x}_2 \\ \dot{x}_3 \end{bmatrix} = \begin{bmatrix} -1 & 1 & 0 & 0 & 0 & \vdots & 0 & 0 & 0 \\ 0 & -1 & 1 & 0 & 0 & \vdots & 1 & 0 & 0 \\ 0 & 0 & 0 & -1 & 0 & \vdots & 0 & -1 & 0 \\ 0 & -R & 0 & 1 & -1 & \vdots & 0 & 0 & 0 \\ 0 & 0 & 0 & 0 & 1 & \vdots & 0 & 0 & -1 \\ \cdots & \cdots & \cdots & \cdots & \cdots & \cdots & \cdots & \cdots & \cdots \\ 0 & 0 & 0 & 0 & \frac{1}{L} & \vdots & 0 & 0 & 0 \\ \frac{1}{C_1} & 0 & 0 & 0 & 0 & \vdots & 0 & 0 & 0 \\ 0 & 0 & \frac{1}{C_2} & 0 & 0 & \vdots & 0 & 0 & 0 \end{bmatrix} \begin{bmatrix} z_1 \\ z_2 \\ z_3 \\ z_4 \\ z_5 \\ \cdots \\ x_1 \\ x_2 \\ x_3 \end{bmatrix} + \begin{bmatrix} 0 & 0 \\ 0 & 0 \\ 1 & 0 \\ 0 & 0 \\ 0 & 0 \\ \cdots & \cdots \\ 0 & -\frac{1}{L} \\ 0 & 0 \\ 0 & 0 \end{bmatrix}.$$

Now, writing the outputs, let $y_1 = i_3$ and $y_2 = i_5$

$$\text{so} \quad \begin{bmatrix} y_1 \\ y_2 \end{bmatrix} = \begin{bmatrix} 0 & 1 & 0 & 0 & 0 & \vdots & 0 & 0 & 0 \\ 0 & 0 & 0 & 0 & 0 & \vdots & 1 & 0 & 0 \end{bmatrix} \begin{bmatrix} z \\ x \end{bmatrix}.$$

Applying the Gaussian elimination method gives

$$A = \begin{bmatrix} 0 & 0 & 1/L \\ 0 & -1/(RC_1) & -1/(RC_1) \\ -1/C_2 & -1/(RC_2) & -1/(RC_2) \end{bmatrix}; \qquad B = \begin{bmatrix} 0 & -1/L \\ 1/(RC_1) & 0 \\ 1/(RC_2) & 0 \end{bmatrix}$$

$$C = \begin{bmatrix} 0 & -1/R & -1/R \\ 1 & 0 & 0 \end{bmatrix}; \qquad D = \begin{bmatrix} 1/R & 0 \\ 0 & 0 \end{bmatrix}$$

as before.

2.3 State differential equations of mechanical systems

Just as we did for electrical circuits, we will show how translational mechanical systems can be described by state differential equations in normal form. The basic elements of these systems are springs, viscous dampings, and masses as shown in Figure 2.4. The assumed positive direction for displacements, velocities, and forces is indicated by the arrow. These basic passive mechanical components are analogous to the basic components of electrical circuits. The damping element (this can be thought of as a shock absorber) dissipates mechanical energy and thus corresponds to the resistor which dissipates electrical energy. The spring which stores potential energy and the mass which stores kinetic energy correspond to the inductor and capacitor which both store electrical energy. It is not surprising therefore that here too there is a "natural" choice of state variables.

The terminal relation for a spring comes from Hooke's law, which states that $f_K(t) = K(t)[z_2(t) - z_1(t)]$ where $z_i(t)$ denotes the displacement of end i of the spring from its equilibrium position (i.e., not stretched or compressed), $f_K(t)$ is the force applied, and $K(t)$ is the spring constant.

The terminal relation for a damping element is given by $f_D(t) = D(t)[v_2(t) - v_1(t)]$ where $v_i(t)$ denotes the velocity of end i of the damping element, $f_D(t)$ is the force applied, and $D(t)$ is the damping coefficient.

The relation for a mass is found from Newton's second law which states that the rate of change of momentum is equal to the net force acting on the mass. Since

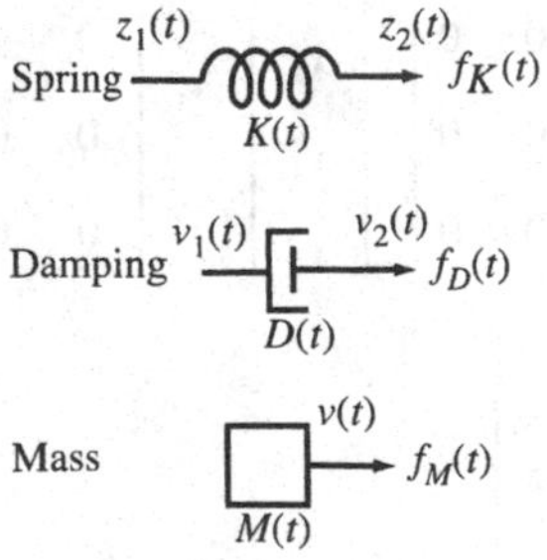

Fig. 2.4.

momentum $= M(t)v(t)$, we see that

$$\frac{d[M(t)v(t)]}{dt} = \frac{dM(t)}{dt}v(t) + M(t)\frac{dv(t)}{dt} = f_M(t)$$

or

$$M(t)\frac{dv(t)}{dt} = f_M(t) - \frac{dM(t)}{dt}v(t). \tag{2.12}$$

When the mass does not change with time (i.e., M is a constant) then $\dot{M} = 0$, so (2.12) becomes

$$M\frac{dv(t)}{dt} = f_M(t). \tag{2.13}$$

Newton's second law also prescribes how the basic element relations are combined because it says that

$$\frac{d[M(t)v(t)]}{dt} = \sum \text{ all forces acting on } M(t). \tag{2.14}$$

Therefore by writing an equation of motion like (2.14) for every point that can move independently, we get a description of the entire mechanical system.

Potential energy is stored in the springs when they are either stretched or compressed and kinetic energy is stored in moving masses. Since these energies cannot change instantaneously, the variables associated with them are natural initial conditions that describe the state of the system. Thus the differences in displacement of the ends of springs from their equilibrium positions and the velocities of masses are "natural" choices for the state variables of mechanical translational systems.

Rotational systems are treated in exactly the same way. The only difference is that the moment of inertia $J(t)$ replaces the mass $M(t)$, and we have angular positions $\theta(t)$ and velocities $\omega(t)$ instead of translational positions and velocities.

Example 2.3

For the translational system illustrated in Figure 2.5, the force $f(t)$ is the input and the output $y(t) = z_1(t)$.

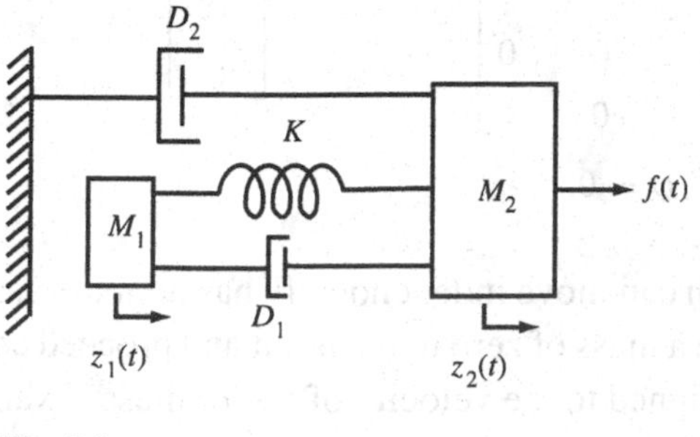

Fig. 2.5.

The equations of motion are

$$M_1 \frac{dv_1(t)}{dt} = K[z_2(t) - z_1(t)] + D_1[v_2(t) - v_1(t)] \tag{2.15}$$

$$M_2 \frac{dv_2(t)}{dt} = f(t) - K[z_2(t) - z_1(t)] - D_2 v_2(t) - D_1[v_2(t) - v_1(t)] \tag{2.16}$$

where the velocity $v_i(t)$ is the derivative of displacement position $z_i(t)$ so $v_i(t) = \dot{z}_i(t)$.

We choose the natural state variables $x_1(t) = v_1(t)$, $x_2(t) = v_2(t)$, and $x_3(t) = z_2(t) - z_1(t)$. Note that since there are three elements that store energy, there are three state variables. We see therefore from (2.15) that

$$\dot{x}_1(t) = \frac{1}{M_1}\dot{v}_1(t) = \frac{Kx_3(t) + D_1[x_2(t) - x_1(t)]}{M_1}.$$

From (2.16) we get

$$\dot{x}_2(t) = \frac{1}{M_2}\dot{v}_2(t) = \frac{f(t) - Kx_3(t) - D_2 x_2(t) - D_1[x_2(t) - x_1(t)]}{M_2}.$$

The third equation comes from

$$\dot{x}_3(t) = \dot{z}_2(t) - \dot{z}_1(t) = x_2(t) - x_1(t).$$

These three equations are the state differential equations in standard form. Since the output here is the position $z_1(t)$, we must introduce an additional state variable in order to generate the output. That is to say, this output cannot be obtained as a linear combination of the state variables, x_1, x_2, x_3. We therefore choose $x_4(t) = z_1(t)$. The output equation is then simply $y(t) = x_4(t)$, and the additional state equation is $\dot{x}_4(t) = x_1(t)$. (The fourth state variable is just the integral of the first state variable.) The A, B, C, D matrices for this system are

$$A = \begin{bmatrix} -\dfrac{D_1}{M_1} & \dfrac{D_2}{M_1} & \dfrac{K}{M_1} & 0 \\ \dfrac{D_1}{M_2} & -\dfrac{D_1 + D_2}{M_2} & -\dfrac{K}{M_2} & 0 \\ -1 & 1 & 0 & 0 \\ -1 & 1 & 0 & 0 \end{bmatrix}; \qquad B = \begin{bmatrix} 0 \\ \dfrac{1}{M_2} \\ 0 \\ 0 \end{bmatrix};$$

$$C = \begin{bmatrix} 0 & 0 & 0 & 1 \end{bmatrix}; \qquad D = 0.$$

It can happen that a point which can move independently has no mass associated with it. In this situation just assign a mass of zero to the point and proceed as before. Of course, no state variable is assigned to the velocity of a zero mass. Example 2.4 will illustrate this point.

Example 2.4

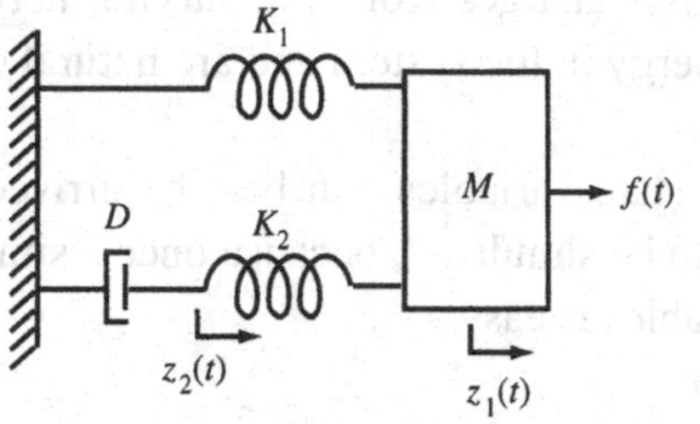

Fig. 2.6.

In Figure 2.6, the point between the damping D and the spring K_2 has no mass but can move independently of mass M.

The equations of motion are

$$M\dot{v}_1(t) = f(t) - K_1 z_1(t) - K_2[z_1(t) - z_2(t)] \tag{2.17}$$

$$0 \cdot \dot{v}_2(t) = K_2[z_1(t) - z_2(t)] - Dv_2(t). \tag{2.18}$$

Note that (2.18) says that the velocity $v_2(t)$ of the point $z_2(t)$ is proportional to $z_1(t) - z_2(t)$. Choosing the state variables to be $x_1(t) = v_1(t)$, $x_2(t) = z_1(t)$, and $x_3 = z_1(t) - z_2(t)$, we get the state differential equations

$$\dot{x}_1(t) = \frac{f(t) - K_1 x_2(t) - K_2 x_3(t)}{M}$$

$$\dot{x}_2(t) = x_1(t)$$

$$\dot{x}_3(t) = v_1(t) - v_2(t) = x_1(t) - \frac{K_2}{D} x_3(t).$$

Note that $v_2(t)$ was eliminated from the right-hand side of the third state equation by using (2.18).

2.4 Choice of state variables

We should be aware of the fact that the choice of the state variables is somewhat free. Suppose $\boldsymbol{x}_a(t)$ denotes one valid set of state variables: then if $\boldsymbol{x}_b(t)$ can be obtained from $\boldsymbol{x}_a(t)$ via a nonsingular transformation T (i.e., for an $n \times n$ matrix T for which $\det T \neq 0$) we can write $\boldsymbol{x}_b(t) = T\boldsymbol{x}_a(t)$ and $\boldsymbol{x}_a(t) = T^{-1}\boldsymbol{x}_b(t)$. Taking the derivative with respect to time

$$\dot{\boldsymbol{x}}_b = T[A(t)\boldsymbol{x}_a + B(t)\boldsymbol{u}] = TA(t)T^{-1}\boldsymbol{x}_b + TB(t)\boldsymbol{u} \tag{2.19}$$

and from $\boldsymbol{y}(t) = C_a(t)\boldsymbol{x}_a(t) + D(t)\boldsymbol{u}(t)$, using $C_b(t) = C_a(t)T^{-1}$, we can get

$$\boldsymbol{y}(t) = C_b(t)\boldsymbol{x}_b(t) + D(t)\boldsymbol{u}(t). \tag{2.20}$$

Thus the state equations in terms of the state $\boldsymbol{x}_b$ can be obtained.

We see therefore that the choice of state variables is fairly wide. For circuits, it is most convenient to choose inductor currents and capacitor voltages (for fixed networks) or inductor fluxes and capacitor charges (for time-varying networks). These represent the sources of stored energy in the system and are natural choices for the state variables.

For other systems a valid choice of state variables can best be arrived at by considering that the dynamic system is to be simulated, because once a simulation has been achieved a choice of state variables is easy.

2.5 Simulation diagrams

A simulation of a linear dynamic system consists of an interconnection of three basic components in such a way that the (differential or difference) dynamical equations of the simulator are identical to those describing the system. The basic elements of simulation are the following.

Basic simulator elements
    1 Dynamic element
        a For analog systems this is an integrator with the input-output relation

$$u(t) \longrightarrow \boxed{\int} \longrightarrow \qquad y(t) = y(t_0) + \int_{t_0}^{t} u(\tau)\, d\tau$$

        denoted by $y = D^{-1}u$.
        b For discrete-time systems this is a delay

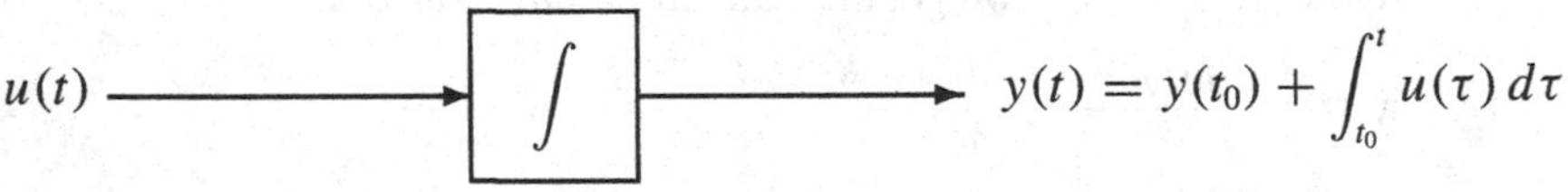

$$u(t_k) \longrightarrow \boxed{\nabla} \longrightarrow \qquad y(t_k) = u(t_{k-1})$$

        with the input-output relation denoted by $y = E^{-1}u$.
    2 Summing element (adder)

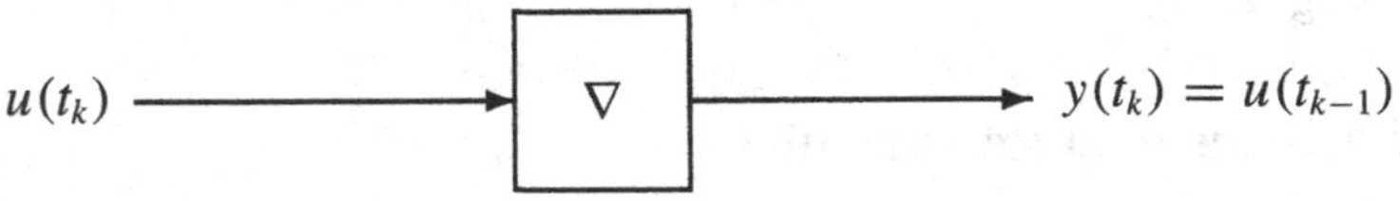

$$y(t) = \sum_{i=1}^{r} u_i(t).$$

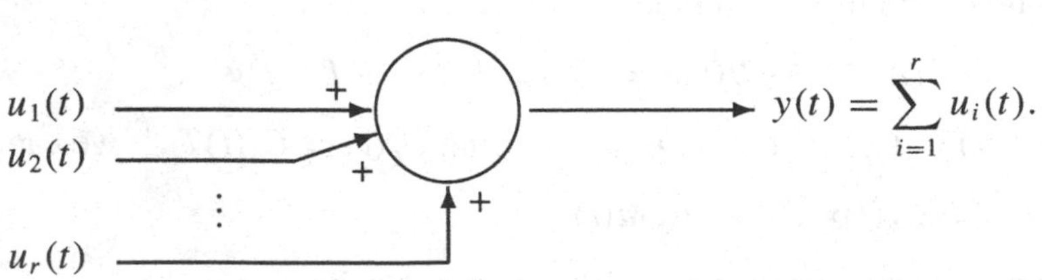

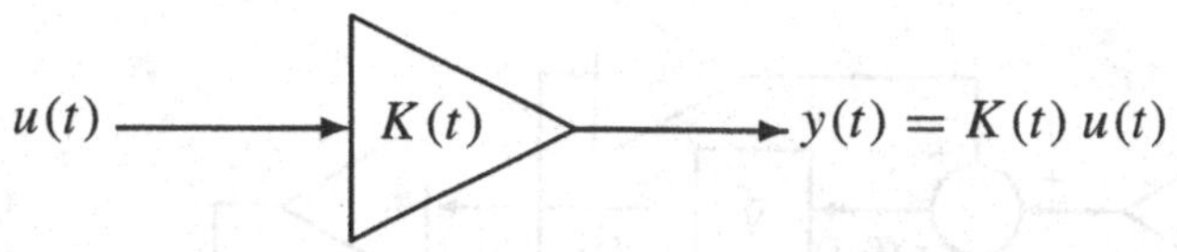

3 Scaling element (amplifier or attenuator)

Note: We specify integrators (not differentiators) because

1 Good physical approximations to differentiators are hard to achieve.
2 Rapidly changing signals cause saturation and overloading.
3 Differentiators are very susceptible to noise disturbances.

Also delays (*not advances*) are used because the simulation should be physically realizable.

We always choose our state variables to be the outputs of the dynamic elements. When this is done, writing the state equations from the simulation diagram is easy.

The fewest number of dynamic elements should be used. The number of dynamic elements is the order of the simulated system. Using the fewest possible number of dynamic elements gives what is called a *minimal realization.* The procedures presented here will not always result in a minimal realization, but we must be sure not to introduce any obviously unnecessary dynamics. In Chapter 8 we define the concepts of controllability and observability. It turns out that the minimal realization of a system is both controllable and observable.

Example 2.5

Consider the simulation in Figure 2.7.

Choose x_1 and x_2 as the outputs of the integrators. Then

$$\dot{x}_1(t) = x_2(t) + b_1(t)u(t)$$

$$\dot{x}_2(t) = -a_1(t)x_1(t) - a_2(t)x_2(t) + u(t)$$

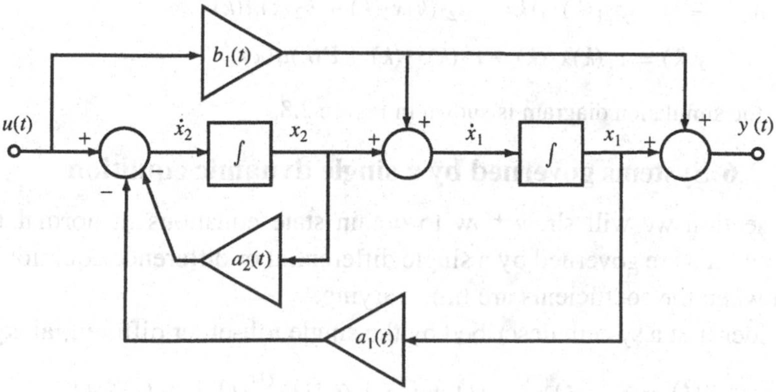

Fig. 2.7.

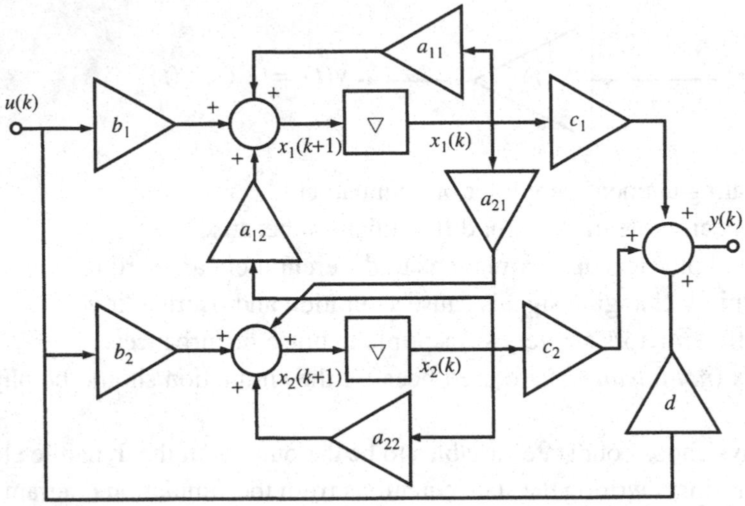

Fig. 2.8.

and

$$y(t) = x_1(t) + b_1(t)u(t).$$

These are the state equations which give

$$A(t) = \begin{bmatrix} 0 & 1 \\ -a_1(t) & -a_2(t) \end{bmatrix}; \qquad B(t) = \begin{bmatrix} b_1(t) \\ 1 \end{bmatrix};$$

$$C(t) = [1 \quad 0]; \qquad D(t) = b_1(t).$$

Example 2.6
For the discrete state equations

$$x_1(k+1) = a_{11}(k)x_1(k) + a_{12}(k)x_2(k) + b_1(k)u(k)$$

$$x_2(k+1) = a_{21}(k)x_1(k) + a_{22}(k)x_2(k) + b_2(k)u(k)$$

$$y(k) = c_1(k)x_1(k) + c_2(k)x_2(k) + d(k)u(k)$$

the simulation diagram is shown in Figure 2.8.

2.6 Systems governed by a single dynamic equation

In this section we will show how to obtain state equations in normal form for systems which are governed by a single differential or difference equation of order n, even when the coefficients are time-varying.

Consider first a system described by the single nth-order differential equation

$$\begin{aligned} y^{(n)}(t) + \alpha_{n-1}(t)y^{(n-1)}(t) + \cdots + \alpha_1(t)y^{(1)}(t) + \alpha_0(t)y(t) \\ = \beta_n(t)u^{(n)}(t) + \beta_{n-1}(t)u^{(n-1)}(t) + \cdots + \beta_1(t)u^{(1)}(t) + \beta_0(t)u(t) \end{aligned} \tag{2.21}$$

where $y^{(j)}$ denotes the jth derivative, and the coefficient of the highest derivative of $y(t)$ has been normalized to 1. Furthermore, notice that the highest order derivative of $u(t)$ cannot exceed the highest order derivative of $y(t)$. This is to ensure that the system can be put into normal form. If an input derivative order exceeds the highest output derivative order then differentiators would be needed for the simulation.

Procedure for obtaining a simulation

Step 1 Place the highest order derivative of $y(t)$ on the left and all other terms on the right. It is convenient to use the notation D^j to denote the jth derivative. Applied to our system this gives

$$\begin{aligned} D^n y(t) = \beta_n(t)D^n u(t) + \beta_{n-1}(t)D^{n-1}u(t) \\ - \alpha_{n-1}(t)D^{n-1}y(t) + \cdots + \beta_1(t)Du(t) \\ - \alpha_1(t)Dy(t) + \beta_0(t)u(t) - \alpha_0(t)y(t). \end{aligned} \tag{2.22}$$

Step 2 Integrate both sides n times. For the system (2.22), this gives

$$\begin{aligned} y(t) = D^{-n}[\beta_n(t)D^n u(t) + \beta_{n-1}(t)D^{n-1}u(t) \\ - \alpha_{n-1}(t)D^{n-1}y(t) + \cdots + \beta_1(t) \times Du(t) \\ - \alpha_1(t)Dy(t) + \beta_0(t)u(t) - \alpha_0(t)y(t)]. \end{aligned} \tag{2.23}$$

Step 3 Integrate each term by parts to eliminate all derivatives on $u(t)$ and $y(t)$. (This is the most difficult part for time-varying systems.) This results in

$$\begin{aligned} y(t) = b_n(t)u(t) + D^{-1}\{b_{n-1}(t)u(t) - a_{n-1}(t)y(t) + D^{-1}[b_{n-2}(t)u(t) \\ - a_{n-2}(t)y(t) + D^{-1}(\cdots + D^{-1}(b_0(t)u(t) - a_0(t)y(t))\ldots)]\}. \end{aligned} \tag{2.24}$$

Note: When all the coefficients α_i and β_i are constant then $b_i(t) = \beta_i$ and $a_i(t) = \alpha_i$, but otherwise they come from the integration by parts.

Step 4 Draw the simulation diagram from (2.24). This gives the system shown in Figure 2.9. We see that $y = x_n + b_n(t)u$.

Step 5 Label the output of each integrator as a state variable and write the state equations

$$\begin{aligned} \dot{x}_1 &= b_0(t)u - a_0(t)(x_n + b_n(t)u) - a_0(t)x_n + (b_0(t) - a_0(t)b_n(t))u \\ \dot{x}_{i+1} &= b_i(t)u - a_i(t)(x_n + b_n(t)u) + x_i \\ &= x_i - a_i(t)x_n + (b_i(t) - a_i(t)b_n(t))u \quad \text{for } i = 1, \ldots, n-1. \end{aligned} \tag{2.25}$$

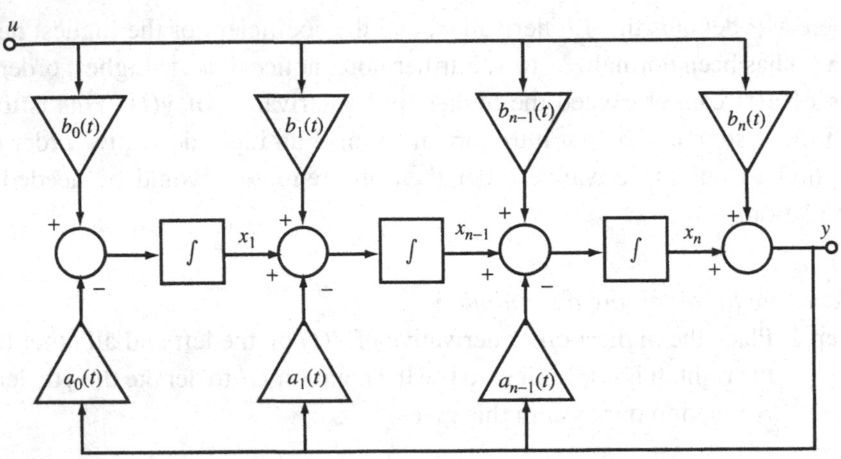

Fig. 2.9.

This gives

$$\begin{bmatrix} \dot{x}_1 \\ \dot{x}_2 \\ \dot{x}_3 \\ \vdots \\ \dot{x}_n \end{bmatrix} = \underbrace{\begin{bmatrix} 0 & 0 & \cdots & 0 & -a_0 \\ 1 & 0 & \cdots & 0 & -a_1 \\ 0 & 1 & 0 & \cdots & -a_2 \\ \vdots & \ddots & \ddots & \ddots & \vdots \\ 0 & \cdots & \cdots & 1 & -a_{n-1} \end{bmatrix}}_{A(t)} \begin{bmatrix} x_1 \\ x_2 \\ x_3 \\ \vdots \\ x_n \end{bmatrix} + \underbrace{\begin{bmatrix} b_0 - a_0 b_n \\ b_1 - a_1 b_n \\ b_2 - a_2 b_n \\ \vdots \\ b_{n-1} - a_{n-1} b_n \end{bmatrix}}_{B(t)} u$$

$$C = \begin{bmatrix} 0 & \cdots & 0 & 1 \end{bmatrix} \qquad D = b_n. \tag{2.26}$$

Note: If $b_n = 0$ then $D = 0$. This means that all forward paths from the input to output must pass through at least one integrator.

When the system is governed by a single difference equation,

$$\begin{aligned} &y(k+n) + \alpha_{n-1}(k)y(k+n-1) + \cdots + \alpha_1(k)y(k+1) + \alpha_0(k)y(k) \\ &\quad = \beta_n(k)u(k+n) + \beta_{n-1}(k)u(k+n-1) + \cdots + \beta_1(k)u(k+1) \\ &\quad + \beta_0(k)u(k). \end{aligned} \tag{2.27}$$

Using the notation $Ef(k) = f(k+1)$ we see that we can write the equation as

$$\begin{aligned} &E^n y(k) + \alpha_{n-1}(k)E^{n-1}y(k) + \cdots + \alpha_1(k)Ey(k) + \alpha_0(k)y(k) \\ &\quad = \beta_n(k)E^n u(k) + \beta_{n-1}(k)E^{n-1}u(k) + \cdots + \beta_1(k)Eu(k) + \beta_0(k)u(k). \end{aligned} \tag{2.28}$$

Note: The equation when written using the operator notation is seen to be identical to the differential equation except that E (the advance operator) has replaced D (the derivative operator). For constant coefficients α_i and β_i the identical procedure can be used except we end up with delay elements instead of integrators. For time-varying coefficients it is easier to apply the procedure to discrete systems because there is no integration by parts. Just use

$$E^{-1}\{a(k)Ef(k)\} = a(k-1)f(k) \quad \text{but} \quad D^{-1}\{a(t)Df(t)\} = a(t)f(t) - D^{-1}\{\dot{a}(t)f(t)\}.$$

Example 2.7

Consider

$$y^{(4)}(t) + 3ty^{(3)}(t) + 4y^{(2)}(t) + 2y^{(1)}(t) + \alpha(t)y(t) = u^{(2)}(t) + e^{-t}u^{(1)}(t) + u(t).$$

After Step 2 we have

$$y(t) = D^{-4}[-3tD^3y + D^2(u-4y) + e^{-t}Du - 2Dy + u - \alpha(t)y].$$

Integration by parts gives

$$D^{-3}[3tD^3y] = 3ty - D^{-1}(9y)$$

and

$$D^{-1}[e^{-t}Du] = e^{-t}u + D^{-1}(e^{-t}u)$$

so

$$y(t) = D^{-1}\{-3ty + D^{-1}[(u+5y) + D^{-1}(e^{-t}u - 2y) + D^{-1}(e^{-t}u + u - \alpha(t)y)]\}.$$

The simulation diagram is Figure 2.10 from which the state equations can be obtained easily.

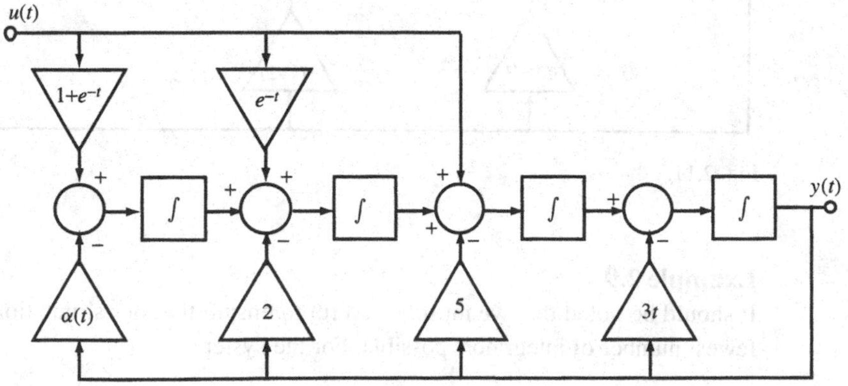

Fig. 2.10.

Example 2.8
For the difference equation

$$y(k+3) + 3ky(k+2) + e^{-k}y(k+1) + y(k) = k^2u(k+3) - (k+1)u(k+1)$$

after Step 2 we have

$$y(k) = E^{-3}\{k^2E^3u(k) - 3kE^2y(k) - (k+1)Eu(k) - e^{-k}Ey(k) - y(k)\}.$$

Now

$$\begin{aligned} E^{-3}\{k^2E^3u(k)\} &= (k-3)^2u(k)\\ E^{-2}\{kE^2y(k)\} &= (k-2)y(k)\\ E^{-1}\{(k+1)Eu(k)\} &= ku(k)\\ E^{-1}\{e^{-k}Ey(k)\} &= e^{-(k-1)}y(k)\end{aligned}$$

so

$$\begin{aligned} y(k) = (k-3)^2u(k) - E^{-1}\{&3(k-2)y(k) + E^{-1}[ku(k)\\ &+ e^{-(k-1)}y(k) + E^{-1}(y(k))]\}.\end{aligned}$$

The simulation diagram is shown in Figure 2.11, and the state equations can now be easily written.

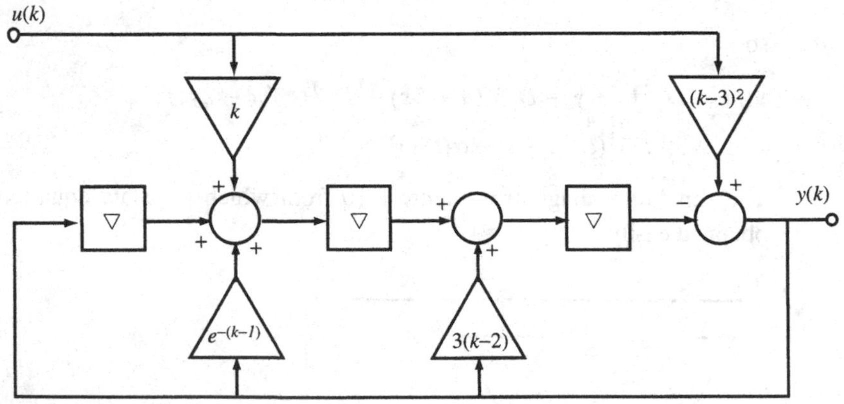

Fig. 2.11.

Example 2.9
It should be noted that we must be careful to ensure that our simulation uses the fewest number of integrators possible. For the system

$$\dddot{y} + \ddot{y} - \dot{y} - y = \ddot{u} - u,$$

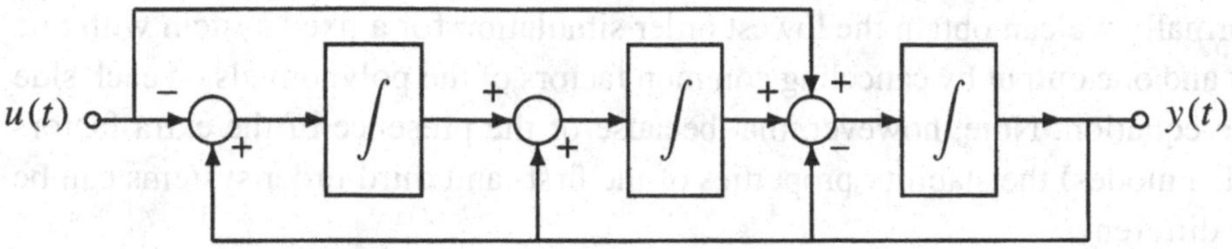

Fig. 2.12.

if we blindly follow the procedure outlined we get

$$
\begin{aligned}
y &= D^{-3}\{D^2u - D^2y + Dy + y - u\} \\
&= D^{-1}\{u - y + D^{-1}[y + D^{-1}(y - u)]\}.
\end{aligned}
$$

This has the simulation shown in Figure 2.12 and which uses 3 integrators. Note, however, that the first-order system

$$\dot{y} + y = u$$

satisfies the given equation, since if this equation is differentiated twice and its negative added to the result we obtain

$$\dddot{y} + \ddot{y} - \dot{y} - y = \ddot{u} - u.$$

Thus the first-order system of Figure 2.13 simulates the original equation. This could have been discovered by writing the original equation as

$$
\begin{aligned}
& D^3y + D^2y - Dy - y = D^2u - u \\
\text{or} \quad & (D^3 + D^2 - D - 1)y = (D^2 - 1)u \\
\text{or} \quad & (D^2 - 1)(D + 1)y = (D^2 - 1)u.
\end{aligned}
$$

We see that $(D^2 - 1)$ is a factor on both sides of the equation. Even though this is operator notation, it means that if the first-order equation $(D + 1)y = u$ is operated on by the operator $(D^2 - 1)$ (i.e., take 2 derivatives and then subtract the equation from the result) we can obtain the desired equation. Therefore the first-order equation satisfies the desired third-order equation, and the system can be simulated by the first-order equation.

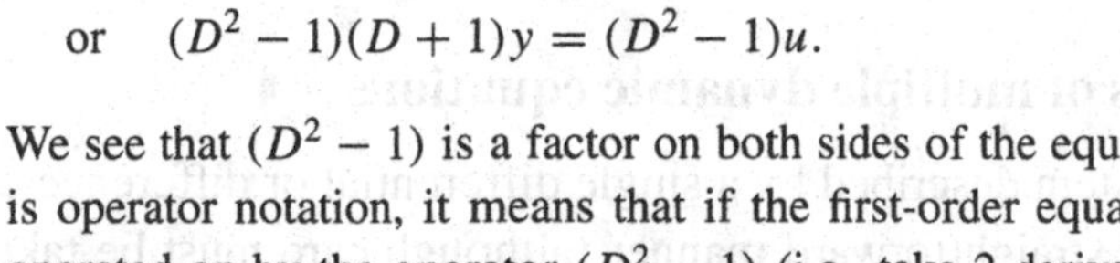

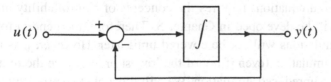

Fig. 2.13.

Formally we can obtain the lowest order simulation for a fixed system with one input and one output by canceling common factors of the polynomials on each side of the equation. Note, however, that because of the presence of the extra factors (hidden modes) the stability properties of the first- and third-order systems can be quite different.

Special case
We can obtain the state equations with considerably less work for the special case of a system governed by the nth-order equation

$$y^{(n)}(t) + \alpha_{n-1}(t)y^{(n-1)}(t) + \cdots + \alpha_1(t)y^{(1)}(t) + \alpha_0(t)y(t) = \beta_0(t)u(t) \tag{2.29}$$

which contains no input derivatives. Even though the coefficients are time-varying, we can avoid integration by parts in this case because we can choose the state variables to be $x_1(t) = y(t)$, $x_2(t) = \dot{y}(t)$, $x_2(t) = \ddot{y}(t)$, $x_n(t) = y^{(n-1)}(t)$. The state equations then become

$$\begin{aligned}
\dot{x}_1 &= x_2 \\
\dot{x}_2 &= x_3 \\
&\vdots \\
\dot{x}_{n-1} &= x_n \\
\dot{x}_n &= -a_0(t)x_1 - a_1(t)x_2 - \cdots - a_{n-1}(t)x_n + b_0(t)u \\
y &= x_1.
\end{aligned}$$

2.7 Systems of multiple dynamic equations

We showed how a system described by a single differential or difference equation can be simulated in a straightforward manner (although care must be taken if we want to ensure that it is a minimum-order simulation). When a system is described by a set of differential equations then a number of approaches can be used to obtain a simulation. None of them, however, gives a straightforward approach for obtaining the lowest order simulation. The problem of obtaining the lowest order simulation for a set of differential equations requires the concepts of controllability and observability which will be developed in Chapter 8. Therefore, the general topic of minimum-order simulations will not be covered until later. However, it is necessary to have some simulation (even if not of the lowest order) since the reduction method starts with an unreduced simulation. We will illustrate two procedures with a particular example.

Example 2.10

$$\ddot{y}_1 + \dot{y}_1 + \dot{y}_2 + y_2 = u$$
$$\dot{y}_1 + y_1 - \ddot{y}_2 - \dot{y}_2 = \ddot{u} - \dot{u} + u.$$

Procedure 1

Begin by placing the highest order derivative of a different output variable on the left in each of the equations

$$D^2 y_1 = -D(y_1 + y_2) + u - y_2$$
$$D^2 y_2 = -D^2 u + D(u + y_1 - y_2) + (y_1 - u).$$

The remaining steps are the same as for a single equation:

$$y_1 = D^{-1}\{-y_1 - y_2 + D^{-1}(u - y_2)\}$$
$$y_2 = -u + D^{-1}\{u + y_1 - y_2 + D^{-1}(y_1 - u)\}.$$

This is simulated as shown in Figure 2.14 which is a fourth-order system. Note that the first-order system

$$\dot{x} = -x + u$$
$$\begin{bmatrix} y_1 \\ y_2 \end{bmatrix} = \begin{bmatrix} 1 \\ 2 \end{bmatrix} x + \begin{bmatrix} 0 \\ -1 \end{bmatrix} u$$

satisfies the original equations:

First equation $\quad \ddot{y}_1 + \dot{y}_1 + \dot{y}_2 + y_2 = \ddot{x} + \dot{x} + 2\dot{x} - \dot{u} + 2x - u = u.$

Second equation $\quad \dot{y}_1 + y_1 - \ddot{y}_2 - \dot{y}_2 = \dot{x} + x - 2\ddot{x} + \ddot{u} - 2\dot{x} + \dot{u}$
$$= \ddot{u} - \dot{u} + u.$$

So the equations can actually be represented by a first-order system.

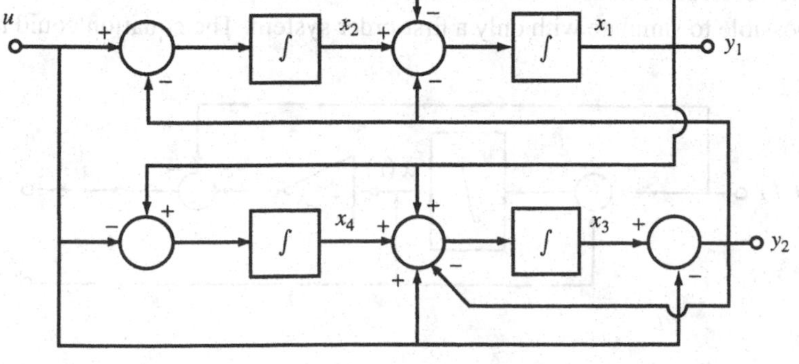

Fig. 2.14.

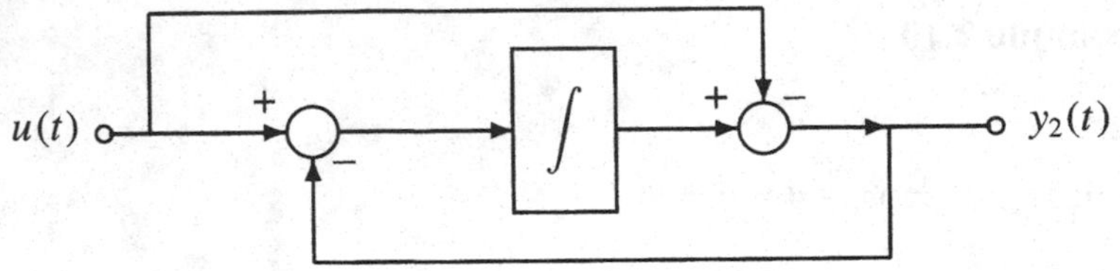

Fig. 2.15.

Procedure 2
We can frequently get a clue to the order by trying to write a single equation in terms of one of the output variables. In Example 2.10 we write

$$D^2y_1 + Dy_1 + Dy_2 + y_2 = u$$
$$Dy_1 + y_1 - D^2y_2 - Dy_2 = D^2u - Du + u.$$

Differentiate the first equation and add to the second to get $D^3y_1 + D^2y_1 + Dy_1 + y_1 = D^2u + u$. This can be expressed as

$$(D^2 + 1)(D + 1)y_1 = (D^2 + 1)u. \tag{2.30}$$

This shows that when y_1 satisfies $Dy_1 + y_1 = u$ it also satisfies (2.18), and we conclude that y_1 can be simulated by the system shown in Figure 2.13.

If we do the same for y_2 (differentiate the second equation and subtract from the first) we get $D^3y_2 + D^2y_2 + Dy_2 + y_2 = -D^3u + D^2u - Du + u$, which can be written as

$$(D^2 + 1)(D + 1)y_2 = (D^2 + 1)(-D + 1)u. \tag{2.31}$$

Thus if y_2 satisfies $Dy_2 + y_2 = -Du + u$ it also satisfies (2.31). We can then write $y_2 = -u + D^{-1}(u - y_2)$. We conclude that y_2 can be simulated as shown in Figure 2.15.

We see that the equation can certainly be simulated by a second-order system. However, since in this case both characteristic roots are the same, it turns out to be possible to simulate with only a first-order system. The equation could have been

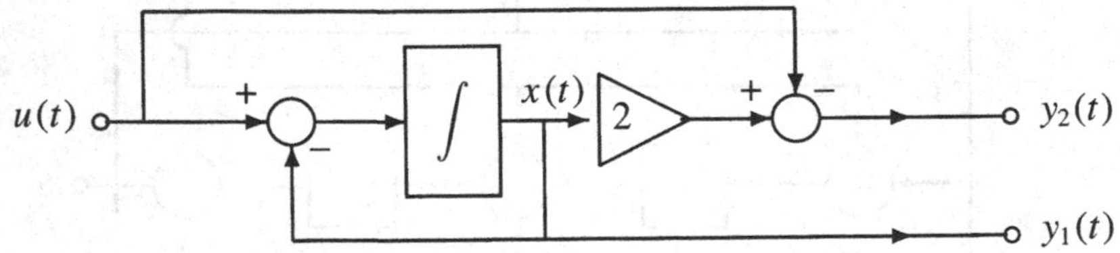

Fig. 2.16.

written as

$$y_2 = -u + 2D^{-1}(u/2 - y_2/2)$$

$$\text{or} \quad y_2 + u = 2D^{-1}\left[u - \frac{y_2 + u}{2}\right].$$

By defining $x = (y_2 + u)/2 = y_1$ we see that $x = D^{-1}(u - x)$ which has the simulation shown in Figure 2.16.

Problems

2.1 Write the state differential equations and output equations for each of the following circuits. Be sure to indicate the A, B, C, and D matrices.

a)

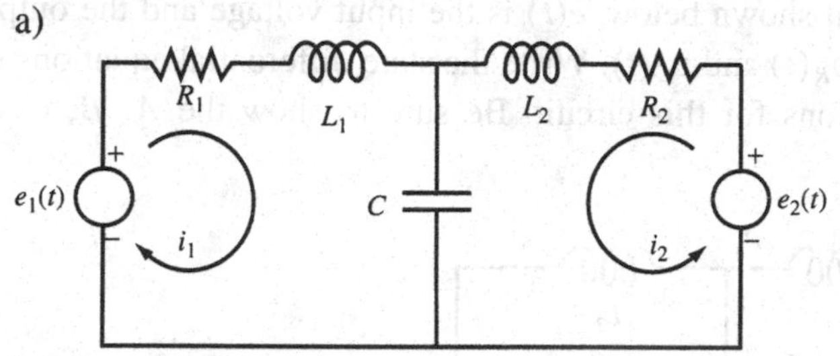

i_1 and i_2 are outputs.

b)

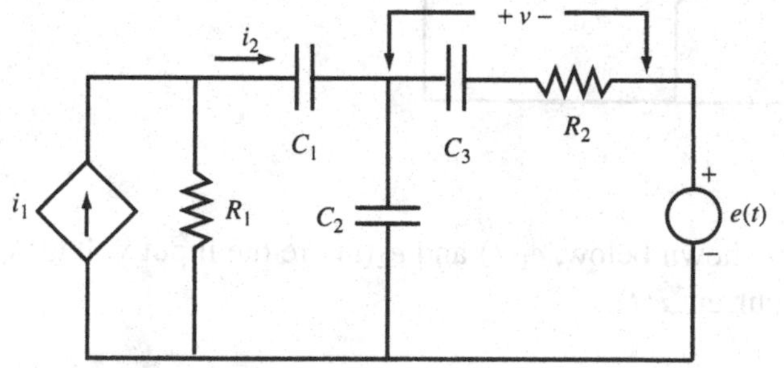

v and i_2 are outputs.

c)

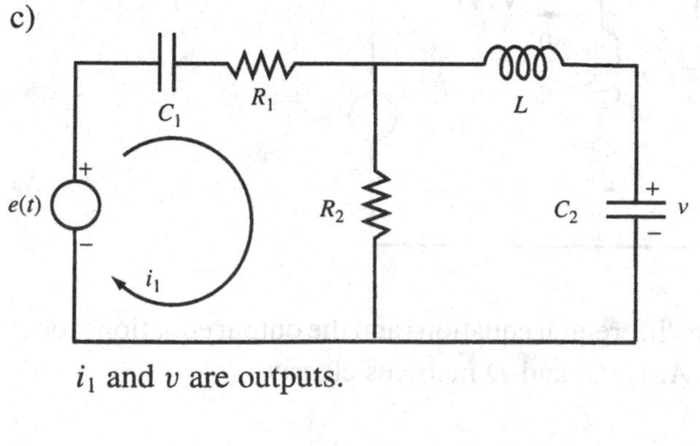

i_1 and v are outputs.

2.2 Write the state differential equations for the circuit below.

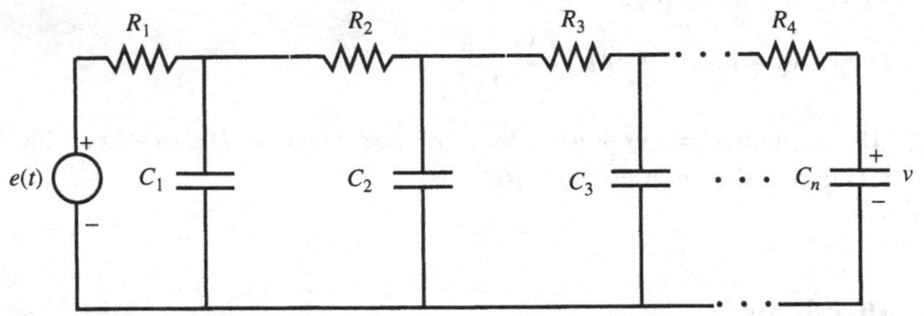

2.3 For the circuit shown below, $e(t)$ is the input voltage and the outputs are the voltages $v_R(t)$ and $v_C(t)$. Write the state differential equations and the output equations for this circuit. Be sure to show the A, B, C, and D matrices.

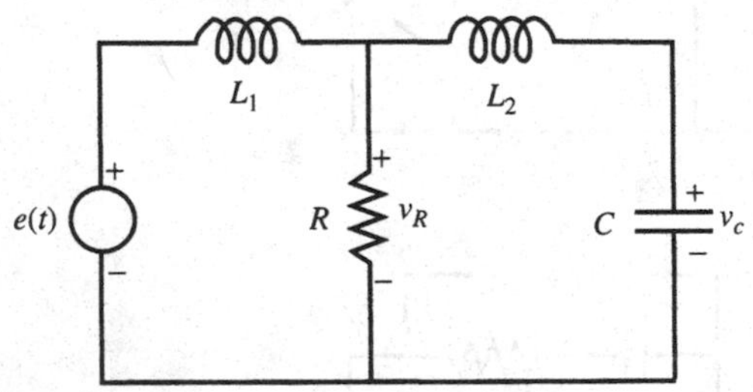

2.4 For the circuit shown below, $e_1(t)$ and $e_2(t)$ are the input voltages and the output is the current $i(t)$.

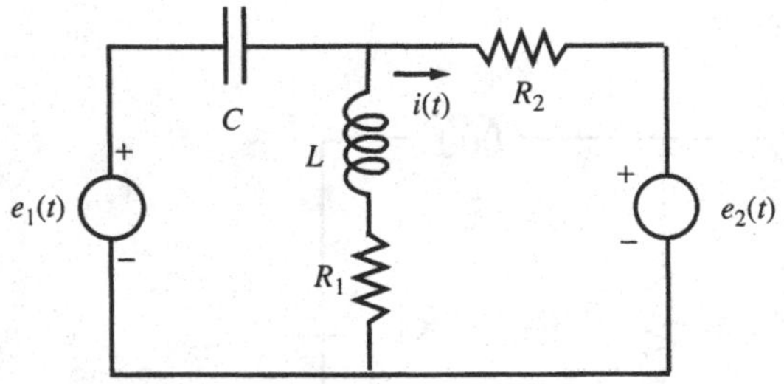

Write the state differential equations and the output equations for this circuit and show the A, B, C, and D matrices clearly.

2.5 In the circuit shown below, the inputs are the applied voltages $e_1(t)$ and $e_2(t)$. The output is the voltage $v(t)$.

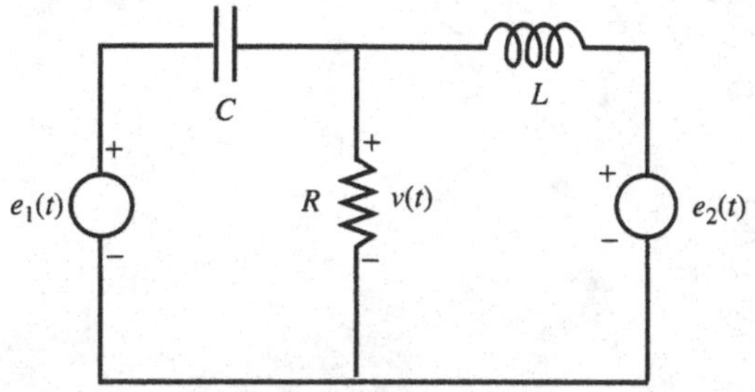

Write the state equations and output equations for this circuit.

2.6 For the circuit shown below, write the state differential equations and the output equations and show the A, B, C, and D matrices. As shown, $e(t)$ is

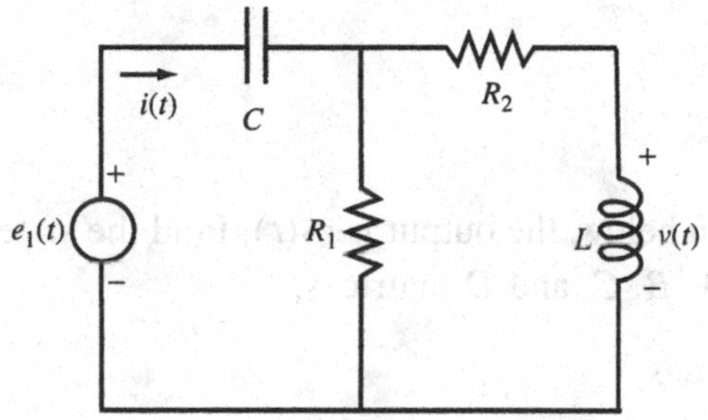

the input voltage, and the outputs are the voltage $v(t)$ and the current $i(t)$.

2.7 The equivalent circuit of a three-stage amplifier is shown below where i_4 is the output.

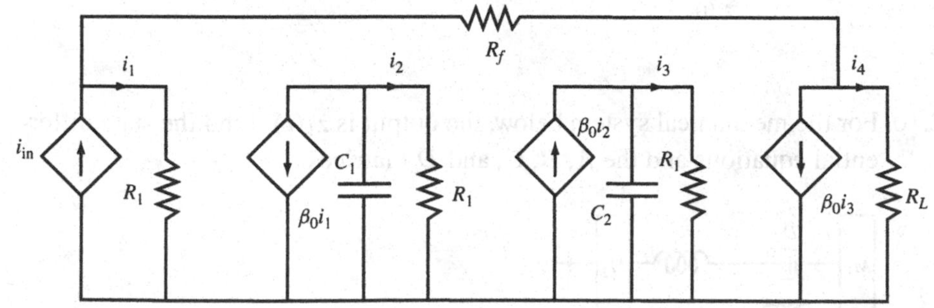

Write the state differential equations and the A, B, C, and D matrices.

2.8 Find the state differential equations and the A, B, C, and D matrices for each of the mechanical systems in Fig. P 2.8. The input is $f(t)$ and the output is $z_1(t)$ for both systems.

a)

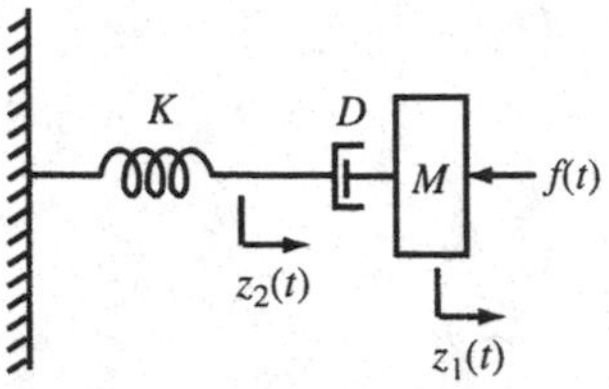

b)

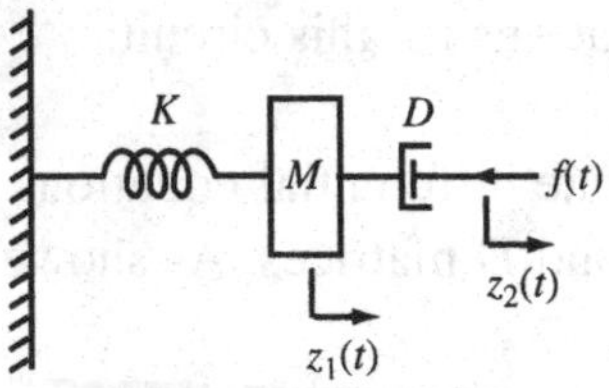

Fig. P 2.8.

2.9 For the mechanical system below, the output is $z_2(t)$. Find the state differential equations and the A, B, C, and D matrices.

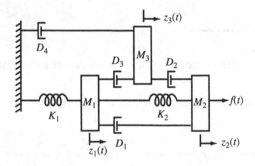

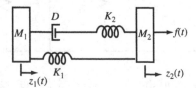

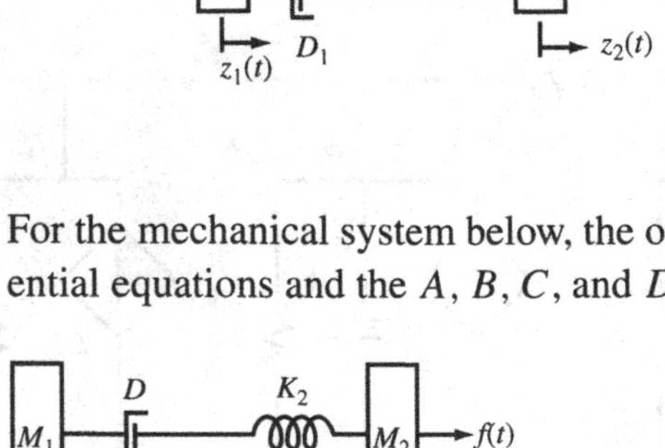

2.10 For the mechanical system below, the output is $z_1(t)$. Find the state differential equations and the A, B, C, and D matrices.

2.11 Find the simulation diagram and the A, B, C, and D matrices for the analog systems:

a) $\dddot{y}(t) + 2\ddot{y}(t) - \dot{y}(t) + 3y(t) = 4\dot{u}(t) + u(t)$
b) $5t^2\ddot{y}(t) + (t-1)\dot{y}(t) + t^2 y(t) = \ddot{u}(t) + tu(t)$.

2.12 For the system described by the differential equation

$$\ddot{y}(t) + t\dot{y}(t) + 3ty(t) = 2\ddot{u}(t) + 2t\dot{u}(t) - t^2u(t)$$

find the state equations showing the A, B, C, and D matrices.

2.13 Write the state differential equations and show the A, B, C, and D matrices for the system

$$\ddot{y}(t) + 2e^{-2t}\dot{y}(t) + t^2y(t) = 2\ddot{u}(t) + u(t).$$

2.14 Write the state differential equations and the A, B, C, and D matrices for the system described by

$$\ddot{y}(t) + 2\cos(t)\dot{y}(t) + ty(t) = t\ddot{u}(t) + e^{-t}\dot{u}(t) - u(t).$$

2.15 Write the state differential equations showing the A, B, C, and D matrices for each of the systems

a) $\ddot{y}(t) + \frac{2}{t+2}\dot{y}(t) + y(t) = u(t)$
b) $\ddot{y}(t) + \frac{2}{t+2}\dot{y}(t) + y(t) = 2\ddot{u}(t) + u(t)$.

2.16 Draw the simulation diagram and show the A, B, C, and D matrices for the discrete systems

a) $y(k+1) + 2y(k) + y(k-1) = u(k+1) + u(k)$
b) $y(k+1) + k^2y(k) + ky(k-1) = u(k+1) - e^{-k}u(k-1)$.

2.17 Find the A, B, C, and D matrices of the system described by

$$ky(k+2) - 6k^2y(k+1) + 11y(k) - 6ky(k-1) = u(k+1) - 4k^2u(k-1).$$

2.18 Find the A, B, C, and D matrices of the state equations for the system

$$y(k) + k^2y(k-1) + ky(k-2) = u(k) + k^2u(k-1) + ku(k-2).$$

2.19 Write the state difference equations and the A, B, C, and D matrices for the discrete system described by

$$y(k+2) + k^2y(k) = u(k+1) + 2ku(k-1).$$

2.20 Write the state difference equations and the A, B, C, and D matrices for the system

$$\frac{1}{k+1}y(k+2) + k^2 y(k) = u(k+2) + ku(k+1) + 2u(k).$$

2.21 Find the simulation diagram and the A, B, C, and D matrices of the system described by the coupled equations

$$e^{-t}\dot{y}_1(t) + y_1(t) + \ddot{y}_2(t) + y_2(t) = tu(t)$$
$$\dot{y}_1(t) + \dot{y}_2(t) + ty_2(t) = \dot{u}(t) + t^2 u(t).$$

2.22 Find the state equations and the A, B, C, and D matrices for the analog systems

a) $\ddot{y}_1(t) + 4\dot{y}_1(t) + 3y_2(t) = u_1(t) + u_2(t)$
$\dot{y}_1(t) + y_1(t) + \ddot{y}_2(t) + 5\dot{y}_2(t) = u_2(t)$

b) $\dot{y}_1(t) + t\dot{y}_2(t) = u(t)$; $\quad \ddot{y}_1(t) + y_2(t) = tu(t)$.

2.23 Find the A, B, C, and D matrices for the discrete systems

a) $y_1(k) + y_2(k+2) = u_1(k)$
$y_1(k+1) + 3y_1(k) - 2y_2(k) = u_2(k)$

b) $y_1(k+1) + ky_2(k+1) = u(k)$
$y_1(k+2) + y_2(k+1) = ku(k)$.

2.24 A discrete system is described by the equations

$$ky_1(k+1) + e^{-k}y_2(k+1) = u(k)$$
$$y_1(k+1) + ky_2(k) = k^2 u(k).$$

Write the state equations of this system and show the A, B, C, and D matrices.

2.25 Write the state equations for the system governed by

$$y_1(k+1) + ky_2(k+1) = u(k)$$
$$y_1(k+2) + y_2(k) = ku(k).$$

2.26 Show that the system governed by

$$y_1(k+1) + y_2(k+1) = u(k)$$
$$y_1(k+2) + y_2(k) = u(k)$$

can be described by second-order state equations. Write the A, B, C, and D matrices for this second-order realization.

3

Principles of time-domain analysis

3.1 Introduction

As we indicated earlier, in the indirect approach to the analysis of linear systems, when we seek the response of some linear system S to a general input function $u(t)$, we express $u(t)$ as a linear combination of elementary functions, we find the response of the system to each elementary function, and we then express the response as a linear combination of these elementary responses. The utility of this approach lies in the fact that the responses to each elementary function are related in a simple fashion, so, knowing the response to one typical elementary function, we can express the response to any other very easily.

3.2 Elementary analog signals

The basic elementary functions used in time-domain analysis of analog systems are defined by

$$\mu_{i+1}(t-\tau) = \int_{-\infty}^{t} \mu_i(\lambda - \tau)\, d\lambda \quad \text{for } i = 0, 1, 2, \ldots. \tag{3.1}$$

These are the so-called singularity functions, the two most important of which are the unit impulse $\underset{\sim}{\delta}_\tau = \{\delta(t-\tau)\} \triangleq \{\mu_0(t-\tau)\}$ and the unit-step function

$$\underset{\sim}{1}_\tau = \{1(t-\tau)\} \triangleq \{\mu_1(t-\tau)\} \quad \text{where } 1(t) = \begin{cases} 1 & \text{for } t > 0 \\ 0 & \text{for } t < 0. \end{cases} \tag{3.2}$$

The entire family can thus be generated by integrating the member with the next lowest index when the unit impulse $\underset{\sim}{\mu}_0 = \underset{\sim}{\delta}$ is defined.

3.3 Impulses and the impulse response

The unit impulse or δ-function can be defined by the two properties

$$\int_{-\infty}^{t} f(\lambda)\delta(\lambda-\tau)\,d\lambda = \begin{cases} 0 & \text{for } t<\tau \\ f(\tau) & \text{for } t>\tau \end{cases} \tag{3.3}$$

and $\qquad \delta(t)=\delta(-t) \qquad$ for all t.

From this, the usual properties are obtained:

$$1 \quad \int_{-\infty}^{t} \delta(\lambda-\tau)\,d\lambda = \left\{\begin{matrix} 0 & \text{for } t<\tau \\ 1 & \text{for } t>\tau \end{matrix}\right\} \triangleq 1(t-\tau) \tag{3.4}$$

$$2 \quad \int_{\tau-}^{\tau+} \delta(\lambda-\tau)\,d\lambda = 1\text{: the area under the impulse is 1}$$

$$3 \quad f(\tau) = \int_{-\infty}^{\infty} f(\lambda)\delta(\tau-\lambda)\,d\lambda\text{: the sifting property.} \tag{3.5}$$

Property 3 can be viewed as expressing a general function $\underset{\sim}{f}$ as a linear combination of functions $\underset{\sim}{\delta}_\lambda$ which occur at different time instants λ.

As we showed earlier, the zero-state response to any input function $u(t)$ is thus

$$\begin{aligned} y(t) = S\{u(t)\} &= S\left\{\int_{-\infty}^{\infty} u(\lambda)\delta(t-\lambda)\,d\lambda\right\} \\ &= \int_{-\infty}^{\infty} u(\lambda)S\{\delta(t-\lambda)\}\,d\lambda = \int_{-\infty}^{\infty} u(\lambda)h(t,\lambda)\,d\lambda \end{aligned} \tag{3.6}$$

where $h(t,\lambda) \triangleq$ the zero-state response of the system observed at time t to a unit impulse that occurs at time λ. This is called the impulse response function.

3.4 Step function and step response

An arbitrary signal can also be considered to be composed of unit-step functions. Observe that

$$f(t) = f(t_0) + \int_{t_0}^{t} \dot{f}(\lambda)\,d\lambda \qquad \text{for } t \geq t_0 \qquad \text{for any function } \underset{\sim}{f}$$

so

$$f(t) = f(t_0)1(t-t_0) + \int_{t_0}^{\infty} \dot{f}(\lambda)1(t-\lambda)\,d\lambda \qquad \text{for } t \geq t_0. \tag{3.7}$$

This expression gives any $\{f(t)\}$ as a linear combination of unit-step functions. Just as in the case of resolution into impulses, the response to an arbitrary input can be written as

$$y(t) = S\{u(t)\} = u(t_0)S\{1(t-t_0)\} + \int_{t_0}^{\infty} \dot{u}(\lambda)S\{1(t-\lambda)\}\,d\lambda. \tag{3.8}$$

Thus, for $t \geq t_0$,

$$y(t) = u(t_0)a(t, t_0) + \int_{t_0}^{\infty} \dot{u}(\lambda)a(t, \lambda)\, d\lambda \tag{3.9}$$

where $a(t, \tau) \triangleq$ the zero-state response observed at time t to a unit step that starts at time τ. This is called the unit-step response function or just the step response.

3.5 Relation between the step and impulse response

Since

$$a(t, \tau) = S\{1(t - \tau)\} = \int_{-\infty}^{\infty} 1(\lambda - \tau)h(t, \lambda)\, d\lambda = \int_{\tau}^{\infty} h(t, \lambda)\, d\lambda \tag{3.10}$$

which shows that the step response is the integral on the second variable of the impulse response, or by taking the partial with respect to τ

$$\frac{\partial a(t, \tau)}{\partial \tau} = \frac{\partial}{\partial \tau}\left[-\int_{\infty}^{\tau} h(t, \eta)\, d\eta\right] = -h(t, \tau) \tag{3.11}$$

the impulse response is thus the negative derivative of the step response with respect to the second variable.

Similarly, we can relate the responses of a linear system to any of the singularity functions $\{\mu_i(t)\}$. Since

$$\mu_{i+1}(t - \tau) = \int_{-\infty}^{t} \mu_i(\lambda - \tau)\, d\lambda \tag{3.12}$$

then

$$S\{\mu_{i+1}(t - \tau)\} = S\left\{\int_{-\infty}^{t} \mu_i(\lambda - \tau)\, d\lambda\right\}. \tag{3.13}$$

Let us change variables to $\xi = \lambda - t$ in order to get fixed limits. Then for

$$m_i(t, \tau) \triangleq S\{\mu_i(t - \tau)\}$$

$$m_{i+1}(t, \tau) = S\{\mu_{i+1}(t - \tau)\} = \int_{-\infty}^{0} S\{\mu_i(\xi + t - \tau)\}\, d\xi$$

$$= \int_{-\infty}^{0} m_i(t, \tau - \xi)\, d\xi$$

and for $\eta = \tau - \xi$

$$m_{i+1}(t, \tau) = \int_{\tau}^{\infty} m_i(t, \eta)\, d\eta \tag{3.14}$$

which is the integral over the second variable of $m_i(t, \eta)$ or

$$m_i(t, \tau) = -\frac{\partial m_{i+1}(t, \tau)}{\partial \tau}. \tag{3.15}$$

Now, if a system is causal then

$$h(t, \tau) = 0 \quad \text{for } t < \tau \quad \text{and} \quad a(t, \tau) = 0 \quad \text{for } t < \tau \tag{3.16}$$

since the response to an impulse or step input cannot occur before the input is applied.

If a system is fixed then

$$h(t, \tau) = h(t + T, \tau + T) \quad \text{and} \quad a(t, \tau) = a(t + T, \tau + T) \quad \text{for any } T \tag{3.17}$$

since the response observed at t to an input applied at τ is identical to that observed at $t + T$ to the same input applied at $\tau + T$. In particular, for $T = -\tau$,

$$h(t, \tau) = h(t - \tau, 0) \quad \text{and} \quad a(t, \tau) = a(t - \tau, 0). \tag{3.18}$$

In other words, the responses depend only on the time elapsed between application of input and observation of response. With only a minor abuse of our notation, we will write

$$\left.\begin{array}{l} h(t-\tau) \text{ to mean } h(t-\tau, 0) \\ a(t-\tau) \text{ to mean } a(t-\tau, 0) \end{array}\right\} \text{this is standard usage.}$$

Thus for causal systems, the response to an arbitrary input $u(t)$ has the form

$$y(t) = \int_{-\infty}^{t} u(\lambda) h(t, \lambda)\, d\lambda \tag{3.19}$$

or

$$y(t) = u(t_0) a(t, t_0) + \int_{t_0}^{t} \dot{u}(\lambda) a(t, \lambda)\, d\lambda \tag{3.20}$$

and

$$a(t, \tau) = \int_{\tau}^{t} h(t, \eta)\, d\eta. \tag{3.21}$$

Note: If we consider inputs that are initially zero (i.e., $\lim_{t_0 \to -\infty} u(t_0) = 0$), then the superposition integral for step responses looks like

$$y(t) = \int_{-\infty}^{\infty} \dot{u}(\lambda) a(t, \lambda)\, d\lambda \tag{3.22}$$

which is similar to the form for impulse responses.

For fixed systems

$$y(t) = \int_{-\infty}^{\infty} u(\lambda)h(t-\lambda)\,d\lambda = \int_{-\infty}^{\infty} u(t-\xi)h(\xi)\,d\xi \tag{3.23}$$

or

$$y(t) = \int_{-\infty}^{\infty} \dot{u}(\lambda)a(t-\lambda)\,d\lambda = \int_{-\infty}^{\infty} \dot{u}(t-\xi)a(\xi)\,d\xi \tag{3.24}$$

and

$$a(t-\tau) = \int_{\tau}^{\infty} h(t-\lambda)\,d\lambda. \tag{3.25}$$

For $\tau = 0$ (3.26) becomes

$$a(t) = \int_{0}^{\infty} h(t-\lambda)\,d\lambda = \int_{-\infty}^{t} h(\eta)\,d\eta \tag{3.26}$$

(i.e., the step response is the integral of the impulse response) and equivalently

$$h(t) = \frac{da(t)}{dt}. \tag{3.27}$$

If the input starts at a time t_0 (i.e., $u(t) = 0$ for $t \leq t_0$), then

$$y(t) = \int_{t_0}^{\infty} u(\lambda)h(t,\lambda)\,d\lambda = \int_{t_0}^{\infty} \dot{u}(\lambda)a(t,\lambda)\,d\lambda. \tag{3.28}$$

Note: If $u(t)$ is discontinuous at t_0, we will always use the convention that integration starts from t_0^- and the derivative of the step discontinuity at t_0 (i.e., the δ-function at t_0) is included in $\dot{u}(\lambda)$.

The form of the superposition integral, usually written as

$$y(t) = \int_{0}^{t} u(\lambda)h(t-\lambda)\,d\lambda \quad \text{or} \quad y(t) = \int_{0}^{t} \dot{u}(\lambda)a(t-\lambda)\,d\lambda \tag{3.29}$$

called the convolution integral, is valid only for the very special situation of a fixed, causal system for which the input starts at time zero.

3.6 Graphical interpretation of the convolution integral

Consider a general form of the convolution integral

$$f(t) = \int_{a}^{b} f_1(\lambda) f_2(t-\lambda)\,d\lambda.$$

Suppose $\{f_1(\lambda)\}$ and $\{f_2(\lambda)\}$ are as shown in the figure.

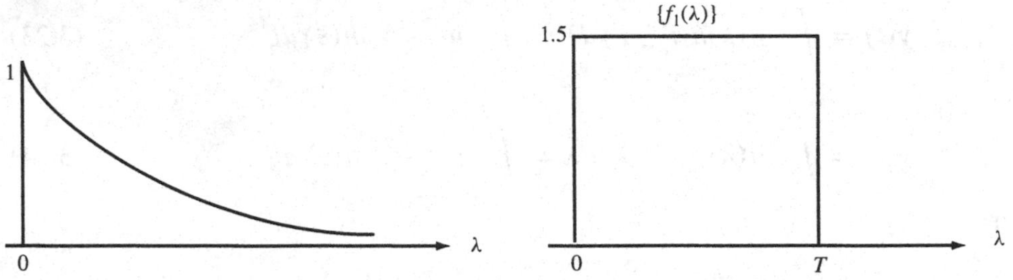

Then if we plot $\{f_2(-\lambda)\}$ against λ, this looks like

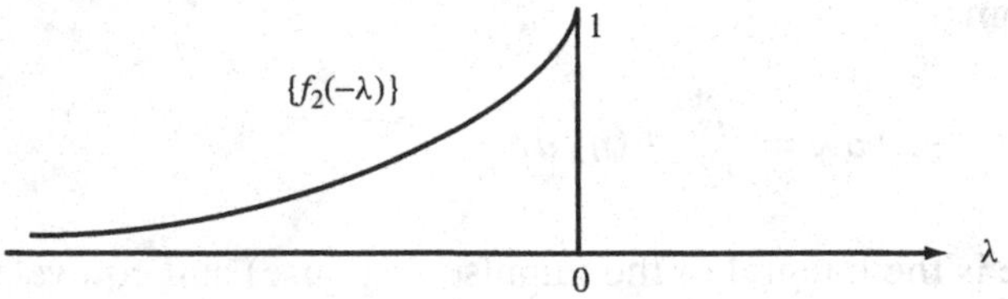

Now, shift by an amount t_1, so $\{f_2(t_1 - \lambda)\}$ is

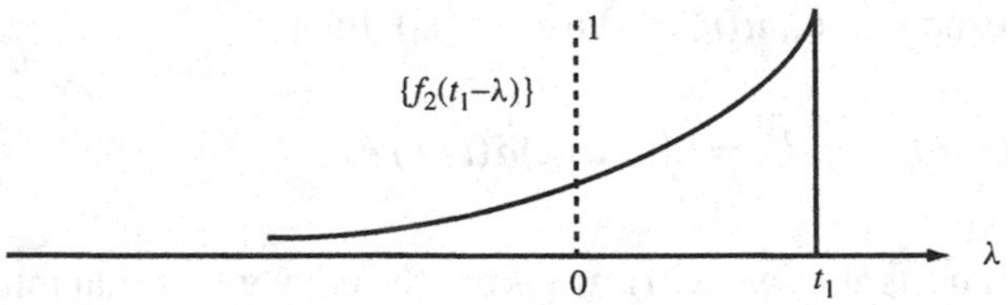

and the product function $\{f_1(\lambda)\ f_2(t_1 - \lambda)\}$ versus λ is

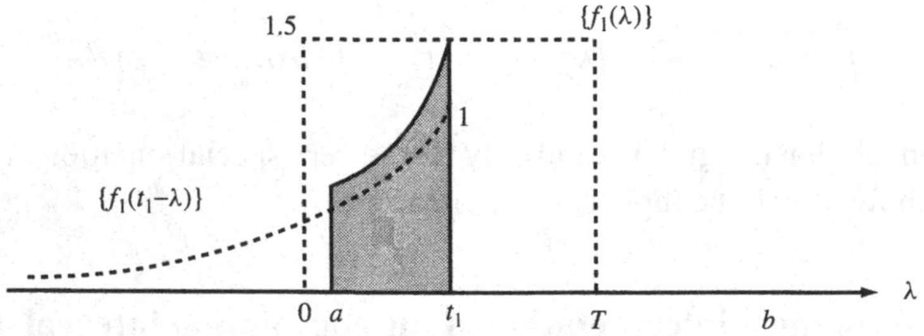

and the function $\{f(t)\}$ evaluated at t_1, (i.e., $f(t_1)$) is the area under this curve between a and b. To get $f(t)$ at another time t_2, we would shift $f_2(-\lambda)$ by t_2, take the product, and get the area under the curve between a and b. This process is repeated for all t.

Of course, this is just a graphical interpretation of the process of convolution. To evaluate, we actually integrate the functions inside the integral.

Example 3.1

A fixed linear system has the step response

$$a(t) = [1 - e^{-2t}]\,1(t).$$

Find the zero-state response to the input $u(t) = e^{-t}\,1(t)$.

Since neither the input nor the step response can be resolved into a finite sum of singularity functions, we use the superposition integral. The output is

$$y(t) = \int_{-\infty}^{\infty} h(t-\tau)u(\tau)\,d\tau$$

where

$$h(t) = \frac{da(t)}{dt} = 2e^{-2t}\,1(t) + [1 - e^{-2t}]\delta(t) = 2e^{-2t}\,1(t)$$

because $f(t)\delta(t) = f(0)\delta(t)$. Therefore

$$\begin{aligned} y(t) &= 2\int_{-\infty}^{\infty} e^{-2(t-\tau)}1(t-\tau)e^{-\tau}1(\tau)\,d\tau \\ &= 2e^{-2t}\underbrace{\int_0^t e^{\tau}1(t-\tau)1(\tau)\,d\tau}_{\text{if } t<0 \text{ this is zero}} = 2\int_0^t e^{\tau}\,d\tau e^{-2t}1(t) \\ &= 2e^{-2t}[e^t - 1]1(t) = 2[e^{-t} - e^{-2t}]1(t). \end{aligned}$$

In terms of the step response,

$$y(t) = u(t_0)a(t-t_0) + \int_{t_0}^{\infty} \dot{u}(\tau)a(t-\tau)\,d\tau$$

with

$$\dot{u}(t) = -e^{-t}1(t) + \delta(t).$$

Here $t_0 = 0$, so if we use $t_0^- = 0^-$

$$\begin{aligned} y(t) &= \int_{0^-}^{\infty} [\delta(\tau) - e^{-\tau}1(\tau)]\left[1 - e^{-2(t-\tau)}\right]1(t-\tau)\,d\tau \\ &= \int_{0^-}^{\infty} \delta(\tau)\left[1 - e^{-2(t-\tau)}\right]1(t-\tau)\,d\tau - \int_{0^-}^{t} e^{-\tau}\left(1 - e^{-2(t-\tau)}\right)1(t-\tau)\,d\tau \\ &= \{[1 - e^{-2t}] + [e^{-t} + e^{-t} - 1 - e^{-2t}]\}1(t) \\ &= 2[e^{-t} - e^{-2t}]1(t). \end{aligned}$$

Example 3.2

Suppose we want to find the zero-state response of the system of Example 3.1 to the input $u(t) = 5\ 1(t) - 5\ 1(t-1)$. Since the system is linear,

$$S\{5\ 1(t) - 5\ 1(t-1)\} = 5S\{1(t)\} - 5S\{1(t-1)\}$$

and since it is fixed,

$$S\{1(t-1)\} = \left[1 - e^{-2(t-1)}\right] 1(t-1).$$

Thus the response to $u(t)$ is

$$S\{u(t)\} = 5[1 - e^{-2t}]1(t) - 5\left[1 - e^{-2(t-1)}\right] 1(t-1).$$

To find the response to an input $\hat{u}(t) = t\ 1(t)$ (i.e., $\underset{\sim}{\hat{u}} = \underset{\sim}{\mu_2}$) we use

$$m_2(t) = \int_{-\infty}^{t} a(\tau)\, d\tau$$

which, for fixed systems, is obtained from (3.14) exactly as (3.26) was obtained from (3.10). Thus

$$m_2(t) = \int_{-\infty}^{t} \left[1 - e^{-2\tau}\right] 1(\tau)\, d\tau = \int_{-\infty}^{t} \left[1 - e^{-2\tau}\right] d\tau\ 1(t)$$
$$= \left[t - \frac{1}{2} + \frac{e^{-2t}}{2}\right] 1(t).$$

Example 3.3

The step response of a fixed linear system is shown in Figure 3.1. Find the response to $u(t) = t\,1(t-1) - (t-2)\,1(t-2)$. There are a number of approaches, but the easiest is to decompose $u(t)$ into the singularity functions. That is, express $u(t)$ as

$$u(t) = (t-1)\,1(t-1) + 1(t-1) - (t-2)1(t-2)$$
$$= \mu_2(t-1) + \mu_1(t-1) - \mu_2(t-2).$$

From the given sketch of $a(t)$, we can write

$$a(t) = \mu_2(t-1) - 2\mu_2(t-2) + \mu_2(t-3).$$

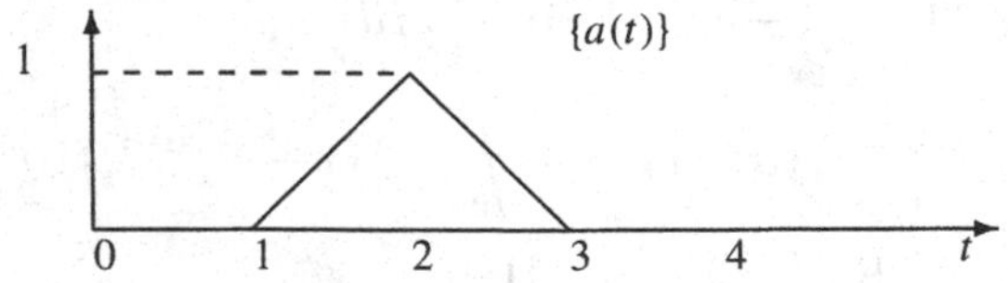

Fig. 3.1.

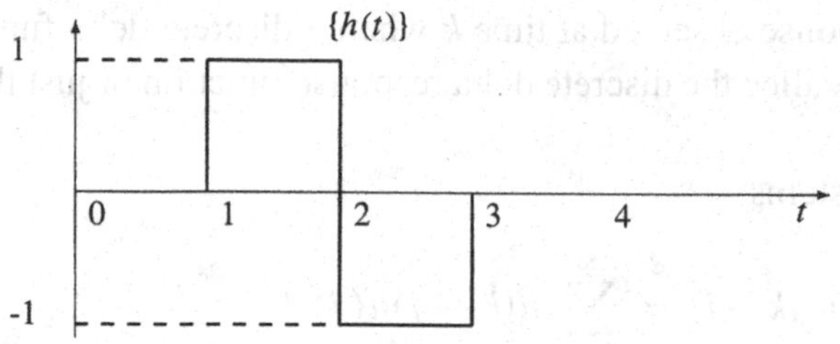

Fig. 3.2.

Now, $a(t) = S\{\mu_1(t)\}$, and we know that

$$m_2(t) = S\{\mu_2(t)\} = \int_{-\infty}^{t} a(\tau)\, d\tau = \mu_3(t-1) - 2\mu_3(t-2) + \mu_3(t-3)$$

where $\mu_3(t) = (t^2/2)\, 1(t)$. Since the system is linear and fixed,

$$\begin{aligned} y(t) = S\{u(t)\} &= S\{\mu_2(t-1)\} + S\{\mu_1(t-1)\} - S\{\mu_2(t-2)\} \\ &= m_2(t-1) + a(t-1) - m_2(t-2) \end{aligned}$$

where $a(t)$ and $m_2(t)$ are given above.

If we wanted to find $h(t)$ (see Figure 3.2), then

$$\begin{aligned} h(t) = \frac{da(t)}{dt} &= \mu_1(t-1) - 2\mu_1(t-2) + \mu_1(t-3) \\ &= 1(t-1) - 2\,1(t-2) + 1(t-3). \end{aligned}$$

3.7 The discrete delta and delta response

For discrete-time systems, the elementary function of major interest is the discrete delta function or Kronecker delta:

$$\delta(k - k_0) = \delta_{kk_0} = \begin{cases} 0 & \text{for } k \neq k_0 \\ 1 & \text{for } k = k_0. \end{cases} \tag{3.30}$$

Any discrete function can be written as

$$\underset{\sim}{f} = \{f(k)\} = \sum_{i=-\infty}^{\infty} f(i)\{\delta(k-i)\}. \tag{3.31}$$

Proceeding as in the continuous-time case,

$$\begin{aligned} y(k) &= S\left\{\sum_{i=-\infty}^{\infty} u(i)\delta(k-i)\right\} \\ &= \sum_{i=-\infty}^{\infty} u(i)S\{\delta(k-i\} = \sum_{i=-\infty}^{\infty} u(i)h(k,i) \end{aligned} \tag{3.32}$$

where $h(k, i)$ = the response observed at time k when a discrete delta function is applied at time i. This is called the discrete delta response function or just the delta response.

As before for fixed systems

$$y(k) = \sum_{i=-\infty}^{\infty} u(i)h(k-i) = \sum_{j=-\infty}^{\infty} u(k-j)h(j) \tag{3.33}$$

for causal systems

$$y(k) = \sum_{i=-\infty}^{\infty} u(i)h(k,i) \tag{3.34}$$

and if the input starts at time k_0 then

$$y(k) = \sum_{i=k_0}^{\infty} u(i)h(k,i). \tag{3.35}$$

It is of interest to note that for discrete systems, we can express the discrete delta function $\{\delta(k)\}$ as

$$\{\delta(k)\} = \{1(k)\} - \{1(k-1)\}.$$

Thus, $S\{\delta(k-i)\} = S\{1(k-i)\} - S\{(1(k-i-1)\}$. So by the linearity property,

$$h(k,i) = a(k,i) - a(k,i+1) \tag{3.36}$$

for discrete-time systems. That is to say, we can find the delta response from the discrete step response. Analogously, since

$$1(k) = \sum_{j=0}^{\infty} \delta(k-j)$$

we find that

$$\begin{aligned} a(k,i) = S\{1(k-i)\} &= S\left\{\sum_{j=0}^{\infty} \delta(k-i-j)\right\} \\ &= \sum_{j=0}^{\infty} S\{\delta(k-i-j)\} = \sum_{j=0}^{\infty} h(k,i+j) \end{aligned} \tag{3.37}$$

and for causal systems

$$a(k,i) = \sum_{j=0}^{k-i} h(k,i+j) = \sum_{\ell=i}^{k} h(k,\ell)$$

so the discrete step response can be found from the discrete delta response.
Note: It is not always easy to find a closed form expression from this.

3.8 Systems with multiple inputs and outputs

When a linear system has many inputs and outputs (see Figure 3.3; remember, we are only concerned with the zero-state response $A(\theta; \underset{\sim}{u}) = S\{\underset{\sim}{u}\}$) then we can define the following:

$h_{ij}(t, \tau) \triangleq$ the zero-state response observed at time t at the ith output when the jth input $\underset{\sim}{u}_j = \{\delta(t - \tau)\}$ and all other inputs are zero.

Therefore, when $\underset{\sim}{u}_j$ is some arbitrary input function and all other inputs are zero (i.e., $\underset{\sim}{u}_k = \underset{\sim}{0}$ for $k \neq j$), we obtain

$$y_i(t) = \int_{-\infty}^{\infty} h_{ij}(t, \lambda) u_j(\lambda)\, d\lambda$$

so when all inputs are present,

$$y_i(t) = \int_{-\infty}^{\infty} \sum_{j=1}^{r} h_{ij}(t, \lambda) u_j(\lambda)\, d\lambda \tag{3.38}$$

and this is true for $i = 1, 2, \ldots, m$. This entire set of equations may be expressed in vector-matrix notation as

$$\boldsymbol{y}(t) = \int_{-\infty}^{\infty} H(t, \lambda) \boldsymbol{u}(\lambda)\, d\lambda \tag{3.39}$$

where $\boldsymbol{y}$ is an m-vector, H is an $m \times r$ matrix, and $\boldsymbol{u}$ is an r-vector. Thus, $H(t, \lambda)$ is the matrix with elements $h_{ij}(t, \lambda)$, and it is called the impulse response matrix (i.e., $H(t, \lambda) = [h_{ij}(t, \lambda)]$).

Similarly, for discrete-time systems, let $h_{ij}(k, \ell) =$ the zero-state response at output i observed at time k to a discrete delta function applied to jth input at time

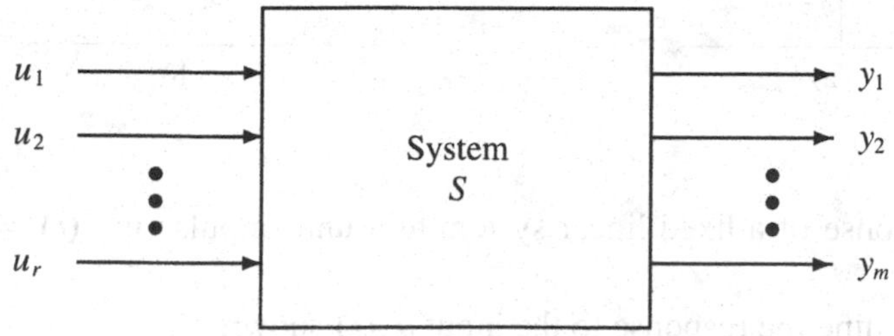

Fig. 3.3.

ℓ with all other inputs zero. Then

$$y(k) = \sum_{\ell=-\infty}^{\infty} H(k, \ell)u(\ell) \tag{3.40}$$

where $H(k, \ell)$ is the $m \times r$ matrix with elements $h_{ij}(k, \ell)$, and it is called the delta response matrix.

Problems

3.1 For the function shown below:

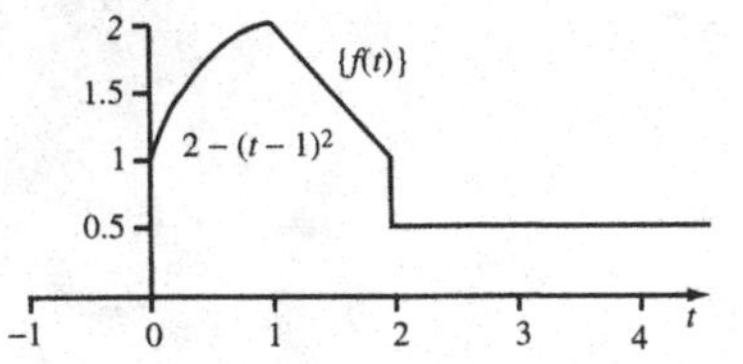

a) Find an expression in terms of the elementary functions $\{\mu_i\}$.
b) Sketch the derivative function, and find an expression for it in terms of elementary functions.
c) Sketch the integral function, and find an analytic expression for it in terms of elementary functions.

3.2 Find the response of each of the networks below to a unit ramp, a unit step, and a unit impulse.

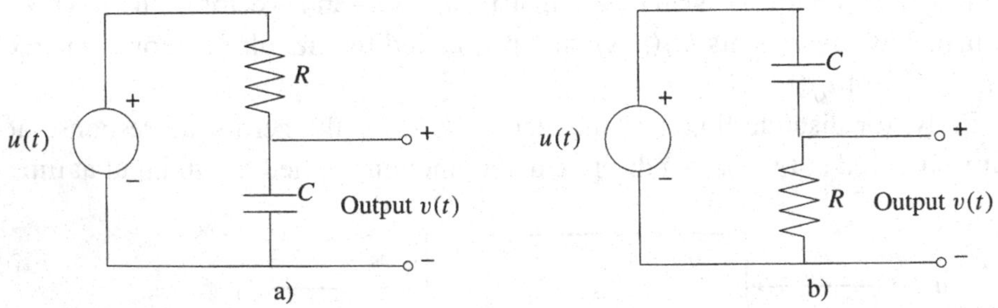

3.3 The response of a fixed linear system to a unit impulse is $h(t) = 1(t) - 1(t-1)$.
a) Determine the response to the input $u_a(t)$ shown.
b) Determine the response to the input $u_b(t)$ shown.

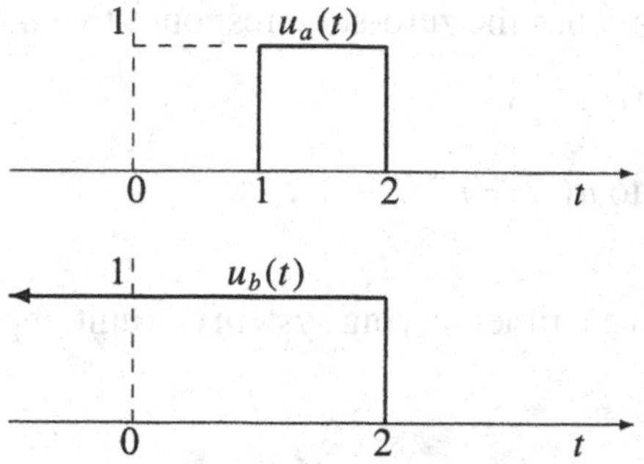

3.4 A fixed, linear system has the unit-step response $a(t) = te^{-t}\,1(t)$.
a) Find the impulse response $h(t) = S\{\delta(t)\}$.
b) Find the response $m_2(t)$ to the unit ramp function $\mu_2(t)$.
c) Find the response $y(t)$ to the function $u(t)$ shown below.

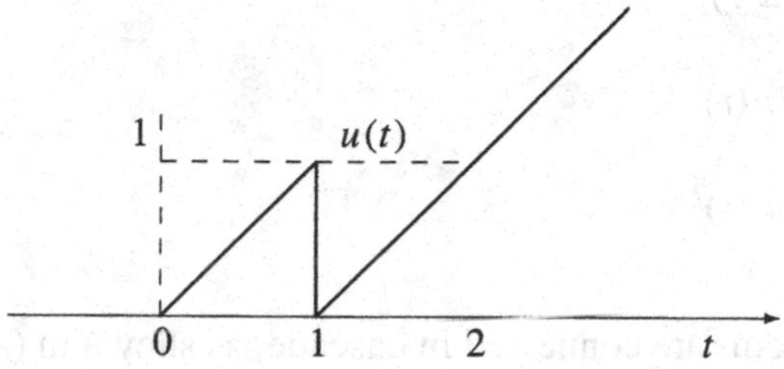

3.5 A linear fixed system has the response to a unit impulse applied at $t = 0$ given by $h(t) = 1(t-2)$. Find the response of this system to the input $u(t)$ shown.

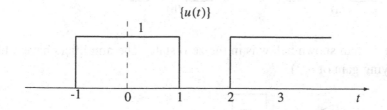

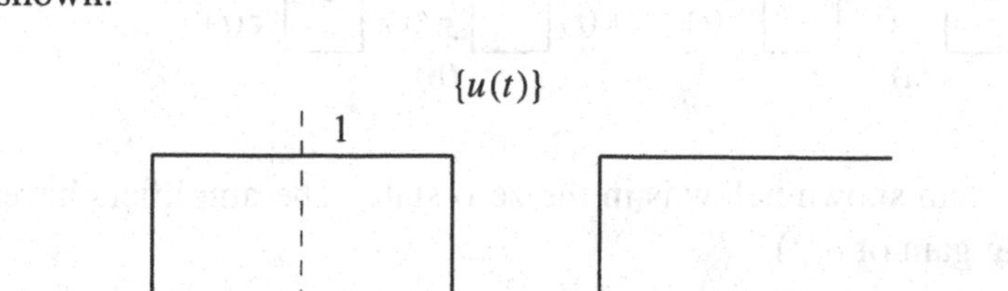

3.6 A fixed linear system has the impulse response $h(t) = te^{-t}\,1(t)$. Find the response of the system to the input $u(t) = 1(t) - 1(t-1)$.

3.7 The response of a time-varying linear system at time t to a unit step applied at time τ is $a(t, \tau) = \tau(1-e^{-t})\,1(t-\tau)$. Find the impulse response $h(t, \tau)$.

3.8 A linear time-varying system S has the zero-state response to a unit step

$$a(t, t_0) = S\{1(t - t_0)\} = t_0\ 1(t - t_0).$$

Find the zero-state response to $u(t) = t\ 1(t - 1)$.

3.9 The zero-state response of a linear time-varying system to a unit step applied at time τ is

$$a(t, \tau) = \tau t\ 1(t - \tau).$$

a) Find the impulse response $h(t, \tau)$.
b) Find the zero-state response to the input $x(t) = e^{-t} 1(t - 1)$.

3.10 Find the step response for each linear system below:

a) $\dfrac{dy(t)}{dt} + \dfrac{2}{t} y(t) = u(t)$

b) $\dfrac{dy(t)}{dt} + 2y(t) = tu(t)$

c) $\dfrac{dy(t)}{dt} + 2ty(t) = u(t)$

3.11 Two fixed, linear systems are connected in cascade, as shown in (a). If the two systems are interchanged as in (b), show that the response of the overall system to any input is unchanged.

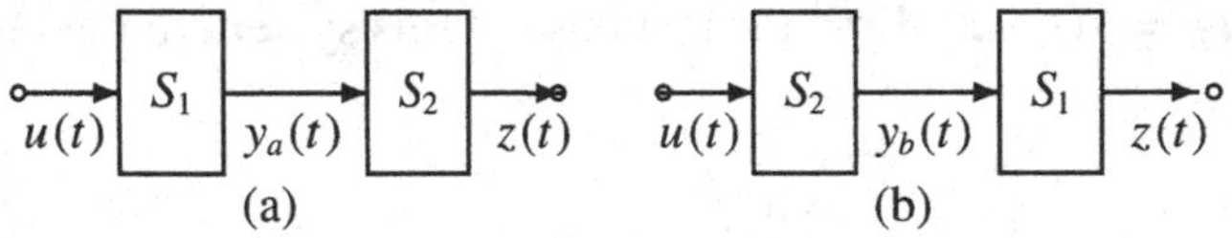

3.12 The system shown below is in the zero state. The amplifiers have a time-varying gain of $\alpha(t)$.

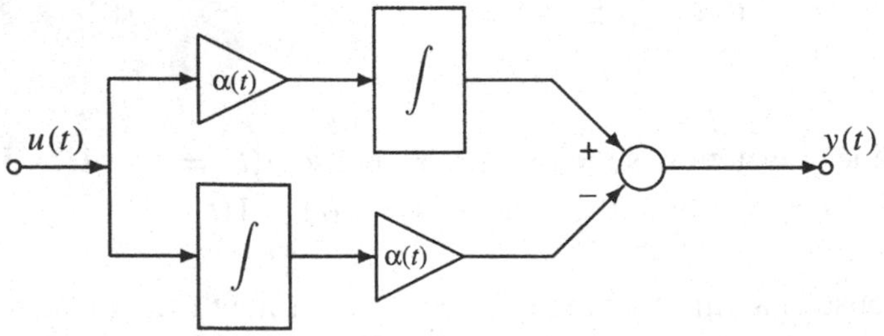

a) Find the impulse response $h(t, t_0)$ of this system.
b) When $\alpha(t) = t$, is the system fixed or time-varying?
c) For $\alpha(t) = t$, calculate the response of the system to the input:

$$u(t) = \begin{cases} 2 & \text{for } 0 < t < 1 \\ 0 & \text{otherwise.} \end{cases}$$

3.13 A time-invariant linear system has the impulse response and input as shown.

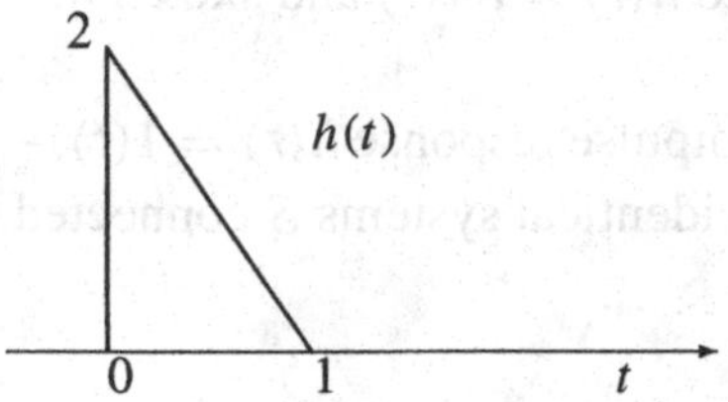

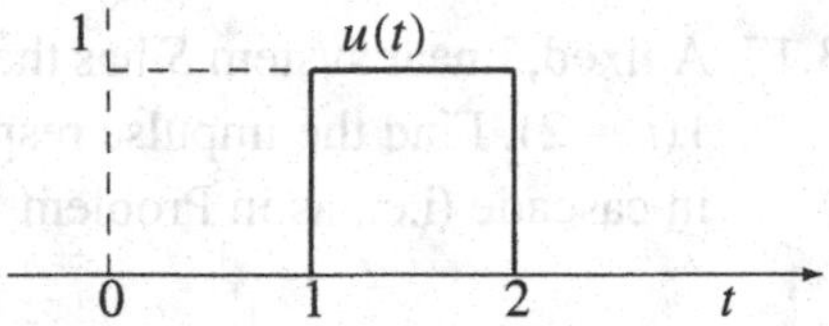

a) Find an expression for the step response $a(t)$ and sketch it.
b) Find an expression for the zero-state response $y(t)$ to $u(t)$ and sketch it.

3.14 For the time-varying system described by the state equations below:

$$\dot{x}(t) = u(t)$$
$$y(t) = -tx(t) + t^2 u(t)$$

a) Find the unit impulse response $h(t, \tau)$; i.e., the zero-state response to the unit impulse $\delta(t - \tau)$.
b) Find the zero-state response to the input $u(t) = e^{-t} 1(t - 1)$.

3.15 A fixed linear system has the impulse response $h(t)$ shown below.

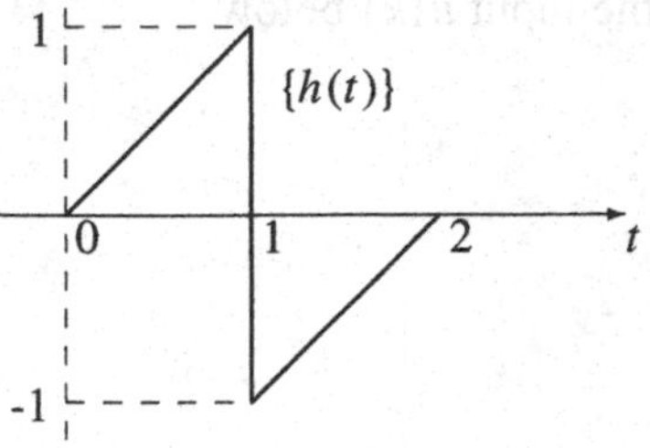

a) Write an expression for this $h(t)$.
b) Find an expression for the step response $a(t)$.
c) Sketch the step response $a(t)$.

3.16 A fixed, linear system has the impulse response shown below.

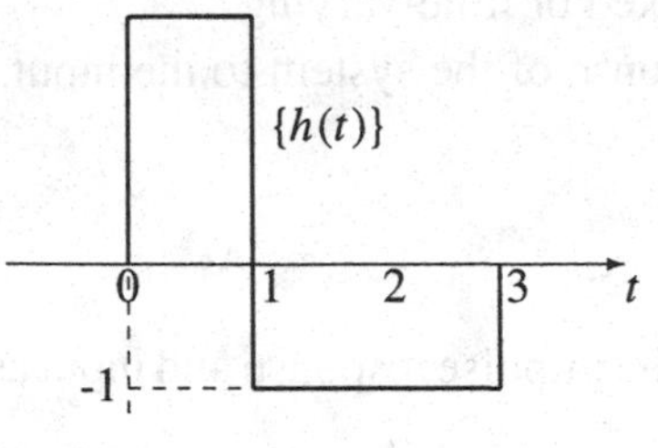

Find the zero-state response of the system to $u(t) = t\ 1(t)$ and sketch it.

3.17 A fixed, linear system S has the zero-state impulse response $h(t) = 1(t) - 1(t-2)$. Find the impulse response of two identical systems S connected in cascade (i.e., as in Problem 11).

3.18 A fixed, linear system S has the zero-state step response shown below.

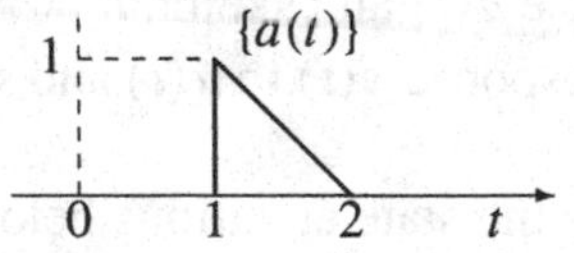

Find the impulse response of two identical systems S connected in cascade

3.19 A fixed, linear, discrete system has the zero-state delta response
$h(k) = k1(k) - 2(k-2)1(k-2) + (k-4)1(k-4)$.

Find the zero-state response to $u(k) = k1(k)$.

3.20 A fixed, linear, discrete system has the zero-state step response $a(k)$ as shown. Find the response to the input $u(k)$ below.

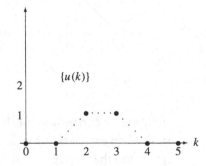

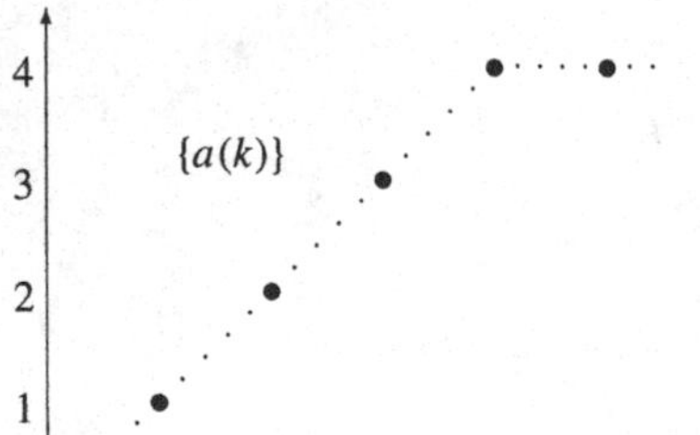

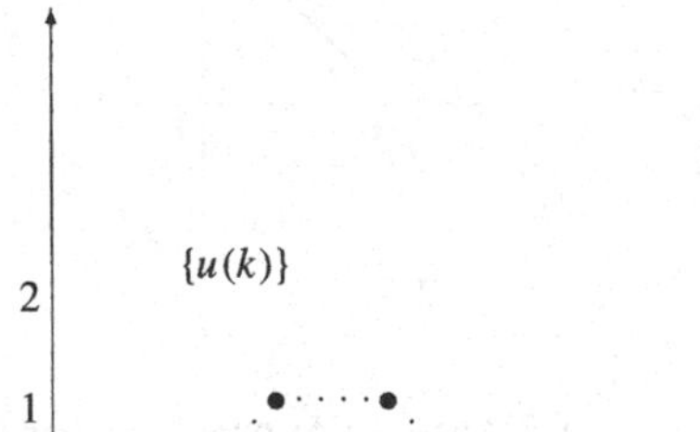

3.21 A fixed, linear, discrete system has the delta response $h(k) = \delta(k) - \delta(k-1)$.
   a) Find the step response $a(k)$.
   b) Find the zero-state response $y(k)$ when the input is $u(k) = k^2\, 1(k)$.

3.22 A fixed, linear, discrete system, which is in the zero state, is observed to have the output $y(k) = 1(k)$ when the input is $u(k) = \delta(k) - \delta(k-1)$. Find the delta response $h(k)$ of this system.

4

Solution of the dynamic state equations

4.1 Introduction

As mentioned earlier, an important aspect of analyzing systems is to find the response of systems to arbitrary inputs. Chapter 2 showed how the dynamic state equations could be obtained for linear systems. Chapter 3 showed that the zero-state response of a linear system to arbitrary input signals could be expressed in terms of a superposition integral (for analog systems) or superposition summation (for discrete systems).

In this chapter, the solutions of the dynamic state equations will be found and expressed in terms of an important matrix called the *transition matrix*. It will also be shown that the impulse (or delta) response matrix can be expressed directly in terms of the transition matrix and the B, C, and D matrices of the state equations. Therefore, in view of the results of Chapter 3, the response of a linear dynamic system to an arbitrary input can be determined directly once the transition matrix is known.

4.2 Analog systems

For a system described by the state equations

$$\dot{\boldsymbol{x}}(t) = A(t)\boldsymbol{x}(t) + B(t)\boldsymbol{u}(t) \tag{4.1a}$$

$$\boldsymbol{y}(t) = C(t)\boldsymbol{x}(t) + D(t)\boldsymbol{u}(t) \tag{4.1b}$$

the problem is to obtain a general solution to these equations for an arbitrary initial condition $\boldsymbol{x}(t_0) = \boldsymbol{x}_0$ and an arbitrary input function $\{\boldsymbol{u}(t)\}$.

First consider the unforced system

$$\dot{\boldsymbol{x}}(t) = A(t)\boldsymbol{x}(t) \quad \text{with} \quad \boldsymbol{x}(t_0) = \boldsymbol{x}_0. \tag{4.2}$$

Let $Q(t)$ be an $n \times n$ matrix of time functions that satisfies the matrix differential equation

$$\dot{Q}(t) = A(t)Q(t), \tag{4.3}$$

and the condition $Q(\hat{t}) = I$ at some finite time $\hat{t}$.
Note: $\dot{Q}(t)$ denotes a matrix whose i, j elements are the derivatives of the i, j elements of $Q(t)$.

Theorem 4.1 *If $A(t)$ is bounded (i.e., $|a_{ij}(t)| < \infty$ for all t on some interval $\mathcal{I}$, which contains $\hat{t}$ and t_0), then $Q(t)$ is nonsingular for all finite t on $\mathcal{I}$, and the solution to the unforced system is*

$$\boldsymbol{x}(t) = Q(t)Q^{-1}(t_0)\boldsymbol{x}_0 \triangleq \phi(t, t_0)\boldsymbol{x}_0. \tag{4.4}$$

The matrix $Q(t)Q^{-1}(t_0) \triangleq \phi(t, t_0)$ is called the *transition matrix* of the system. In some of the older literature this was called the matrizant (Frazer, Duncan, and Collar 1938: 53–6) or matricant (Gantmacher 1959: vol. 2, 125–31), but transition matrix is more widely used.

Proof: For $A(t)$ bounded on an interval $\mathcal{I}$, we see that if $\boldsymbol{x}(t) = \boldsymbol{0}$ for any finite time t on $\mathcal{I}$, then the solution to $\dot{\boldsymbol{x}}(t) = A(t)\boldsymbol{x}(t)$ is identically zero for all t on $\mathcal{I}$ because $\dot{\boldsymbol{x}}(t)$ will be identically zero. Suppose that at some finite $t_a \in \mathcal{I}$, $Q(t_a)$ is singular, then the columns of $Q(t_a)$ are linearly dependent. This means that there exists an $\boldsymbol{\alpha} \neq \boldsymbol{0}$ such that $\tilde{\boldsymbol{x}}(t_a) \triangleq Q(t_a)\boldsymbol{\alpha} = \boldsymbol{0}$. Therefore $\tilde{\boldsymbol{x}}(t) = \boldsymbol{0}$ for all $t \in \mathcal{I}$. However, at $\hat{t} \in \mathcal{I}$ we know that $\tilde{\mathbf{x}}(\hat{t}) = Q(\hat{t})\boldsymbol{\alpha} = \boldsymbol{\alpha} \neq \boldsymbol{0}$. This contradiction shows that the assumption that $Q(t)$ could be singular was wrong. Thus $Q(t)$ cannot be singular for any finite $t \in \mathcal{I}$, so $Q^{-1}(t)$ exists for all finite $t \in \mathcal{I}$.

The second part of the theorem is easily verified since

$$\dot{\boldsymbol{x}}(t) = \dot{Q}(t)Q^{-1}(t_0)\boldsymbol{x}_0 = A(t)Q(t)Q^{-1}(t_0)\boldsymbol{x}_0 = A(t)\boldsymbol{x}(t)$$

so $\boldsymbol{x}(t)$ satisfies (4.2) and

$$\boldsymbol{x}(t_0) = Q(t_0)Q^{-1}(t_0)\boldsymbol{x}_0 = \boldsymbol{x}_0.$$

Thus $\boldsymbol{x}(t)$ satisfies the initial condition, and we see that if we can obtain the transition matrix $\phi(t, t_0)$ or any fundamental matrix $Q(t)$ that satisfies the matrix differential equation, we can get a general solution to the unforced system equations. QED.

Properties of $\phi(t, t_0)$
We can establish the following properties of $\phi(t, t_0)$:

1 $\phi(t_0, t_0) = I$ for any finite t_0. This comes directly from $\phi(t_0, t_0) = Q(t_0)Q^{-1}(t_0) = I$.

2 $\phi(t_2, t_1)\phi(t_1, t_0) = \phi(t_2, t_0)$ for all finite t_0, t_1, and t_2 on $\mathcal{I}$. This follows from

$$\underbrace{Q(t_2)Q^{-1}(t_1)}_{\phi(t_2,t_1)}\underbrace{Q(t_1)Q^{-1}(t_0)}_{\phi(t_1,t_0)} = Q(t_2)Q^{-1}(t_0) = \phi(t_2, t_0).$$

Note: We do not require $t_2 \geq t_1 \geq t_0$.

3 $\phi(t_1, t_2) = \phi^{-1}(t_2, t_1)$ for all finite t_1, t_2 on $\mathcal{I}$. This follows directly from the two preceding properties or from $[Q(t_2)Q^{-1}(t_1)]^{-1} = Q(t_1)Q^{-1}(t_2) = \phi(t_1, t_2)$. *Note:* The $\phi(t_2, t_1)$ is nonsingular for all finite t_2, t_1 on $\mathcal{I}$.

4 $\dot{\phi}(t, t_0)|_{t_0=t} = A(t)$. This property follows from $\dot{Q}(t)Q^{-1}(t_0) = A(t)Q(t)Q^{-1}(t_0)$ evaluated at $t_0 = t$. It is a useful but not conclusive check on the results of the evaluation methods for $\phi(t, t_0)$ discussed later in this chapter.

Now we are ready to consider the forced system (4.1a) (i.e., with an input present)

$$\dot{\boldsymbol{x}}(t) = A(t)\boldsymbol{x}(t) + B(t)\boldsymbol{u}(t). \tag{4.1a}$$

For any input $\boldsymbol{u}(t)$, we wish to obtain a solution to this set of differential equations that satisfies the initial condition $\boldsymbol{x}(t_0) = \boldsymbol{x}_0$. That is, we want a complete solution.

The method used is called the variation of the parameter. Since we know that the unforced solution is $\boldsymbol{x}(t) = \phi(t, t_0)\boldsymbol{x}_0$, we assume that a complete solution has the form

$$\boldsymbol{x}(t) = \phi(t, t_0)\boldsymbol{f}(t) \tag{4.5}$$

and we seek the conditions that $\boldsymbol{f}(t)$ must satisfy to make this assumption true. Thus,

$$\begin{aligned}\dot{\boldsymbol{x}}(t) &= \dot{\phi}(t, t_0)\boldsymbol{f}(t) + \phi(t, t_0)\dot{\boldsymbol{f}}(t)\\ &= A(t)\phi(t, t_0)\boldsymbol{f}(t) + \phi(t, t_0)\dot{\boldsymbol{f}}(t)\\ &= A(t)\boldsymbol{x}(t) + \phi(t, t_0)\dot{\boldsymbol{f}}(t).\end{aligned} \tag{4.6}$$

Now equating this to $\dot{\boldsymbol{x}}(t)$ in (4.1a), we see that

$$\phi(t, t_0)\dot{\boldsymbol{f}}(t) = B(t)\boldsymbol{u}(t)$$

so

$$\dot{\boldsymbol{f}}(t) = \phi^{-1}(t, t_0)B(t)\boldsymbol{u}(t) = \phi(t_0, t)B(t)\boldsymbol{u}(t). \tag{4.7}$$

Integrating gives

$$\boldsymbol{f}(t) = \boldsymbol{f}(t_0) + \int_{t_0}^{t} \phi(t_0, \lambda)B(\lambda)\boldsymbol{u}(\lambda)\,d\lambda \tag{4.8}$$

and substituting (4.8) into (4.5) gives

$$\boldsymbol{x}(t) = \phi(t, t_0)\boldsymbol{f}(t_0) + \int_{t_0}^{t} \phi(t, t_0)\phi(t_0, \lambda)B(\lambda)\boldsymbol{u}(\lambda)\,d\lambda$$

so

$$x(t) = \phi(t, t_0)x_0 + \int_{t_0}^{t} \phi(t, \lambda)B(\lambda)u(\lambda)d\lambda. \tag{4.9}$$

The expression for $y(t)$ is obtained by substituting (4.9) into the output equation (4.1b) to give

$$y(t) = \underbrace{C(t)\phi(t, t_0)x_0}_{\text{Zero-input response}} + \underbrace{\int_{t_0}^{t} C(t)\phi(t, \lambda)B(\lambda)u(\lambda)\,d\lambda + D(t)u(t)}_{\text{Zero-state response}}. \tag{4.10}$$

This can be related to the form of the superposition integral from time-domain analysis by noting that with $x_0 = \mathbf{0}$ (i.e., the system is initially in the zero state) and $t_0 = -\infty$, the zero-state response is

$$y(t) = \int_{-\infty}^{t} C(t)\phi(t, \lambda)B(\lambda)u(\lambda)\,d\lambda + D(t)u(t). \tag{4.11}$$

By inserting a unit step $1(t - \lambda)$ into the integral, we can change the upper limit to $+\infty$. Then, using the property $u(t) = \int_{-\infty}^{\infty} u(\lambda)\delta(t - \lambda)\,d\lambda$ gives

$$y(t) = \int_{-\infty}^{\infty} [C(t)\phi(t, \lambda)B(\lambda)1(t - \lambda) + D(t)\delta(t - \lambda)]u(\lambda)\,d\lambda \tag{4.12}$$

which corresponds to the expression we found earlier for the zero-state response of multiple-input multiple-output systems when we identify the impulse response matrix as

$$H(t, \lambda) = C(t)\phi(t, \lambda)B(\lambda)1(t - \lambda) + D(t)\delta(t - \lambda). \tag{4.13}$$

This relates $H(t, \lambda)$ to the matrices of the state differential and output equation. For fixed systems, since A, B, C, and D are constant matrices, we find

$$H(t - \lambda) = C\phi(t - \lambda)B1(t - \lambda) + D\delta(t - \lambda).$$

Note: When $D(t) \neq 0$ (i.e., there is a direct connection from input to output) then an impulse appears in $H(t, \lambda)$.

4.3 The transition matrix

In the preceding section, we saw that the complete solution of the state differential equations as well as the impulse response matrix can be determined by finding the transition matrix $\phi(t, t_0)$.

4.3.1 The physical meaning of $\phi(t, t_0)$

One approach to finding $\phi(t, t_0)$ is from an interpretation of the physical meaning of the transition matrix.

If we take the expression (4.4), for the vector solution to the unforced system, and expand it for one component $x_i(t)$, we see that the solution for a single state variable can be written as

$$x_i(t) = \phi_{i1}(t, t_0)x_{10} + \phi_{i2}(t, t_0)x_{20} + \cdots + \phi_{in}(t, t_0)x_{n0}. \quad (4.14)$$

This shows that if $x_{j0} = 1$ and all other state variables are zero at $t = t_0$, then the solution for $x_i(t)$ is

$$x_i(t) = \phi_{ij}(t, t_0) \quad \text{when} \quad x_{j0} = 1 \quad \text{and} \quad x_{k0} = 0 \quad \text{for} \quad k \neq j. \quad (4.15)$$

Thus in terms of a simulation diagram, we can visualize the i, jth element of the transition matrix $\phi(t, t_0)$ as just the response observed at the output of the ith integrator at time t when a unit initial condition (voltage) is placed on the jth integrator at $t = t_0$ and all other integrators have zero initial conditions. All inputs are, of course, zero.

Therefore, according to (4.15), one method of finding the matrix $\phi(t, t_0)$ is to determine the responses of the simulation to unit initial conditions applied to each integrator output separately.

Example 4.1

The state equations for the unforced system (see Figure 4.1) are

$$\dot{x}_1 = x_2$$
$$\dot{x}_2 = -\frac{1}{t+1}x_2 \quad (\text{i.e., } \underset{\sim}{u} = \underset{\sim}{0}).$$

Now

1 $\phi_{21}(t, t_0)$ is the response observed at $x_2(t)$ when $x_1(t_0) = 1$ and $x_2(t_0) = 0$. We see that this gives $x_2(t) = \phi_{21}(t, t_0) = 0$ for $t \geq t_0 > -1$.

2 $\phi_{11}(t, t_0)$ is the response observed at $x_1(t)$ when $x_1(t_0) = 1$ and $x_2(t_0) = 0$. We see from (1) that since $x_2(t) \equiv 0$ then $x_1(t) = \phi_{11}(t, t_0) = 1$ for $t \geq t_0 > -1$.

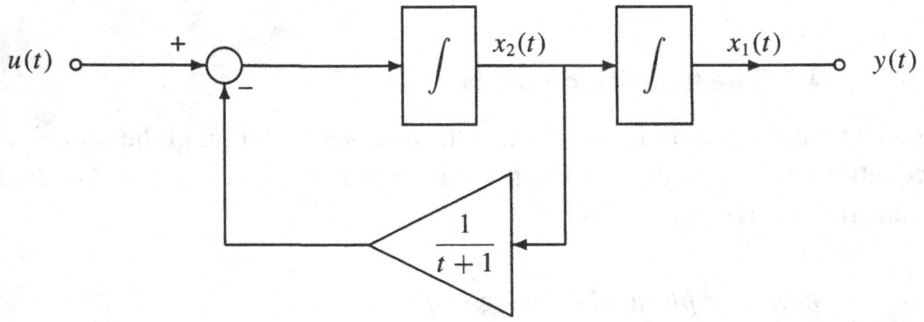

Fig. 4.1. System of Example 4.1

3 $\phi_{22}(t, t_0)$ is the response observed at $x_2(t)$ when $x_2(t_0) = 1$ and $x_1(t_0) = 0$. To get this, we must solve the first-order differential equation

$$\dot{x}_2(t) = -\frac{1}{t+1}x_2(t) \quad \text{subject to } x_2(t_0) = 1.$$

This can be done by direct integration of

$$\frac{dx_2}{x_2} = -\frac{dt}{t+1}.$$

Then

$$\int_{x_2(t_0)}^{x_2(t)} \frac{dx}{x} = -\int_{t_0}^{t} \frac{d\tau}{\tau+1}$$

so

$$\ln[x_2(t)] - \ln[x_2(t_0)] = -[\ln(t+1) - \ln(t_0+1)]$$

or

$$\ln x_2(t) = \ln x_2(t_0) + \ln(t_0+1) - \ln(t+1)$$

so

$$x_2(t) = \frac{t_0+1}{t+1}x_2(t_0)$$

and for $x_2(t_0) = 1$ we get

$$\phi_{22}(t, t_0) = \frac{t_0+1}{t+1}.$$

4 $\phi_{12}(t, t_0)$ is the response observed at $x_1(t)$ when $x_2(t_0) = 1$ and $x_1(t_0) = 0$. Since $\dot{x}_1 = x_2$ and we just found x_2 for these initial conditions, so

$$\phi_{12}(t, t_0) = x_1(t) = \int_{t_0}^{t} \frac{t_0+1}{\tau+1}\,d\tau$$

$$= (t_0+1)[\ln(t+1) - \ln(t_0+1)] = (t_0+1)\ln\left(\frac{t+1}{t_0+1}\right).$$

Thus,

$$\phi(t, t_0) = \begin{bmatrix} 1 & (t_0+1)\ln\left(\dfrac{t+1}{t_0+1}\right) \\ 0 & \dfrac{t_0+1}{t+1} \end{bmatrix}.$$

Note: $\phi(t_0, t_0) = I$ and $\det\phi(t, t_0) = \frac{t_0+1}{t+1} \neq 0$ for all $-1 < t_0 \leq t < \infty$. (Verify that this $\phi(t, t_0)$ satisfies the other transition matrix properties.) The impulse response is found from (4.13) to be

$$h(t, t_0) = [1\ 0]\phi(t, t_0) \begin{bmatrix} 0 \\ 1 \end{bmatrix} 1(t - t_0) = \phi_{12}(t, t_0) 1(t - t_0)$$

$$= (t_0 + 1) \ln \left[\frac{(t_0 + 1)}{t + 1} \right] 1(t - t_0).$$

4.3.2 $\phi(t)$ for constant system matrix A

To obtain the transition matrix using the physical meaning of $\phi(t, t_0)$, we must be able to integrate the unforced system equations. For time-varying systems, this cannot always be done in closed form. However, when the system matrix A is fixed then there are methods for finding $\phi(t, t_0)$ that do not require solving differential equations.

When A is not a function of time, we can seek a solution to

$$\dot{Q}(t) = AQ(t) \quad \text{satisfying } Q(0) = I. \tag{4.16}$$

We recall that a first-order differential equation $\dot{x}(t) = ax(t)$ has the solution $x(t) = e^{at}$ which satisfies $x(0) = 1$, where the exponential function can be expressed as the infinite series

$$e^{at} = 1 + at + \frac{a^2 t^2}{2!} + \cdots = \sum_{i=0}^{\infty} \frac{a^i t^i}{i!}.$$

It turns out that the infinite series of matrices

$$I + At + A^2 \frac{t^2}{2!} + \cdots = \sum_{i=0}^{\infty} A^i \frac{t^i}{i!} \triangleq e^{At}$$

can be shown to converge absolutely for any finite t, so we can call it e^{At}.

Claim: The $Q(t)$ satisfying (4.16) is

$$Q(t) = e^{At} \triangleq \sum_{i=0}^{\infty} A^i \frac{t^i}{i!}. \tag{4.17}$$

Proof: Setting $t = 0$, we see that $e^{A0} = I$ so the initial condition is satisfied. Also, since the series converges absolutely, it can be differentiated term by term to give

$$\frac{de^{At}}{dt} = \frac{d}{dt}\left[I + At + A^2 \frac{t^2}{2!} + A^3 \frac{t^3}{3!} + \cdots \right] = A + A^2 t + A^3 \frac{t^2}{2!} + \cdots$$

$$= A \left[I + At + A^2 \frac{t^2}{2!} + \cdots \right] = Ae^{At}.$$

QED.

Presumably, therefore, to get $\phi(t, t_0)$, we need

$$\phi(t, t_0) = e^{At}[e^{At_0}]^{-1}.$$

However, the situation is really simpler than this.

Claim: For a constant matrix A, the solution to the unforced system

$$\dot{\boldsymbol{x}}(t) = A\boldsymbol{x}(t) \quad \text{which satisfies } \boldsymbol{x}(t_0) = \boldsymbol{x}_0 \tag{4.18}$$

is

$$\boldsymbol{x}(t) = e^{A(t-t_0)}\boldsymbol{x}_0. \tag{4.19}$$

Proof: Since $Q(0) = e^{A0} = I$, when $t = t_0$ in (4.19), we see that $\boldsymbol{x}(t_0) = \boldsymbol{x}_0$ so the initial condition is satisfied. Also, since

$$\dot{\boldsymbol{x}}(t) = \frac{d}{dt}e^{A(t-t_0)}\boldsymbol{x}_0 = Ae^{A(t-t_0)}\frac{d(t-t_0)}{dt}\boldsymbol{x}_0 = A\boldsymbol{x}(t)$$

so (4.18) is satisfied. QED.

In other words, we see that $\phi(t, t_0) = e^{A(t-t_0)}$, so we will change notation to write $\phi(t - t_0) = e^{A(t-t_0)}$. This is consistent with our usual practice for fixed systems (i.e., since $\phi(t + T, t_0 + T) = \phi(t, t_0)$ for any T, by choosing $T = -t_0$ we get $\phi(t - t_0, 0)$, which is what we really mean when we write $\phi(t - t_0)$). Note also that $[e^{At}]^{-1} = e^{-At}$ (verify this from the series definition), so this is again consistent with

$$\phi(t - t_0) = e^{At}\left[e^{At_0}\right]^{-1} = e^{At}e^{-At_0} = e^{A(t-t_0)}.$$

4.3.3 Calculating e^{At}

The infinite series definition of e^{At} is a very useful form for proving properties of e^{At}, but it is not very useful for actually calculating the matrix, except in very special cases, because it is not easy to see the closed form from the infinite series form. If a computer is available to perform the operations, then values of the matrix e^{At} can be obtained for various values of t by programming a suitable algorithm. To get an analytical expression, however, we must use a different approach.

Method 1

In finding solutions to the unforced system $\dot{\boldsymbol{x}}(t) = A\boldsymbol{x}(t)$ satisfying $\boldsymbol{x}(0) = \boldsymbol{x}_0$ the classical approach is to try a solution of the form $\boldsymbol{x}(t) = e^{\lambda t}\boldsymbol{\xi}$, where $\boldsymbol{\xi}$ is a constant vector, and then determine whether a solution of this form can exist. Proceeding in this way, we find that

$$\dot{\boldsymbol{x}}(t) = \lambda e^{\lambda t}\boldsymbol{\xi} = Ae^{\lambda t}\boldsymbol{\xi} = e^{\lambda t}A\boldsymbol{\xi}. \tag{4.20}$$

Since $e^{\lambda t} \neq 0$ for finite t, we can divide and get

$$\lambda \boldsymbol{\xi} = A\boldsymbol{\xi} \quad \text{or equivalently} \quad [\lambda I - A]\boldsymbol{\xi} = \mathbf{0}.$$

Note that $\boldsymbol{\xi} = \mathbf{0}$ always satisfies, but this is not of interest (it is called a trivial solution), so we ask under what conditions does a solution exist for $\boldsymbol{\xi} \neq \mathbf{0}$.

By a well-known result of linear algebra (see Appendix, Section A.7), the system of n linear algebraic equations in n unknowns with zero right-hand side has a nonzero solution if and only if the coefficient matrix has a zero determinant. In this case then, for a solution $\boldsymbol{\xi} \neq \mathbf{0}$ to exist, we require that $\det[\lambda I - A] = 0$. This equation is called the characteristic equation of the matrix A. When $\det[\lambda I - A]$ is expanded out, we obtain an nth-degree polynomial $p(\lambda)$ in λ called the *characteristic polynomial*

$$p(\lambda) = \det[\lambda I - A] = \lambda^n + a_{n-1}\lambda^{n-1} + a_{n-2}\lambda^{n-2} + \cdots + a_1\lambda + a_0. \tag{4.21}$$

Thus the characteristic equation

$$p(\lambda) = 0 \tag{4.22}$$

is an nth-degree polynomial equation that has n roots (i.e., n solutions for λ). The n roots are the only values of λ that make $\det[\lambda I - A] = 0$ and thus are the only values of λ for which nonzero vectors $\boldsymbol{\xi}$ can be found. These roots denoted as $\lambda_1, \lambda_2, \ldots, \lambda_n$ are called the eigenvalues or characteristic roots of A.

For $\lambda = \lambda_i$, the equation $[\lambda_i I - A]\boldsymbol{\xi}_i = \mathbf{0}$ has a nontrivial solution $\boldsymbol{\xi}_i$. These $\boldsymbol{\xi}_i$ are called the *eigenvectors* of A associated with the eigenvalues λ_i.

Note: In general the eigenvalues need not be distinct (i.e., a root could be repeated), but things are much simpler when they are.

Example 4.2

$$A = \begin{bmatrix} 0 & 1 \\ -2 & -3 \end{bmatrix} \quad \text{so} \quad \det[\lambda I - A] = \det \begin{bmatrix} \lambda & -1 \\ 2 & \lambda + 3 \end{bmatrix} = \lambda(\lambda + 3) + 2 = 0$$

or $\lambda^2 + 3\lambda + 2 = (\lambda + 1)(\lambda + 2) = 0$.

Thus $\lambda = \lambda_1 = -1$ and $\lambda = \lambda_2 = -2$ are the two eigenvalues. For $\lambda = -1$, the eigenvector is found by solving the algebraic equations

$$\begin{bmatrix} -1 & -1 \\ 2 & 2 \end{bmatrix} \begin{bmatrix} \xi_1 \\ \xi_2 \end{bmatrix} = \begin{bmatrix} 0 \\ 0 \end{bmatrix} \quad \text{which gives} \quad \xi_1 = -\xi_2.$$

The vector $\boldsymbol{\xi}_1 = k_1 \begin{bmatrix} 1 \\ -1 \end{bmatrix}$ for any $k_1 \neq 0$ is thus an eigenvector associated with $\lambda = -1$. For $\lambda = -2$, we solve

$$\begin{bmatrix} -2 & -1 \\ 2 & 1 \end{bmatrix} \begin{bmatrix} \xi_1 \\ \xi_2 \end{bmatrix} = \begin{bmatrix} 0 \\ 0 \end{bmatrix} \quad \text{which gives} \quad 2\xi_1 = -\xi_2.$$

Thus $\boldsymbol{\xi}_2 = k_2 \begin{bmatrix} -1 \\ 2 \end{bmatrix}$ for any $k_2 \neq 0$ is an eigenvector associated with $\lambda = -2$.

Suppose that the λ_i are distinct (i.e., $\lambda_i \neq \lambda_j$ for $i \neq j$), and consider the vector

$$\boldsymbol{x}(t) = k_1 e^{\lambda_1 t} \boldsymbol{\xi}_1 + k_2 e^{\lambda_2 t} \boldsymbol{\xi}_2 + \cdots + k_n e^{\lambda_n t} \boldsymbol{\xi}_n. \tag{4.23}$$

This vector satisfies the equation $\dot{\boldsymbol{x}} = A\boldsymbol{x}$ since

$$\begin{aligned} \dot{\boldsymbol{x}}(t) &= k_1 \lambda_1 e^{\lambda_1 t} \boldsymbol{\xi}_1 + k_2 \lambda_2 e^{\lambda_2 t} \boldsymbol{\xi}_2 + \cdots + k_n \lambda_n e^{\lambda_n t} \boldsymbol{\xi}_n \\ &= k_1 e^{\lambda_1 t} A\boldsymbol{\xi}_1 + k_2 e^{\lambda_2 t} A\boldsymbol{\xi}_2 + \cdots + k_n e^{\lambda_n t} A\boldsymbol{\xi}_n \\ &= A\boldsymbol{x}(t). \end{aligned}$$

Furthermore, *any* solution can be written in this form (due to the linear independence of the $\boldsymbol{\xi}_i$; see Appendix, Theorem A.14).

In order to satisfy the initial condition $\boldsymbol{x}(0) = \boldsymbol{x}_0$, we must choose the k_i properly, that is,

$$\begin{aligned} \boldsymbol{x}(0) &= k_1 \boldsymbol{\xi}_1 + k_2 \boldsymbol{\xi}_2 + \cdots + k_n \boldsymbol{\xi}_n \\ &= [\boldsymbol{\xi}_1, \boldsymbol{\xi}_2, \ldots, \boldsymbol{\xi}_n] \begin{bmatrix} k_1 \\ k_2 \\ \vdots \\ k_n \end{bmatrix} \triangleq T\boldsymbol{k} = \boldsymbol{x}_0 \end{aligned} \tag{4.24}$$

where T is an $n \times n$ matrix whose columns are $\boldsymbol{\xi}_1, \ldots, \boldsymbol{\xi}_n$.

A theorem of matrix theory tells us that the eigenvectors associated with distinct eigenvalues are linearly independent (see Appendix, Theorem A.14), and another theorem states that an $n \times n$ matrix with n linearly independent columns is nonsingular (i.e., its inverse exists; see Appendix, Theorem A.6). Thus T^{-1} exists, so we must choose $\boldsymbol{k} = T^{-1}\boldsymbol{x}_0$ to satisfy the initial condition.

Substitute this result into (4.23) to get

$$\begin{aligned} \boldsymbol{x}(t) &= \left[e^{\lambda_1 t}\boldsymbol{\xi}_1, e^{\lambda_2 t}\boldsymbol{\xi}_2, \ldots, e^{\lambda_n t}\boldsymbol{\xi}_n\right] \boldsymbol{k} \\ &= [\boldsymbol{\xi}_1, \boldsymbol{\xi}_2, \ldots, \boldsymbol{\xi}_n] \begin{bmatrix} e^{\lambda_1 t} & 0 & \cdots & 0 \\ 0 & e^{\lambda_2 t} & \cdots & 0 \\ \vdots & \vdots & \ddots & \vdots \\ 0 & 0 & \cdots & e^{\lambda_n t} \end{bmatrix} \boldsymbol{k}. \end{aligned}$$

So

$$\boldsymbol{x}(t) = T \begin{bmatrix} e^{\lambda_1 t} & 0 & \cdots & 0 \\ 0 & e^{\lambda_2 t} & \cdots & 0 \\ \vdots & \vdots & \ddots & \vdots \\ 0 & 0 & \cdots & e^{\lambda_n t} \end{bmatrix} T^{-1}\boldsymbol{x}_0 \triangleq T e^{\Lambda t} T^{-1} \boldsymbol{x}_0 \tag{4.25}$$

where $\Lambda = \text{diag}(\lambda_1 \cdots \lambda_n)$ and $e^{\Lambda t} = \text{diag}(e^{\lambda_1 t} \cdots e^{\lambda_n t})$. Comparing this with our previous result (4.19), we see that

$$e^{At} = \phi(t) = Te^{\Lambda t}T^{-1} \tag{4.26}$$

where T is a matrix whose n columns are the eigenvectors of A.

Method 2

An alternate approach to obtaining (4.26) is to observe that if a matrix T exists such that

$$T^{-1}AT = \Lambda = \begin{bmatrix} \lambda_1 & 0 & \cdots & 0 \\ 0 & \lambda_2 & \cdots & 0 \\ \vdots & \vdots & \ddots & \vdots \\ 0 & 0 & \cdots & \lambda_n \end{bmatrix}$$

(i.e., A can be diagonalized) then for *any* function $f(\lambda)$ which can be expanded in a power series

$$f(\lambda) = \sum_{i=0}^{\infty} a_i \lambda^i$$

and which is defined at the eigenvalues λ_i, we can write

$$f(\Lambda) = \sum_{i=0}^{\infty} a_i \Lambda^i = \begin{bmatrix} \sum_{i=0}^{\infty} a_i \lambda_1^i & 0 & \cdots & 0 \\ 0 & \sum_{i=0}^{\infty} a_i \lambda_2^i & \cdots & 0 \\ \vdots & \vdots & \ddots & \vdots \\ 0 & 0 & \cdots & \sum_{i=0}^{\infty} a_i \lambda_n^i \end{bmatrix}$$

$$= \begin{bmatrix} f(\lambda_1) & 0 & \cdots & 0 \\ 0 & f(\lambda_2) & \cdots & 0 \\ \vdots & \vdots & \ddots & \vdots \\ 0 & 0 & \cdots & f(\lambda_n) \end{bmatrix} \tag{4.27}$$

which defines that function $f(\Lambda)$ for the diagonal matrix Λ because

$$\Lambda^i = \begin{bmatrix} \lambda_1^i & 0 & \cdots & 0 \\ 0 & \lambda_2^i & \cdots & 0 \\ \vdots & \vdots & \ddots & \vdots \\ 0 & 0 & \cdots & \lambda_n^i \end{bmatrix}.$$

Since we can write $A = T\Lambda T^{-1}$ which gives $\Lambda = T^{-1}AT$, then substituting for Λ in (4.27) and using $(T^{-1}AT)^i = T^{-1}A^iT$ gives

$$f(\Lambda) = \sum_{i=0}^{\infty} a_i (T^{-1}AT)^i = T^{-1}\left[\sum_{i=0}^{\infty} a_i A^i\right] T.$$

We can now define the function $f(A)$ for the matrix A as

$$f(A) \triangleq \sum_{i=0}^{\infty} a_i A^i = Tf(\Lambda)T^{-1}. \tag{4.28}$$

Thus if we find T and Λ, we easily get any function $f(A)$ for which $f(\lambda)$ is defined at the eigenvalues of A. In particular we can use (4.28) to get the function e^{At}.

Finding the matrix T (which is formed from the eigenvectors of A) can actually be avoided because for matrices A with distinct eigenvalues $\lambda_1, \ldots, \lambda_n$, the matrix $f(\Lambda)$ can be expressed as

$$f(\Lambda) = f(\lambda_1)E_1 + f(\lambda_2)E_2 + \cdots + f(\lambda_n)E_n = \sum_{k=0}^{n} f(\lambda_k)E_k \tag{4.29}$$

where E_i is a matrix with 1 in the i, i position and zeros everywhere else. Thus

$$f(A) = \sum_{k=0}^{n} f(\lambda_k)TE_kT^{-1} = \sum_{k=0}^{n} f(\lambda_k)Z_{k0} \tag{4.30}$$

where $Z_{k0} = TE_kT^{-1}$.

Note that the Z_{k0} matrices depend only on the matrix A and not on the function $f(\lambda)$, so they can be evaluated independently and used for any function. We now show how to find the Z_{k0} matrices without first getting T.

The method of trial functions

When using this method we will choose a number of "trial functions" for which $f(A)$ is easy to calculate directly (i.e., polynomials) and for which only one of the Z_{k0} in (4.30) is nonzero. By choosing n different functions, we can find each Z_{k0}.

Example 4.3

For $A = \begin{bmatrix} 0 & 1 \\ -2 & -3 \end{bmatrix}$ we found $p(\lambda) = (\lambda+1)(\lambda+2)$ so $\lambda_1 = -1$ and $\lambda_2 = -2$.

Choose $f_1(\lambda) = (\lambda+2)$ as the first trial function (i.e., a function that is nonzero at only one of the eigenvalues). Then

$$f_1(A) = [A + 2I] = f_1(-1)Z_{10} + f_1(-2)Z_{20} = Z_{10}$$

$$Z_{10} = [A + 2I] = \begin{bmatrix} 2 & 1 \\ -2 & -1 \end{bmatrix}.$$

Next, choose $f_2(\lambda) = (\lambda + 1)$ which gives

$$f_2(A) = [A + I] = f_2(-1)Z_{10} + f_2(-2)Z_{20} = -Z_{20}$$

$$Z_{20} = -[A + I] = \begin{bmatrix} -1 & -1 \\ 2 & 2 \end{bmatrix}$$

so for any function $f(\lambda)$, which exists at $\lambda = -1$ and -2, we have

$$f(A) = f(-1)\begin{bmatrix} 2 & 1 \\ -2 & -1 \end{bmatrix} + f(-2)\begin{bmatrix} -1 & -1 \\ 2 & 2 \end{bmatrix}$$

and in particular for $f(\lambda) = e^{\lambda t}$, we get

$$e^{At} = e^{-t}\begin{bmatrix} 2 & 1 \\ -2 & -1 \end{bmatrix} + e^{-2t}\begin{bmatrix} -1 & -1 \\ 2 & 2 \end{bmatrix} = \begin{bmatrix} 2e^{-t} - e^{-2t} & e^{-t} - e^{-2t} \\ 2\left(e^{-2t} - e^{-t}\right) & 2e^{-2t} - e^{-t} \end{bmatrix}.$$

We can actually derive a formula for finding the matrices Z_{k0}. When A has distinct eigenvalues, then $p(\lambda)$ can be written in factored form as

$$p(\lambda) = (\lambda - \lambda_1)(\lambda - \lambda_2)\ldots(\lambda - \lambda_n). \tag{4.31}$$

Let us now choose the trial functions as

$$f_k(\lambda) = \prod_{\substack{i=1 \\ i\neq k}}^{n} (\lambda - \lambda_i) \quad \text{for} \quad k = 1\ldots n. \tag{4.32}$$

Note that each function $f_k(\lambda)$ is just $p(\lambda)$ with the factor $(\lambda - \lambda_k)$ removed. This gives

$$f_k(A) = \prod_{\substack{i=1 \\ i\neq k}}^{n} (A - \lambda_i I) \quad \text{for} \quad k = 1\ldots n.$$

and since

$$f_k(\lambda_j) = \begin{cases} 0 & \text{for} \quad j \neq k \\ \prod_{\substack{i=1 \\ i\neq k}}^{n} (\lambda_k - \lambda_i) & \text{for} \quad j = k \end{cases}$$

we get Sylvester's formula

$$Z_{k0} = \prod_{\substack{i=1 \\ i\neq k}}^{n} \left[\frac{A - \lambda_i I}{\lambda_k - \lambda_i}\right]. \tag{4.33}$$

It is usually easier to choose your own trial functions rather than memorize this formula.

Example 4.4

$$A = \begin{bmatrix} 0 & 0 & -2 \\ 0 & -1 & 2 \\ 1 & 0 & 3 \end{bmatrix} \quad \text{so} \quad [\lambda I - A] = \begin{bmatrix} \lambda & 0 & 2 \\ 0 & \lambda+1 & -2 \\ -1 & 0 & \lambda-3 \end{bmatrix}$$

and

$$p(\lambda) = \det[\lambda I - A] = \lambda[(\lambda+1)(\lambda-3)] + 2(\lambda+1) = (\lambda+1)[\lambda^2 - 3\lambda + 2]$$
$$= (\lambda+1)(\lambda-1)(\lambda-2) \quad \text{so} \quad \lambda_1 = -1,\ \lambda_2 = 1,\ \lambda_3 = 2.$$

Using trial functions:

$$f(A) = f(-1)Z_{10} + f(1)Z_{20} + f(2)Z_{30}.$$

Choose $f_1(\lambda) = (\lambda-1)(\lambda-2) = \lambda^2 - 3\lambda + 2$ so $f_1(A) = [A-I][A-2I] = 6Z_{10}$ and $Z_{10} = 1/6[A-I][A-2I]$. For $f_2(\lambda) = (\lambda+1)(\lambda-2)$ we get $f_2(A) = [A+I][A-2I] = -2Z_{20}$, so $Z_{20} = -1/2[A+I][A-2I]$.

For $f_3(\lambda) = (\lambda+1)(\lambda-1)$ we get $f_3(A) = [A-I][A+I] = 3Z_{30}$. The resulting matrices are

$$Z_{30} = \frac{1}{3}\begin{bmatrix} -1 & 0 & -2 \\ 0 & -2 & 2 \\ 1 & 0 & 2 \end{bmatrix}\begin{bmatrix} 1 & 0 & -2 \\ 0 & 0 & 2 \\ 1 & 0 & 4 \end{bmatrix} = \begin{bmatrix} -1 & 0 & -2 \\ 2/3 & 0 & 4/3 \\ 1 & 0 & 2 \end{bmatrix}$$

$$Z_{20} = -\frac{1}{2}\begin{bmatrix} 1 & 0 & -2 \\ 0 & 0 & 2 \\ 1 & 0 & 4 \end{bmatrix}\begin{bmatrix} -2 & 0 & -2 \\ 0 & -3 & 2 \\ 1 & 0 & 1 \end{bmatrix} = \begin{bmatrix} 2 & 0 & 2 \\ -1 & 0 & -1 \\ -1 & 0 & -1 \end{bmatrix}$$

$$Z_{10} = \frac{1}{6}\begin{bmatrix} -1 & 0 & -2 \\ 0 & -2 & 2 \\ 1 & 0 & 2 \end{bmatrix}\begin{bmatrix} -2 & 0 & -2 \\ 0 & -3 & 2 \\ 1 & 0 & 1 \end{bmatrix}\begin{bmatrix} 0 & 0 & 0 \\ 1/3 & 1 & -1/3 \\ 0 & 0 & 0 \end{bmatrix}$$

so

$$e^{At} = e^{-t}Z_{10} + e^{t}Z_{20} + e^{2t}Z_{30}$$
$$= \begin{bmatrix} 2e^t - e^{2t} & 0 & 2(e^t - e^{2t}) \\ \frac{1}{3}e^{-t} - e^t + \frac{2}{3}e^{2t} & e^{-t} & -\frac{1}{3}e^{-t} - e^t + \frac{4}{3}e^{2t} \\ -e^t + e^{2t} & 0 & -e^t + 2e^{2t} \end{bmatrix}.$$

When the matrix A has repeated eigenvalues then $p(\lambda)$ can be written in factored form as

$$p(\lambda) = (\lambda - \lambda_1)^{n_1}(\lambda - \lambda_2)^{n_2}\ldots(\lambda - \lambda_s)^{n_s} \tag{4.34}$$

where s denotes the number of distinct eigenvalues and n_i is the multiplicity of the eigenvalue λ_i. Note that $\sum_{i=1}^{s} n_i = n =$ the order of the system. For $n_i = 1$, the eigenvalue λ_i is called simple. The previous case of all distinct eigenvalues occurs when $n_i = 1$ for all i, so $s = n$.

A derivation somewhat more complicated than for the case of distinct eigenvalues (see Appendix, Section A.20) leads to the result that, for any function $f(\lambda)$ for which the derivative values $\left[\frac{d^j f(\lambda)}{d\lambda^j}\right]_{\lambda=\lambda_i}$ exist for $i = 1, \ldots, s$ and $j = 0, \ldots, n_i - 1$ (these are called the values of f on the *spectrum* of A), a function $f(A)$ of the square matrix A can be obtained from

$$f(A) = \sum_{i=1}^{s} \sum_{j=0}^{n_i-1} \left[\frac{d^j f(\lambda)}{d\lambda^j}\right]_{\lambda=\lambda_i} Z_{ij}. \tag{4.35}$$

Equation (4.30) is a special case of (4.35) for $s = n$ and $n_i = 1$ for all $i = 1, \ldots, n$. The method of trial functions can be used to evaluate the matrices Z_{ij}.

Example 4.5

For $A = \begin{bmatrix} 0 & 0 & -2 \\ 0 & 1 & 0 \\ 1 & 0 & 3 \end{bmatrix}$ find $e^{At} = \phi(t)$.

$$\det[\lambda I - A] = \det \begin{bmatrix} \lambda & 0 & 2 \\ 0 & \lambda - 1 & 0 \\ -1 & 0 & \lambda - 3 \end{bmatrix} = \lambda(\lambda - 1)(\lambda - 3) + 2(\lambda - 1)$$

$$= (\lambda - 1)[\lambda^2 - 3\lambda + 2] = (\lambda - 1)^2(\lambda - 2) = p(\lambda).$$

Note that the root $\lambda_1 = 1$ is repeated (its multiplicity $= 2$); $\lambda_2 = 2$ is a simple root, so $f(A) = f(1)Z_{10} + f'(1)Z_{11} + f(2)Z_{20}$.

Using trial function $f_1(\lambda) = (\lambda - 1)^2$ so $f_1'(\lambda) = 2(\lambda - 1)$, thus

$$f_1(A) = (A - I)^2 = Z_{20} = \begin{bmatrix} -1 & 0 & -2 \\ 0 & 0 & 0 \\ 1 & 0 & 2 \end{bmatrix} \begin{bmatrix} -1 & 0 & -2 \\ 0 & 0 & 0 \\ 1 & 0 & 2 \end{bmatrix} = \begin{bmatrix} -1 & 0 & -2 \\ 0 & 0 & 0 \\ 1 & 0 & 2 \end{bmatrix}.$$

Next choose $f_2(\lambda) = (\lambda - 1)(\lambda - 2)$, so $f_2'(\lambda) = 2\lambda - 3$ which gives

$$f_2(A) = (A - I)(A - 2I) = -Z_{11} = \begin{bmatrix} -1 & 0 & -2 \\ 0 & 0 & 0 \\ 1 & 0 & 2 \end{bmatrix} \begin{bmatrix} -2 & 0 & -2 \\ 0 & -1 & 0 \\ 1 & 0 & 1 \end{bmatrix} = \begin{bmatrix} 0 & 0 & 0 \\ 0 & 0 & 0 \\ 0 & 0 & 0 \end{bmatrix}.$$

Finally we can choose $f_3(\lambda) = (\lambda - 2)$, so $f_3'(\lambda) = 1$ which gives

$$f_3(A) = A - 2I = -Z_{10} + Z_{11} = \begin{bmatrix} -2 & 0 & -2 \\ 0 & -1 & 0 \\ 1 & 0 & 1 \end{bmatrix}.$$

Thus $e^{At} = e^t Z_{10} + te^t Z_{11} + e^{2t} Z_{20}$ is found to be

$$e^{At} = \begin{bmatrix} 2e^t - e^{2t} & 0 & 2e^t - 2e^{2t} \\ 0 & e^t & 0 \\ -e^t + e^{2t} & 0 & -e^t + 2e^{2t} \end{bmatrix}.$$

Example 4.6

For $A = \begin{bmatrix} 0 & 2 & -2 \\ 0 & 1 & 0 \\ 1 & -1 & 3 \end{bmatrix}$ we find $\det[\lambda I - A] = \det \begin{bmatrix} \lambda & -2 & 2 \\ 0 & \lambda - 1 & 0 \\ -1 & 1 & \lambda - 3 \end{bmatrix}$.
Here $p(\lambda) = \lambda(\lambda-1)(\lambda-3)+2(\lambda-1) = (\lambda-1)[\lambda^2-3\lambda+2] = (\lambda-1)^2(\lambda-2)$, so again $\lambda_1 = 1$ has multiplicity 2 and $\lambda_2 = 2$ has multiplicity of 1.

As before for $f_1(\lambda) = (\lambda - 1)^2$, $f_1'(\lambda) = 2(\lambda - 1)$, we get

$$f_1(A) = (A - I)^2 = Z_{20} = \begin{bmatrix} -1 & 2 & -2 \\ 0 & 0 & 0 \\ 1 & -1 & 2 \end{bmatrix} \begin{bmatrix} -1 & 2 & -2 \\ 0 & 0 & 0 \\ 1 & -1 & 2 \end{bmatrix} = \begin{bmatrix} -1 & 0 & -2 \\ 0 & 0 & 0 \\ 1 & 0 & 2 \end{bmatrix}.$$

Choosing $f_2(\lambda) = (\lambda - 1)(\lambda - 2)$, $f_2'(\lambda) = 2\lambda - 3$ gives

$$f_2(A) = (A - I)(A - 2I) = -Z_{11} = \begin{bmatrix} -1 & 2 & -2 \\ 0 & 0 & 0 \\ 1 & -1 & 2 \end{bmatrix} \begin{bmatrix} -2 & 2 & -2 \\ 0 & -1 & 0 \\ 1 & -1 & 1 \end{bmatrix} = \begin{bmatrix} 0 & -2 & 0 \\ 0 & 0 & 0 \\ 0 & 1 & 0 \end{bmatrix}$$

and for $f_3(\lambda) = (\lambda - 2)$ so $f_3'(\lambda) = 1$ we get $f_3(A) = (A - 2I) = -Z_{10} + Z_{11}$ so

$$Z_{10} = Z_{11} - (A - 2I) = \begin{bmatrix} 2 & 0 & 2 \\ 0 & 1 & 0 \\ -1 & 0 & -1 \end{bmatrix}.$$

For $f(\lambda) = e^{\lambda t}$ we get $e^{At} = e^t Z_{10} + te^t Z_{11} + e^{2t} Z_{20}$, which results in

$$e^{At} = \begin{bmatrix} 2e^t - e^{2t} & 2te^t & 2e^t - 2e^{2t} \\ 0 & e^t & 0 \\ e^{2t} - e^t & -te^t & 2e^{2t} - e^t \end{bmatrix}.$$

Method 3

There is a theorem of matrix theory called the Cayley-Hamilton theorem, which states that every square $n \times n$ matrix A satisfies its own characteristic equation (see Appendix, Section A.17). That is, for

$$p(\lambda) = \det(\lambda I - A) = \lambda^n + a_{n-1}\lambda^{n-1} + a_{n-2}\lambda^{n-2} + \cdots + a_1\lambda + a_0$$

the Cayley-Hamilton theorem states that

$$p(A) = A^n + a_{n-1}A^{n-1} + a_{n-2}A^{n-2} + \cdots + a_1 A + a_0 I = \Theta \tag{4.36}$$

where Θ denotes the zero matrix.

From (4.36), we see that the nth power of A can be expressed as a unique linear combination of the powers A^i for $i = 0, 1, \ldots, n-1$. Namely,

$$A^n = -\sum_{i=0}^{n-1} a_i A^i \tag{4.37}$$

thus

$$\begin{aligned} A^{n+1} &= -\sum_{i=0}^{n-1} a_i A^{i+1} = -a_{n-1}A^n - \sum_{j=1}^{n-1} a_{j-1}A^j \\ &= \sum_{i=0}^{n-1} a_{n-1}a_i A^i - \sum_{j=1}^{n-1} a_{j-1}A^j \triangleq \sum_{i=0}^{n-1} \alpha_{n+1,i}A^i. \end{aligned} \tag{4.38}$$

This shows that $A^{(n+1)}$ can also be expressed as a unique linear combination of the powers A^i for $i = 0, \ldots, n-1$. Proceeding in the same way, we can thus show (by induction) that for any $N > n$, we can express A^N as a unique linear combination of the powers A^i for $i = 0, \ldots, n-1$. That is,

$$A^N = \sum_{i=0}^{n-1} \alpha_{N,i}A^i. \tag{4.39}$$

Therefore, any function $f(\lambda)$ that has a convergent power series expansion

$$f(\lambda) = \sum_{i=0}^{\infty} a_i \lambda^i$$

gives the matrix function

$$f(A) = \sum_{i=0}^{\infty} a_i A^i.$$

This function $f(A)$ can therefore be expressed as a *finite* polynomial

$$f(A) = \gamma_{n-1}A^{n-1} + \gamma_{n-2}A^{n-2} + \cdots + \gamma_1 A + \gamma_0 I \triangleq g(A) \tag{4.40}$$

(since the powers A^j for $j \geq n$ can be written in terms of the first $(n-1)$ powers).

To find the coefficients γ_ℓ in (4.40), consider a matrix A that can be diagonalized. In this case there exists a nonsingular matrix T such that $T^{-1}AT = \Lambda$ where $\Lambda =$

$\mathrm{diag}(\lambda_1 \ldots \lambda_n)$. Thus premultiplying (4.38) by T^{-1} and postmultiplying by T gives

$$f(\Lambda) = \begin{bmatrix} f(\lambda_1) & 0 & \ldots & 0 \\ 0 & f(\lambda_2) & \ldots & 0 \\ \vdots & \vdots & \ddots & \vdots \\ 0 & 0 & \ldots & f(\lambda_n) \end{bmatrix} = \sum_{j=0}^{n-1} \gamma_j \Lambda^j = g(\Lambda)$$

$$= \begin{bmatrix} \sum_{j=0}^{n-1} \gamma_j \lambda_1^j & 0 & \ldots & 0 \\ 0 & \sum_{j=0}^{n-1} \gamma_j \lambda_2^j & \ldots & 0 \\ \vdots & \vdots & \ddots & \vdots \\ 0 & 0 & \ldots & \sum_{j=0}^{n-1} \gamma_j \lambda_n^j \end{bmatrix}. \tag{4.41}$$

We see that

$$f(\lambda_i) = \sum_{j=0}^{n-1} \gamma_j \lambda_i^j = g(\lambda_i) \quad \text{for} \quad i = 1, 2, \ldots, n. \tag{4.42}$$

This gives one equation for each λ_i, so there are n equations which can be solved for the n unknowns $\gamma_0, \gamma_1, \ldots, \gamma_{n-1}$. Thus we see that by choosing the $\gamma_0, \gamma_1, \ldots, \gamma_{n-1}$ to satisfy (4.42) (this makes $f(\lambda_i) = g(\lambda_i)$ for all $i = 1, 2, \ldots, n$) we make $f(A) = g(A)$.

This is a special case of a theorem which states that for any two arbitrary polynomial functions $f(\lambda)$ and $g(\lambda)$ such that

$$\frac{d^j f(\lambda_i)}{d\lambda^j} = \frac{d^j g(\lambda_i)}{d\lambda^j} \quad \text{for} \quad i = 1, \ldots, s; \quad j = 0, \ldots, n_{i-1} \tag{4.43}$$

(i.e., when the values of f and g on the spectrum of A are equal), then $f(A) = g(A)$. Equation (4.43) generalizes the result of (4.42) and can be used to find the coefficients γ_ℓ in (4.40) for matrices which cannot be diagonalized (see Appendix, Section A.20).

Example 4.7

$$A = \begin{bmatrix} 0 & 1 \\ -2 & -3 \end{bmatrix} \quad \text{so} \quad p(\lambda) = (\lambda + 1)(\lambda + 2).$$

Applying Method 3, we choose $f(\lambda) = e^{\lambda t}$ and $g(\lambda) = \gamma_0 + \gamma_1 \lambda$, and we equate these on the spectrum of A to give

$$f(-1) = g(-1) \longrightarrow e^{-t} = \gamma_0 - \gamma_1$$

$$f(-2) = g(-2) \longrightarrow e^{-2t} = \gamma_0 - 2\gamma_1$$

so

$$\gamma_1 = e^{-t} - e^{-2t}, \qquad \gamma_0 = 2e^{-t} - e^{-2t}$$

and

$$e^{At} = (2e^{-t} - e^{-2t})\begin{bmatrix}1 & 0\\0 & 1\end{bmatrix} + (e^{-t} - e^{-2t})\begin{bmatrix}0 & 1\\-2 & -3\end{bmatrix}$$

$$= \begin{bmatrix}2e^{-t} - e^{-2t} & e^{-t} - e^{-2t}\\2(e^{-2t} - e^{-t}) & 2e^{-2t} - e^{-t}\end{bmatrix}.$$

Example 4.8

$$A = \begin{bmatrix}0 & 0 & -2\\0 & 1 & 0\\1 & 0 & 3\end{bmatrix}; \quad \text{find} \quad f(A) = e^{At}.$$

We find that $p(\lambda) = (\lambda-1)^2(\lambda-2)$. Using $f(\lambda) = e^{\lambda t}$ and $g(\lambda) = \gamma_0+\gamma_1\lambda+\gamma_2\lambda^2$, we see that $f'(\lambda) = te^{\lambda t}$, $g'(\lambda) = \gamma_1 + 2\gamma_2\lambda$. So from $f(1) = g(1)$, we get $e^t = \gamma_0 + \gamma_1 + \gamma_2$; from $f'(1) = g'(1)$, we get $te^t = \gamma_1 + 2\gamma_2$; and from $f(2) = g(2)$, we get $e^{2t} = \gamma_0 + 2\gamma_1 + 4\gamma_2$. Solve these three equations for $\gamma_0, \gamma_1, \gamma_2$ to get

$$\gamma_0 = e^{2t} - 2te^t, \qquad \gamma_1 = 2e^t + 3te^t - 2e^{2t}, \qquad \gamma_2 = e^{2t} - e^t - te^t.$$

This results in

$$e^{At} = (e^{2t} - 2te^t)I + (2e^t + 3te^t - 2e^{2t})A + (e^{2t} - e^t - te^t)A^2$$

$$= \begin{bmatrix}2e^t - e^{2t} & 0 & 2(e^t - e^{2t})\\0 & e^t & 0\\e^{2t} - e^t & 0 & 2e^{2t} - e^t\end{bmatrix}.$$

Note: When A has repeated eigenvalues, the minimum polynomial $p_m(\lambda)$ (see Appendix, Section A.19) can be of lower degree than the characteristic polynomial $p(\lambda)$. Then equation (4.35) should actually be written as

$$f(A) = \sum_{i=1}^{s}\sum_{j=0}^{m_i-1}\left[\frac{d^j f(\lambda)}{d\lambda^j}\right]_{\lambda=\lambda_i} Z_{ij} \tag{4.44}$$

where m_i is the multiplicity of λ_i in $p_m(\lambda)$. However, (4.44) and (4.35) will actually give the same results, since it turns out that in (4.35) the $Z_{ij} = 0$ for $j \geq m_i$. This occurred in Example 4.5, where $Z_{11} = \Theta$ since $p_m(\lambda) = (\lambda - 1)(\lambda - 2)$.

For Method 3, (4.39) is actually valid for all $N \geq m =$ the degree of the minimum polynomial. Thus $g(A)$ in (4.40) need only be of degree $(m - 1)$, and the values of f and g need only be equal on the spectrum of A up to the multiplicities $(m_i - 1)$ instead of $(n_i - 1)$ in (4.43). That is, (4.43) could actually be written as

$$\frac{d^j f(\lambda_i)}{d\lambda^j} = \frac{d^j g(\lambda_i)}{d\lambda^j} \quad \text{for} \quad i = 1, \ldots, s; \quad j = 0, \ldots, m_{i-1}. \tag{4.45}$$

Here it turns out that when (4.45) is satisfied then (4.43) is also.

However, since it can be laborious to find $p_m(\lambda)$ and the m_is, it is usually easier to use the previous results as stated.

Example 4.9

$$\text{For } A = \begin{bmatrix} 2 & 1 & 0 & 0 \\ 0 & 2 & 1 & 0 \\ 0 & 0 & 2 & 0 \\ 0 & 0 & 0 & 2 \end{bmatrix}$$

we find that $p(\lambda) = (\lambda - 2)^4$ as before but that $p_m(\lambda) = (\lambda - 2)^3$. Thus $f(A) = f(2)Z_{10} + f'(2)Z_{11} + f''(2)Z_{12}$. Using trial functions, let $f_2(\lambda) = (\lambda - 2)^2$, so $f_2(2) = 0$, $f_2'(2) = 0$, and $f_2''(2) = 2$. Thus

$$f_2(A) = (A - 2I)^2 = \begin{bmatrix} 0 & 0 & 1 & 0 \\ 0 & 0 & 0 & 0 \\ 0 & 0 & 0 & 0 \\ 0 & 0 & 0 & 0 \end{bmatrix} = 2Z_{12}.$$

Next, let $f_1(\lambda) = (\lambda - 2)$, so $f_1(2) = 0$, $f_1'(2) = 1$, $f_1''(2) = 0$; thus

$$f_1(A) = (A - 2I) = \begin{bmatrix} 0 & 1 & 0 & 0 \\ 0 & 0 & 1 & 0 \\ 0 & 0 & 0 & 0 \\ 0 & 0 & 0 & 0 \end{bmatrix}.$$

Finally, let $f_0(\lambda) = 1$, so $f_0(2) = 1$, $f_0'(2) = f_0''(2) = 0$; thus $f_0(A) = I = Z_{10}$. This gives

$$e^{At} = e^{2t}I + te^{2t}Z_{11} + t^2e^{2t}Z_{12} = \begin{bmatrix} e^{2t} & te^{2t} & \frac{1}{2}t^2e^{2t} & 0 \\ 0 & e^{2t} & te^{2t} & 0 \\ 0 & 0 & e^{2t} & 0 \\ 0 & 0 & 0 & e^{2t} \end{bmatrix}.$$

Example 4.10

For the same A as in Example 4.9, since $p(\lambda) = (\lambda - 2)^4$, if we use

$$f(A) = f(2)Z_{10} + f'(2)Z_{11} + f''(2)Z_{12} + f'''(2)Z_{13}$$

let us try the trial function $f_3(\lambda) = (\lambda - 2)^3$, which gives $f_3(2) = 0$, $f_3'(2) = 0$, $f_3''(2) = 0$, and $f_3'''(2) = 3! = 6$. Thus $f_3(A) = (A - 2I)^3 = 6Z_{13}$, but we find that $(A - 2I)^3 = \Theta$ (this is because the minimum polynomial actually is $p_m(\lambda) = (\lambda - 2)^3$). We see, therefore, that $Z_{13} = \Theta$ and the remaining $Z_{\ell i}$ are found just as in Example 4.9.

4.3.4 $\phi(t, t_0)$ for time-varying system matrix $A(t)$

We saw above that it was possible to find the transition matrix of a system with a constant A matrix without solving differential equations. We would like to be able to do the same thing when $A(t)$ depends on t. Since the solution to the unforced time-varying scalar differential equation

$$\dot{x}(t) = a(t)x(t) \quad \text{with} \quad x(t_0) = x_0$$

is $x(t) = e^{\int_{t_0}^{t} a(\tau)\,d\tau} x_0$, by analogy with this scalar case, one *might* think that for time-varying systems $\phi(t, t_0)$ is related to $A(t)$ by

$$\phi(t, t_0) = \exp\left[\int_{t_0}^{t} A(\tau)\,d\tau\right]. \tag{4.46}$$

We shall see that for very special $A(t)$ matrices this is true, but (4.46) is *not* true in general.

Let us define

$$F(t, t_0) \triangleq \int_{t_0}^{t} A(\tau)\,d\tau. \tag{4.47}$$

Theorem 4.2 *If $A(t)F(t, t_0) = F(t, t_0)A(t)$ (i.e., if $F(t, t_0)$ and $A(t)$ commute) for all t and t_0 on an interval $\mathcal{I}$, then $\phi(t, t_0) = e^{F(t,t_0)}$ for all t and t_0 on $\mathcal{I}$.*

This proof is left as an exercise for the reader (see Problem 4.14). It involves using the infinite series definition of the matrix exponential.

Theorem 4.3 *$A(t)F(t, t_0) = F(t, t_0)A(t)$ for all t and t_0 on an interval $\mathcal{I}$ if and only if $A(t)A(\tau) = A(\tau)A(t)$ for all t and τ on $\mathcal{I}$.*

In other words, $A(t)$ and $F(t, t_0)$ commute if and only if $A(t)$ and $A(\tau)$ do. The condition of Theorem 4.3 is somewhat easier to check than that of Theorem 4.2.

Proof: Sufficiency (if)
If $A(t)A(\tau) = A(\tau)A(t)$ for all t and τ on $\mathcal{I}$ then integrate on τ to get

$$A(t)\int_{t_0}^{t} A(\tau)\,d\tau = \int_{t_0}^{t} A(\tau)\,d\tau\,A(t)$$

so $A(t)$ and $F(t, t_0)$ commute.
Necessity (only if)
If $A(t)F(t, t_0) = F(t, t_0)A(t)$ then just take the partial derivative of both sides with respect to t_0 and get $-A(t)A(t_0) = -A(t_0)A(t)$ so $A(t)$ and $A(t_0)$ commute also. This shows that if $A(t)A(\tau) \neq A(\tau)A(t)$ then $A(t)$ and $F(t, t_0)$ could not commute. QED.

Thus we see that it is only necessary to check whether $A(t)$ and $A(\tau)$ commute, in order to determine if $\phi(t, t_0) = e^{F(t,t_0)}$. The importance of this result lies in the fact that for matrices $A(t)$ that satisfy the conditions of Theorem 4.3, Methods 2 or 3 can be used to find $\phi(t, t_0) = e^{F(t,t_0)}$ even though $A(t)$ is not constant. We actually will show this for matrices which can be diagonalized and, as before, just state the result for general matrices.

Let us first observe that if $\Lambda(t)$ is a diagonal matrix with diagonal elements $\lambda_i(t)$ then $\Lambda(t)\Lambda(\tau) = \Lambda(\tau)\Lambda(t)$ for all t and τ, so $\phi(t, t_0) = e^{\int_{t_0}^{t} \Lambda(\tau)\,d\tau}$ is a diagonal matrix with diagonal elements $e^{\int_{t_0}^{t} \lambda_i(\tau)\,d\tau}$, and we can always find $\phi(t, t_0)$ for a diagonal matrix $\Lambda(t)$.

Note that our derivation of Method 2 depended on the existence of a constant matrix T which transformed A into a diagonal matrix (actually into Jordan canonical form when A cannot be diagonalized). Thus for a time-varying $A(t)$, if a constant matrix T exists for which $T^{-1}A(t)T = \Lambda(t)$ where $\Lambda(t)$ is a diagonal matrix whose diagonal elements are the eigenvalues $\lambda_i(t)$ of $A(t)$ (when $A(t)$ has constant eigenvectors, then, just as before, T is the matrix whose columns are these eigenvectors), we can show, as before, that for any function $f(\lambda)$ which can be expanded in a power series we can define

$$f(A(t)) = Tf(\Lambda(t))T^{-1}$$

just as in (4.28), so (4.30) also holds with constant Z_{k0}, and Method 2 can be used to find the constant Z_{k0}. Once the Z_{k0} have been found, the only difference here is that $\phi(t, t_0) = e^{F(t,t_0)}$, so we must use $f(\lambda(t)) = e^{\int_{t_0}^{t} \lambda(\tau)\,d\tau}$ instead of $e^{\lambda(t-t_0)}$ since the eigenvalues of $F(t, t_0)$ are $\int_{t_0}^{t} \lambda_i(\tau)\,d\tau$ where the eigenvalues $\lambda_i(\tau)$ of $A(t)$ are not all constant.

Similarly, (4.42) can be used to find the unknowns $\gamma_0, \gamma_1, \ldots, \gamma_{n-1}$ in Method 3. Here too, $f(\lambda(t)) = e^{\int_{t_0}^{t} \lambda(\tau)\,d\tau}$ instead of $e^{\lambda(t-t_0)}$.

For matrices that cannot be diagonalized, but can be put in Jordan canonical form by a *constant* matrix T (i.e., $T^{-1}A(t)T = J(t)$), we can use Method 2 or 3 to find $\phi(t, t_0)$. For Method 2 use (4.35) to find the constant Z_{ij}, and for Method 3 use (4.43) to find the unknowns $\gamma_0, \gamma_1, \ldots, \gamma_{n-1}$ (see Appendix, Section A.20 for a more complete discussion).

Note: Such a constant T exists if and only if the time-varying matrix $A(t)$ has constant eigenvectors and generalized eigenvectors (see Appendix, Sections A.14–A.15). When this happens, $A(t)$ behaves almost like a constant matrix even though it is time-varying. Just remember to use $f(\lambda(t)) = e^{\int_{t_0}^{t} \lambda(\tau)\,d\tau}$ instead of $e^{\lambda(t-t_0)}$.

The simplest application of these results is to time-dependent matrices $A(t) = \alpha(t)\hat{A}$ where $\hat{A}$ is a constant matrix and $\alpha(t)$ is a scalar time function. For such matrices we see that $A(t)A(\tau) = \alpha(t)\alpha(\tau)\hat{A}^2 = A(\tau)A(t)$. Thus $\phi_a(t, t_0) = e^{\beta(t,t_0)\hat{A}}$ where $\beta(t, t_0) = \int_{t_0}^{t} \alpha(\tau)d\tau$.

Example 4.11
For $A(t) = \begin{bmatrix} 0 & t \\ -2t & -3t \end{bmatrix} = t \begin{bmatrix} 0 & 1 \\ -2 & -3 \end{bmatrix}$

we see that $\beta(t, t_0) = \int_{t_0}^{t} \tau d\tau = \frac{1}{2}(t^2 - t_0^2)$ and $\hat{A} = \begin{bmatrix} 0 & 1 \\ -2 & -3 \end{bmatrix}$. We found in Example 4.3 that this $\hat{A}$ has $\lambda_1 = -1$, $\lambda_2 = -2$, and

$$Z_{10} = \begin{bmatrix} 2 & 1 \\ -2 & -1 \end{bmatrix}; \qquad Z_{20} = \begin{bmatrix} -1 & -1 \\ 2 & 2 \end{bmatrix}.$$

Therefore using $f(\lambda) = e^{\beta(t,t_0)\lambda}$ we get

$$\phi(t, t_0) = e^{\beta(t,t_0)\hat{A}} = e^{-\frac{1}{2}\left(t^2-t_0^2\right)} \begin{bmatrix} 2 & 1 \\ -2 & -1 \end{bmatrix} + e^{-(t^2-t_0^2)} \begin{bmatrix} -1 & -1 \\ 2 & 2 \end{bmatrix}$$

$$= \begin{bmatrix} 2e^{-\frac{1}{2}\left(t^2-t_0^2\right)} - e^{-\left(t^2-t_0^2\right)} & e^{-\frac{1}{2}\left(t^2-t_0^2\right)} - e^{-\left(t^2-t_0^2\right)} \\ 2\left(e^{-\left(t^2-t_0^2\right)} - e^{-\frac{1}{2}\left(t^2-t_0^2\right)}\right) & 2e^{-\left(t^2-t_0^2\right)} - e^{-\frac{1}{2}\left(t^2-t_0^2\right)} \end{bmatrix}.$$

The next example is more complicated.

Example 4.12
The matrix

$$A(t) = \begin{bmatrix} -1-2t & 2t-\frac{1}{2} & -\frac{1}{2} \\ 2t & -\frac{1}{2}-2t & \frac{1}{2} \\ 1-2t & \frac{1}{2}-2t & \frac{1}{2}-4t \end{bmatrix}$$

is not in the simple form mentioned above. We first form

$$A(t)A(t_0) = \begin{bmatrix} \frac{1}{2}+2(t+t_0)+8tt_0 & \frac{1}{2}-8tt_0 & 2(t+t_0) \\ \frac{1}{2}-2(t+t_0)-8tt_0 & \frac{1}{2}+8tt_0 & -2(t+t_0) \\ -\frac{1}{2}-2(t+t_0)+8tt_0 & -\frac{1}{2}+8tt_0 & 2(-t-t_0+8tt_0) \end{bmatrix}$$

and we see that if t and t_0 are interchanged in the above equation, we get the same matrix. Therefore $A(t)$ and $A(t_0)$ commute for all t and t_0, so the transition matrix $\phi(t, t_0) = \exp\left[\int_{t_0}^{t} A(\tau)\, d\tau\right]$. We need the eigenvalues of $A(t)$. Taking

$$p(\lambda) = \det[\lambda I - A(t)] = \lambda^3 + (8t+1)\lambda^2 + 8t(1+t)\lambda + 16t^2 = (\lambda+1)(\lambda+4t)^2$$

we see that the eigenvalues are $\lambda_1 = -1$ and $\lambda_2 = -4t$ with multiplicity 2. Using Method 2 to get $\phi(t, t_0)$, we choose as the first trial function $f_1(\lambda) = (\lambda+1)(\lambda+4t)$, so $f_1'(\lambda) = 2\lambda + 4t + 1$. Then $f_1(A) = (A(t) + I)(A(t) + 4tI) = f_1'(\lambda_2)Z_{21} =$

$(1-4t)Z_{21}$, so

$$Z_{21} = \frac{1}{1-4t}\begin{bmatrix} -2t & 2t-\frac{1}{2} & -\frac{1}{2} \\ 2t & \frac{1}{2}-2t & \frac{1}{2} \\ 1-2t & \frac{1}{2}-2t & \frac{3}{2}-4t \end{bmatrix}\begin{bmatrix} 2t-1 & 2t-\frac{1}{2} & -\frac{1}{2} \\ 2t & 2t-\frac{1}{2} & \frac{1}{2} \\ 1-2t & \frac{1}{2}-2t & \frac{1}{2} \end{bmatrix} = \frac{1}{2}\begin{bmatrix} -1 & 0 & -1 \\ 1 & 0 & 1 \\ 1 & 0 & 1 \end{bmatrix}.$$

Next we choose $f_2(\lambda) = (\lambda + 4t)$, so $f_2'(\lambda) = 1$. Then $f_2(A) = (A(t) + 4tI) = f_2(\lambda_1)Z_{10} + f_2'(\lambda_2)Z_{21} = (4t-1)Z_{10} + Z_{21}$, so

$$Z_{10} = \frac{1}{4t-1}\{A(t) + 4tI - Z_{21}\}$$

$$= \frac{1}{4t-1}\left\{\begin{bmatrix} 2t-1 & 2t-\frac{1}{2} & -\frac{1}{2} \\ 2t & 2t-\frac{1}{2} & \frac{1}{2} \\ 1-2t & \frac{1}{2}-2t & \frac{1}{2} \end{bmatrix} - \frac{1}{2}\begin{bmatrix} -1 & 0 & -1 \\ 1 & 0 & 1 \\ 1 & 0 & 1 \end{bmatrix}\right\}$$

$$= \frac{1}{2}\begin{bmatrix} 1 & 1 & 0 \\ 1 & 1 & 0 \\ -1 & -1 & 0 \end{bmatrix}.$$

Finally we can choose $f_3(\lambda) = 1$ so $f_3'(\lambda) = 0$. This gives $f_3(A) = I = Z_{10} + Z_{20}$, so

$$Z_{20} = I - Z_{10} = \frac{1}{2}\begin{bmatrix} 1 & -1 & 0 \\ -1 & 1 & 0 \\ 1 & 1 & 2 \end{bmatrix}.$$

Now using $f(\lambda) = e^{\int_{t_0}^{t} \lambda(\tau)d\tau}$ and since

$$f'(\lambda) = e^{\int_{t_0}^{t} \lambda(\tau)d\tau}\frac{d\left[\int_{t_0}^{t} \lambda(\tau)d\tau\right]}{d\lambda} = e^{\int_{t_0}^{t} \lambda(\tau)d\tau}\int_{t_0}^{t} d\tau = (t-t_0)e^{\int_{t_0}^{t} \lambda(\tau)d\tau}$$

we find

$$\phi(t,t_0) = e^{-(t-t_0)}Z_{10} + e^{-2\left(t^2-t_0^2\right)}Z_{20} + (t-t_0)e^{-2\left(t^2-t_0^2\right)}Z_{21}$$

$$= \frac{1}{2}\begin{bmatrix} e^{-(t-t_0)} + (1-t+t_0)e^{-2\left(t^2-t_0^2\right)} & e^{-(t-t_0)} - e^{-2\left(t^2-t_0^2\right)} & -(t-t_0)e^{-2\left(t^2-t_0^2\right)} \\ e^{-(t-t_0)} + (t-t_0-1)e^{-2\left(t^2-t_0^2\right)} & e^{-(t-t_0)} + e^{-2\left(t^2-t_0^2\right)} & (t-t_0)e^{-2\left(t^2-t_0^2\right)} \\ (t-t_0+1)e^{-2\left(t^2-t_0^2\right)} - e^{-(t-t_0)} & e^{-2\left(t^2-t_0^2\right)} - e^{-(t-t_0)} & (2+t-t_0)e^{-2\left(t^2-t_0^2\right)} \end{bmatrix}.$$

We can verify that this transition matrix is correct by showing that $\dot{\phi}(t,t_0) = A(t)\phi(t,t_0)$ (see Problem 4.18a).

Method 2 worked for Example 4.12 because the constant matrix

$$T = \begin{bmatrix} 1 & -1 & 0 \\ 1 & 1 & 0 \\ -1 & 1 & -2 \end{bmatrix}$$

puts $A(t)$ into Jordan form.

The fact that $A(t)$ and $A(\tau)$ commute does *not* guarantee that such a constant transformation T exists. When the eigenvectors and generalized eigenvectors are not constant vectors, Methods 2 and 3 can still be used, provided the *form* of the characteristic equation of $F(t, t_0)$ does not change (i.e., the multiplicity of all eigenvalues remain constant) even when the eigenvalues themselves change with time. Since $\phi(t, t_0) = e^{F(t,t_0)}$ when $A(t)$ and $A(t_0)$ commute, we apply the methods to the matrix $F(t, t_0)$ in order to evaluate the matrix function $e^{F(t,t_0)}$. The validity of using these methods for time-varying matrices can be established by the approach used by Gantmacher (1959: vol. 1, ch. 5). For constant matrices F, he defines the spectrum of F in terms of the minimum polynomial and shows that the values of a polynomial $g(\lambda)$ on the spectrum of F determine the matrix $g(F)$ completely. He then defines the matrix function $f(F)$ as $g(F)$ for an arbitrary function $f(\lambda)$ which is defined and equal to the values of $g(\lambda)$ on the spectrum of F. This approach carries over directly to time-varying matrices $F(t)$. It also applies when the spectrum of $F(t)$ is defined (as we did above) in terms of the characteristic polynomial instead of the minimum polynomial since two functions that have values equal on the spectrum as defined by the characteristic polynomial are also equal on the spectrum of the minimum polynomial which is of equal or lower order.

The following example illustrates the application to a matrix for which a constant transformation T to Jordan form does not exist.

Example 4.13

Consider the matrix

$$A(t) = \begin{bmatrix} 1 & t^2 & t^4 \\ 0 & 1 & t^2 \\ 0 & 0 & 1 \end{bmatrix}.$$

First we see that

$$\begin{aligned} A(t)A(t_0) &= \begin{bmatrix} 1 & t^2 & t^4 \\ 0 & 1 & t^2 \\ 0 & 0 & 1 \end{bmatrix} \begin{bmatrix} 1 & t_0^2 & t_0^4 \\ 0 & 1 & t_0^2 \\ 0 & 0 & 1 \end{bmatrix} \\ &= \begin{bmatrix} 1 & t^2 + t_0^2 & t^4 + t^2 t_0^2 + t_0^4 \\ 0 & 1 & t^2 + t_0^2 \\ 0 & 0 & 1 \end{bmatrix} = A(t_0)A(t) \end{aligned}$$

so $A(t)$ and $A(t_0)$ commute and therefore $\phi(t, t_0) = e^{F(t,t_0)}$. We find that the characteristic equation is $\det[\lambda I - A(t)] = p(\lambda) = (\lambda - 1)^3$, so $A(t)$ has an eigenvalue at $\lambda = 1$ of multiplicity 3. The eigenvector and generalized eigenvectors are found to be

$$\boldsymbol{\xi}^{(1)} = \begin{bmatrix} 1 \\ 0 \\ 0 \end{bmatrix}; \quad \boldsymbol{\xi}^{(2)} = \begin{bmatrix} 1 \\ \frac{1}{t^2} \\ 0 \end{bmatrix}; \quad \boldsymbol{\xi}^{(1)} = \begin{bmatrix} 0 \\ 0 \\ \frac{1}{t^4} \end{bmatrix}$$

which are not all constant. Let us apply Method 3 to get $\phi(t, t_0) = e^{F(t,t_0)}$. First we find that

$$F(t, t_0) = \int_{t_0}^{t} A(\tau)d\tau = \begin{bmatrix} (t - t_0) & \frac{1}{3}(t^3 - t_0^3) & \frac{1}{5}(t^5 - t_0^5) \\ 0 & (t - t_0) & \frac{1}{3}(t^3 - t_0^3) \\ 0 & 0 & (t - t_0) \end{bmatrix}$$

which has an eigenvalue at $\lambda(t) = (t - t_0)$ of multiplicity 3. Using $f(\lambda) = e^{\lambda}$, $f'(\lambda) = f''(\lambda) = e^{\lambda}$. For $g(\lambda) = \gamma_2\lambda^2 + \gamma_1\lambda + \gamma_0$ we have $g'(\lambda) = 2\gamma_2\lambda + \gamma_1$ and $g''(\lambda) = 2\gamma_2$. Equating $f(\lambda)$ and $g(\lambda)$ on the spectrum of $F(t, t_0)$ gives the equations

$$\begin{aligned} f(t - t_0) &= e^{(t-t_0)} = \gamma_2(t - t_0)^2 + \gamma_1(t - t_0) + \gamma_0 = g(t - t_0) \\ f'(t - t_0) &= e^{(t-t_0)} = 2\gamma_2(t - t_0) + \gamma_1 = g'(t - t_0) \\ f''(t - t_0) &= e^{(t-t_0)} = 2\gamma_2 = g''(t - t_0). \end{aligned} \tag{4.48}$$

Thus

$$e^{F(t,t_0)} = g\left(F(t, t_0)\right) = \gamma_2 F^2(t, t_0) + \gamma_1 F(t, t_0) + \gamma_0 I \tag{4.49}$$

where

$$F^2(t, t_0) = \begin{bmatrix} (t - t_0)^2 & \frac{2}{3}(t - t_0)(t^3 - t_0^3) & \frac{2}{5}(t - t_0)(t^5 - t_0^5) + \frac{1}{9}(t^3 - t_0^3)^2 \\ 0 & (t - t_0)^2 & \frac{2}{3}(t - t_0)(t^3 - t_0^3) \\ 0 & 0 & (t - t_0)^2 \end{bmatrix}.$$

Solving (4.48) for $\gamma_2, \gamma_1, \gamma_0$ and inserting in (4.49) results in

$$\phi(t, t_0) = e^{(t-t_0)} \begin{bmatrix} 1 & \frac{1}{3}(t^3 - t_0^3) & \frac{1}{5}(t^5 - t_0^5) + \frac{1}{18}(t^3 - t_0^3)^2 \\ 0 & 1 & \frac{1}{3}(t^3 - t_0^3) \\ 0 & 0 & 1 \end{bmatrix}.$$

Again, we can verify that this transition matrix is correct by showing that $\dot{\phi}(t, t_0) = A(t)\phi(t, t_0)$ (see Problem 4.18b). Method 2 will give the same result.

4.4 Discrete-time systems

For discrete-time systems, the state equations are

$$\boldsymbol{x}(k + 1) = A(k)\boldsymbol{x}(k) + B(k)\boldsymbol{u}(k) \tag{4.50a}$$

$$\boldsymbol{y}(k) = C(k)\boldsymbol{x}(k) + D(k)\boldsymbol{u}(k) \tag{4.50b}$$

and as in the case of analog systems the problem is to obtain a general solution for an arbitrary initial condition $\boldsymbol{x}(k_0) = \boldsymbol{x}_0$ and an arbitrary input function $\{\boldsymbol{u}(k)\}$.

In this case, a solution can be obtained directly by iterating the equations (4.50a) starting with $k = k_0$. Thus

$$\begin{aligned}
\boldsymbol{x}(k_0+1) &= A(k_0)\boldsymbol{x}_0 + B(k_0)\boldsymbol{u}(k_0) \\
\boldsymbol{x}(k_0+2) &= A(k_0+1)\boldsymbol{x}(k_0+1) + B(k_0+1)\boldsymbol{u}(k_0+1) \\
&= A(k_0+1)A(k_0)\boldsymbol{x}_0 + A(k_0+1)B(k_0)\boldsymbol{u}(k_0) \\
&\quad + B(k_0+1)\boldsymbol{u}(k_0+1) \\
\boldsymbol{x}(k_0+3) &= A(k_0+2)\boldsymbol{x}(k_0+2) + B(k_0+2)\boldsymbol{u}(k_0+2) \\
&= A(k_0+2)A(k_0+1)A(k_0)\boldsymbol{x}_0 \\
&\quad + A(k_0+2)A(k_0+1)B(k_0)\boldsymbol{u}(k_0) \\
&\quad + A(k_0+2)B(k_0+1)\boldsymbol{u}(k_0+1) + B(k_0+2)\boldsymbol{u}(k_0+2).
\end{aligned}$$

Continuing in this way gives

$$\begin{aligned}
\boldsymbol{x}(k_0+j) &= A(k_0+j-1)\ldots A(k_0)\boldsymbol{x}_0 \\
&\quad + A(k_0+j-1)\ldots A(k_0+1)B(k_0)\boldsymbol{u}(k_0) + \cdots \\
&\quad + A(k_0+j-1)B(k_0+j-2)\boldsymbol{u}(k_0+j-2) \\
&\quad + B(k_0+j-1)\boldsymbol{u}(k_0+j-1) \quad \text{for} \quad j \geq 1.
\end{aligned} \tag{4.51}$$

Let us define the transition matrix as

$$\phi(k,i) \triangleq A(k-1)A(k-2)\ldots A(i) \quad \text{for} \quad k \geq i+1 \tag{4.52}$$

and also $\phi(k,k) = I$, which enables us to write (4.51) as

$$\begin{aligned}
\boldsymbol{x}(k_0+j) &= \phi(k_0+j, k_0)\boldsymbol{x}_0 + \phi(k_0+j, k_0+1)B(k_0)\boldsymbol{u}(k_0) + \cdots \\
&\quad + \phi(k_0+j, k_0+j-1)B(k_0+j-2)\boldsymbol{u}(k_0+j-2) \\
&\quad + \phi(k_0+j, k_0+j)B(k_0+j-1)\boldsymbol{u}(k_0+j-1) \quad \text{for } j \geq 1
\end{aligned} \tag{4.53}$$

or

$$\begin{aligned}
\boldsymbol{x}(k_0+j) &= \phi(k_0+j, k_0)\boldsymbol{x}_0 \\
&\quad + \sum_{i=k_0}^{k_0+j-1} \phi(k_0+j, i+1)B(i)\boldsymbol{u}(i) \quad \text{for } j \geq 1.
\end{aligned} \tag{4.54}$$

Let $k = k_0 + j$ in (4.49). This gives

$$x(k) = \phi(k, k_0)x_0 + \sum_{i=k_0}^{k-1} \phi(k, i+1)B(i)u(i) \quad \text{for} \quad k > k_0. \tag{4.55}$$

Inserting this expression into the output equation (4.50b), we get

$$y(k) = \underbrace{C(k)\phi(k, k_0)x_0}_{\text{zero-input response}} + \underbrace{\sum_{i=k_0}^{k-1} C(k)\phi(k, i+1)B(i)u(i) + D(k)u(k)}_{\text{zero-state response}} \quad \text{for} \quad k > k_0. \tag{4.56}$$

We see from (4.56) that discrete-dynamic systems governed by linear difference equations describe a linear system as defined in Chapter 1.

Again this can be related to the form of the superposition summation by taking $k_0 = -\infty$ and $x_0 = \mathbf{0}$ (i.e., consider only the zero-state response). Then

$$y_{zs}(k) = \sum_{i=-\infty}^{k-1} C(k)\phi(k, i+1)B(i)u(i)1(k-1-i) + D(k)u(k). \tag{4.57}$$

The upper limit can be changed to ∞ without altering the summation, and since the input $u(k)$ can be expressed as

$$u(k) = \sum_{i=-\infty}^{\infty} u(i)\delta(k-i)$$

(4.51) becomes

$$y(k) = \sum_{i=-\infty}^{\infty} \underbrace{[C(k)\phi(k, i+1)B(i)1(k-1-i) + D(k)\delta(k-i)]}_{H(k,i)} u(i). \tag{4.58}$$

This corresponds to the expression we found earlier for the zero-state response of multiple-input multiple-output systems, when we identify the delta response matrix as

$$H(k, i) = C(k)\phi(k, i+1)B(i)\; 1(k-i-1) + D(k)\delta(k-i) \tag{4.59}$$

which relates $H(k, i)$ to the matrices of the difference and output equations. For fixed systems, since A, B, C, and D are constant matrices, we find

$$H(k-i) = C\phi(k-i-1)B\; 1(k-i-1) + D\delta(k-i). \tag{4.60}$$

Note: By our definition of $\phi(k, i)$, we see that for fixed systems

$$\phi(k, i+1) = \phi(k-i-1) = A^{(k-i-1)}$$

for $k \geq i+1$ where $A^0 = I$. Equivalently $\phi(k) = A^k$ for $k \geq 0$.

Thus the matrix A^k can be found for any $k \geq 0$ by using our methods for finding functions of a matrix. That is, for Method 2, use $f(\lambda) = \lambda^k$ in (4.35). The Z_{ji} are evaluated as before to give A^k. For Method 3, use $f(\lambda) = \lambda^k$ in (4.40) and evaluate the γ_i by using (4.43).

For discrete systems, the normal modes are discrete functions of the form $\{\lambda^k\}$ when eigenvalues are distinct and $\{k^{\ell_j}\lambda_j^k\}$ when eigenvalues are repeated in the minimum polynomial (where $0 \leq \ell_j \leq m_{j-1}$).

Properties of $\phi(k, k_0)$

1 $\phi(k_0, k_0) = I$ from the definition. This is the same property as in the analog case.

2 $\phi(k_2, k_1)\phi(k_1, k_0) = \phi(k_2, k_0)$ for $k_2 \geq k_1 \geq k_0$. Note that the restriction on the k_i is not present in the analog case. The result follows from

$$\phi(k_2, k_0) = \underbrace{A(k_2-1)\dots A(k_1)}_{\phi(k_2; k_1)}\,\underbrace{A(k_1-1)\dots A(k_0)}_{\phi(k_1, k_0)}.$$

3 If $A(k)$ is nonsingular for all k, then $(\phi(k_2, k_1)^{-1}$ exists. It is given by

$$\phi(k_2, k_1)^{-1} = A(k_1)^{-1}A(k_1+1)^{-1}\dots A(k_2-1)^{-1} \quad \text{for} \quad k_2 > k_1. \tag{4.61}$$

Note: $\phi(k_2, k_1)^{-1}$ for $k_2 > k_1$ is not the same as $\phi(k_1, k_2)$ from (4.52). This is because (4.52) defines $\phi(k_1, k_2)$ only for $k_1 \geq k_2$. When $k_1 < k_2$ and the $A(k)$ matrices are nonsingular, we can use (4.61) to define $\phi(k_1, k_2) \triangleq \phi(k_2, k_1)^{-1}$.

4 $\phi(k+1, k) = A(k)$ follows directly from (4.52). This corresponds to Property 4 for analog transition matrices.

Example 4.14

Consider the system

$$x(k+1) = \begin{bmatrix} -\frac{3}{2} & -1 \\ 1 & 1 \end{bmatrix} x(k) + \begin{bmatrix} 0 \\ 1 \end{bmatrix} u(k)$$
$$y(k) = x_1(k) + x_2(k) + u(k).$$

To find $\phi(k) = A^k$, we take

$$\det \begin{bmatrix} \lambda+2 & 1 \\ -1 & \lambda-1 \end{bmatrix} = \lambda^2 + \frac{1}{2}\lambda - \frac{1}{2} = (\lambda+1)\left(\lambda - \frac{1}{2}\right) = 0.$$

Then using $f(A) = f(-1)Z_{10} + f\left(\frac{1}{2}\right)Z_{20}$ we can choose $f_2(\lambda) = \lambda + 1$, so $f_2(A) = A + I = \frac{3}{2}Z_{20}$ and $f_1(\lambda) = (\lambda - \frac{1}{2})$, and thus $f_1(A) = A - \frac{1}{2}I = -\frac{3}{2}Z_{10}$

which gives

$$Z_{10} = -\frac{2}{3}\begin{bmatrix} -2 & -1 \\ 1 & \frac{1}{2} \end{bmatrix}; \quad Z_{20} = \frac{2}{3}\begin{bmatrix} -\frac{1}{2} & -1 \\ 1 & 2 \end{bmatrix}.$$

Thus

$$\phi(k) = A^k = -\frac{2}{3}(-1)^k \begin{bmatrix} -2 & -1 \\ 1 & \frac{1}{2} \end{bmatrix} + \frac{2}{3}\left(\frac{1}{2}\right)^k \begin{bmatrix} -\frac{1}{2} & -1 \\ 1 & 2 \end{bmatrix}$$

$$= \frac{2}{3}\begin{bmatrix} 2(-1)^k - \frac{1}{2}\left(\frac{1}{2}\right)^k & (-1)^k - \left(\frac{1}{2}\right)^k \\ \left(\frac{1}{2}\right)^k - (-1)^k & 2\left(\frac{1}{2}\right)^k - \frac{1}{2}(-1)^k \end{bmatrix}.$$

By Method 3 for $f(\lambda) = \lambda^k$ and $g(\lambda) = \gamma_0 + \gamma_1\lambda$, we get

$$(-1)^k = \gamma_0 - \gamma_1 \quad \text{and} \quad \left(\frac{1}{2}\right)^k = \gamma_0 + \frac{1}{2}\gamma_1$$

which gives $\gamma_1 = \frac{2}{3}\left[\left(\frac{1}{2}\right)^k - (-1)^k\right]$ and $\gamma_0 = \frac{1}{3}\left[2\left(\frac{1}{2}\right)^k + (-1)^k\right]$. Therefore we get

$$A^k = \gamma_0 I + \gamma_1 A = \begin{bmatrix} \gamma_0 - \frac{3}{2}\gamma_1 & -\gamma_1 \\ \gamma_1 & \gamma_0 + \gamma_1 \end{bmatrix} = \begin{bmatrix} \frac{4}{3}(-1)^k - \frac{1}{3}\left(\frac{1}{2}\right)^k & -\gamma_1 \\ \gamma_1 & \frac{4}{3}\left(\frac{1}{2}\right)^k - \frac{1}{3}(-1)^k \end{bmatrix}$$

which is the same as we found using Method 2.

We can now obtain

$$h(k) = C\phi(k-1)B\,1(k-1) + D\delta(k)$$

$$= \frac{2}{3}[1 \quad 1]\begin{bmatrix} (-1)^{k-1} - \left(\frac{1}{2}\right)^{(k-1)} \\ 2\left(\frac{1}{2}\right)^{k-1} - \frac{1}{2}(-1)^{k-1} \end{bmatrix} 1(k-1) + \delta(k)$$

$$= \frac{2}{3}\left[\frac{1}{2}(-1)^{k-1} + \left(\frac{1}{2}\right)^{(k-1)}\right] 1(k-1) + \delta(k).$$

To find the response to $u(k) = 1$ for all $k > 0$ and $\boldsymbol{x}(0) = \boldsymbol{0}$, we apply the superposition sum to give

$$y(k) = \sum_{i=-\infty}^{\infty} h(k-i)u(i)$$

$$= \sum_{i=0}^{\infty}\left[\frac{2}{3}\left[\frac{1}{2}(-1)^{k-i-1} + \left(\frac{1}{2}\right)^{k-i-1}\right] 1(k-i-1) + \delta(k-i)\right]$$

$$= \frac{2}{3}\sum_{i=0}^{k-1}\left[\frac{1}{2}(-1)^{k-i-1} + \left(\frac{1}{2}\right)^{k-i-1}\right] 1(k-1) + 1(k).$$

We are not familiar with evaluating such sums but note that

$$S1 \triangleq \sum_{i=0}^{k-1}(-1)^{k-i-1} = \sum_{j=0}^{k-1}(-1)^j = \begin{cases} 1 & \text{for } k \text{ odd} \\ 0 & \text{for } k \text{ even} \end{cases}$$

so we can write

$$S1 = \frac{1}{2}\left[1 - (-1)^k\right].$$

Also,

$$S2 \triangleq \sum_{i=0}^{k-1}\left(\frac{1}{2}\right)^{k-i-1} = \sum_{j=0}^{k-1}\left(\frac{1}{2}\right)^{j} = \sum_{j=0}^{\infty}\left(\frac{1}{2}\right)^{j} - \sum_{j=k}^{\infty}\left(\frac{1}{2}\right)^{j}$$

and let $\ell = j - k$, so

$$S2 = 2 - \sum_{\ell=0}^{\infty}\left(\frac{1}{2}\right)^{\ell}\left(\frac{1}{2}\right)^{k} = 2\left[1 - \left(\frac{1}{2}\right)^{k}\right].$$

Using $S1$ and $S2$, we find the step response is

$$y(k) = \frac{2}{3}\left[\frac{1}{4}\left(1 - (-1)^k\right) + 2 - 2\left(\frac{1}{2}\right)^{k}\right] 1(k-1) + 1(k).$$

Example 4.15
Find $\phi(k)$ for

$$A = \begin{bmatrix} 1 & 1 \\ \frac{1}{2} & \frac{1}{2} \end{bmatrix}.$$

Since $p(\lambda) = \det\begin{bmatrix} \lambda - 1 & -1 \\ -\frac{1}{2} & \lambda - \frac{1}{2} \end{bmatrix} = (\lambda - 1)(\lambda - \frac{1}{2}) - \frac{1}{2} = \lambda(\lambda - \frac{3}{2})$, we see that $\lambda_1 = 0$ and $\lambda_2 = \frac{3}{2}$.

Using $f_1(\lambda) = \lambda$, we find $f_1(A) = A = \frac{3}{2}Z_{20}$, and for $f_2(\lambda) = (\lambda - \frac{3}{2})$, we find $f_2(A) = A - \frac{3}{2}I = -\frac{3}{2}Z_{10}$. Thus

$$Z_{10} = I - \frac{3}{2}A = \begin{bmatrix} \frac{1}{3} & -\frac{2}{3} \\ -\frac{1}{3} & \frac{2}{3} \end{bmatrix}; \qquad Z_{20} = \frac{2}{3}A = \begin{bmatrix} \frac{2}{3} & \frac{2}{3} \\ \frac{1}{3} & \frac{1}{3} \end{bmatrix}.$$

But $A^k = \lambda_1^k Z_{10} + \lambda_2^k Z_{20}$ gives

$$A^k = \left(\frac{3}{2}\right)^k \begin{bmatrix} \frac{2}{3} & \frac{2}{3} \\ \frac{1}{3} & \frac{1}{3} \end{bmatrix} \quad \text{for } k > 0.$$

Note: This result is not correct for $k = 0$ since we know that $A^0 = I$. This is because λ_1^0 is not necessarily 0 for $\lambda_1 = 0$. It turns out that $\lambda_1^0 = 1$ gives the correct result $A^0 = Z_{10} + Z_{20} = I$.

4.5 Analog systems with sampled inputs

When an analog system is excited by an analog input which is completely specified by its values at discrete-time instants (such functions are called "sampled functions" or "sampled data functions" and are discussed in greater detail in Chapter 7), then it is much easier to characterize the system as a discrete-time system and obtain its response at the discrete sampling instants than it is to obtain a complete characterization valid for continuous time.

Basically, a sampled function results from the conversion of a discrete function into an analog function as occurs when a discrete function is passed through a discrete to analog (D to A) converter (see Section 7.5). The simplest kind of sampled function is a piecewise constant function characterized by

$$\boldsymbol{u}_a(t) = \boldsymbol{u}(kT) \quad \text{for } kT \leq t < (k+1)T. \tag{4.62}$$

Therefore if this function $u_a(t)$ is applied as the input to the analog system governed by the state equations (4.1a), we know from (4.9) that the state transition for $kT \leq t < (k+1)T$ can be expressed as

$$\boldsymbol{x}(t) = \phi(t, kT)\boldsymbol{x}(kT) + \int_{kT}^{t} \phi(t, \tau)B(\tau)\mathbf{u}_a(\tau)d\tau. \tag{4.63}$$

Since $\boldsymbol{u}_a(t)$ is piecewise constant, using (4.62) in (4.63) gives

$$\boldsymbol{x}((k+1)T) = \phi((k+1)T, kT)\boldsymbol{x}(kT) + \int_{kT}^{(k+1)T} \phi((k+1)T, \tau)B(\tau)d\tau \boldsymbol{u}(kT) \tag{4.64}$$

Therefore if we define

$$A_d(k) \triangleq \phi((k+1)T, kT) \quad \text{and} \quad B_d(k) \triangleq \int_{kT}^{(k+1)T} \phi((k+1)T, \tau)B(\tau)d\tau \tag{4.65}$$

we get the state equation for the discrete system

$$\boldsymbol{x}(k+1) = A_d(k)\boldsymbol{x}(k) + B_d(k)\boldsymbol{u}(k). \tag{4.66}$$

This equation together with (4.1b) evaluated at $t = kT$ describes the original analog system at the discrete times $t = kT$.

Note:

1 If the A and B matrices are constant, the A_d and B_d will also be constant matrices.

2 These results can be extended to more general sampled functions characterized by

$$u_i(t) = p_i(t - kT)x_i(kT) \quad \text{for} \quad kT \leq t < (k+1)T$$

where $p_i(\tau)$ is some function that is defined only over the interval $0 \leq \tau < T$.

Problems

4.1 Consider the RC circuit shown below, in which the capacitance $C(t)$ changes with time, where $R_1 = 1\Omega$, $R_2 = 0.5\Omega$, and $C(t) = (t+1)f$.

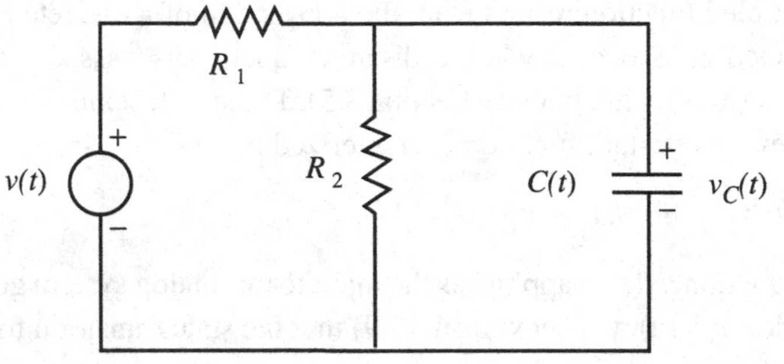

a) Obtain a state differential equation for this circuit using the voltage $v_c(t)$ across the capacitor as the state variable.
b) Find the transition matrix $\phi(t, t_0)$.
c) Find $v_c(t)$ for $t \geq 0$, when $v_c(0) = 0$ and the applied voltage $v(t) = (t+1)1(t)$.

4.2 For each of the matrices below:

1 $$\phi_1(t,\tau) = \begin{bmatrix} e^{-(t-\tau)} & (t-\tau) \\ 0 & 1+t(t-\tau) \end{bmatrix}$$

2 $$\phi_2(t,\tau) = \begin{bmatrix} e^{-(t-\tau)} & e^{-(t-\tau)} - \frac{t+1}{\tau+1} \\ 0 & \frac{t+1}{\tau+1} \end{bmatrix}$$

3 $$\phi_3(t,\tau) = \begin{bmatrix} e^{-(t-\tau)} & e^{-(t-\tau)} \\ 0 & 0 \end{bmatrix}$$

4 $$\phi_4(t,\tau) = \begin{bmatrix} e^{-(t-\tau)} & (t-\tau) \\ 0 & e^{(t-\tau)} \end{bmatrix}$$

a) Determine which of the matrices *are* transition matrices and which are not.
b) Find the $A(t)$ matrix of the system for each transition matrix you found in part (a).

4.3 Find $\phi(t, t_0)$ for the system shown in Problem 3.12, using the physical interpretation of $\phi_{ij}(t, t_0)$ (i.e., the response to initial conditions).

4.4 Find $\phi(t, t_0)$ for the system shown below.

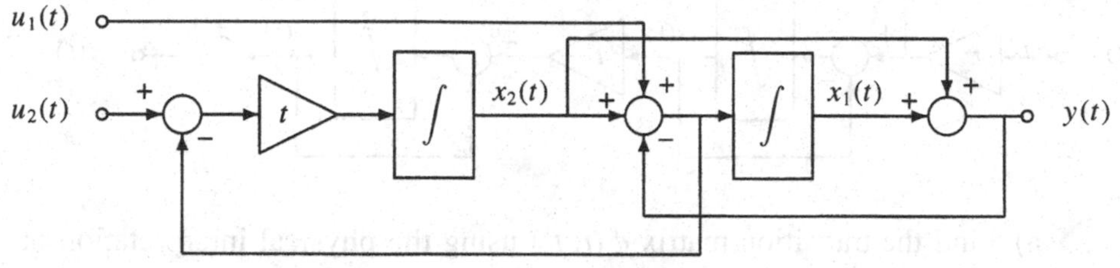

4.5 Find $\phi(t)$ for the system described by

$$\dot{x}_1(t) = x_1(t) + u(t)$$
$$\dot{x}_2(t) = 2x_1(t) - 3x_2(t) + u(t)$$
$$y(t) = x_1(t) - x_2(t) + u(t)$$

using the physical meaning of $\phi_{ij}(t)$ (i.e., the response to initial conditions).

4.6 Use the physical meaning of $\phi_{ij}(t)$ to find $\phi(t)$ for each of the systems below.

a)

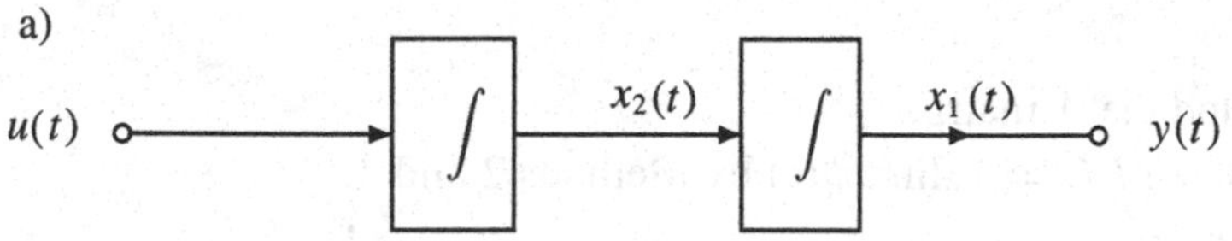

b)

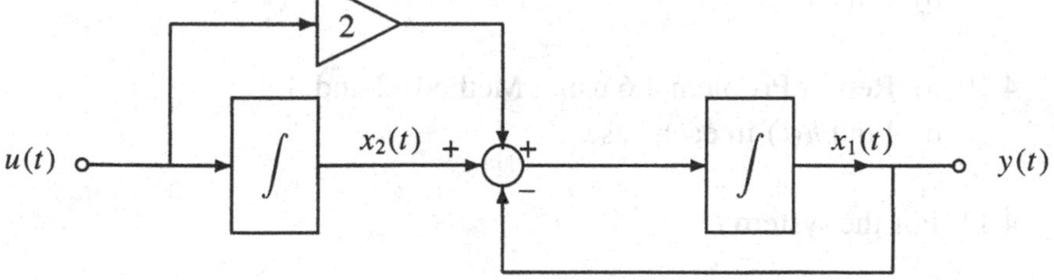

c)

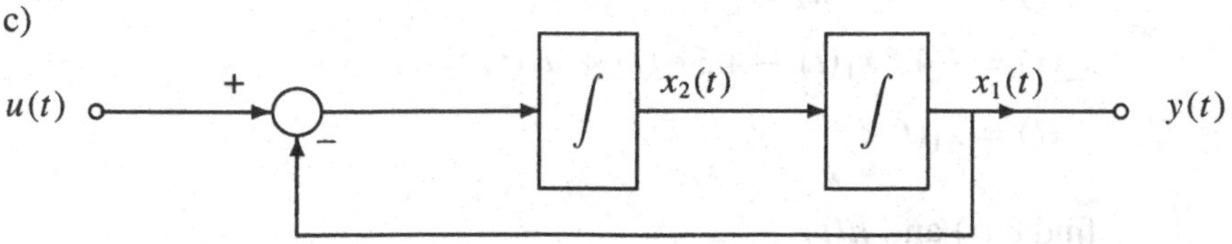

4.7 For the system below

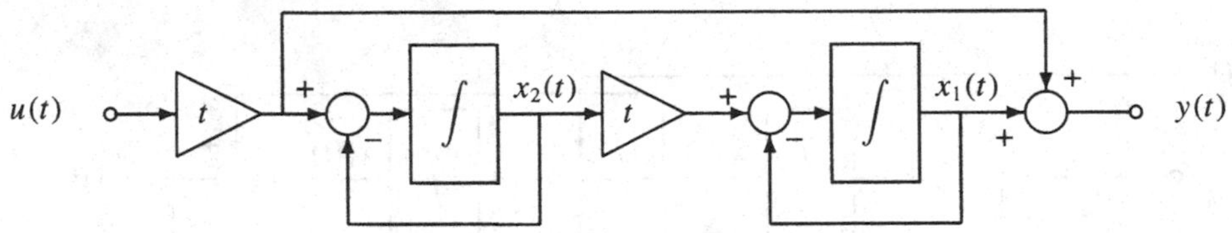

a) Find the transition matrix $\phi(t, t_0)$ using the physical interpretation of $\phi_{ij}(t, t_0)$.
b) Find the impulse response $h(t, t_0)$.

4.8 For the network below

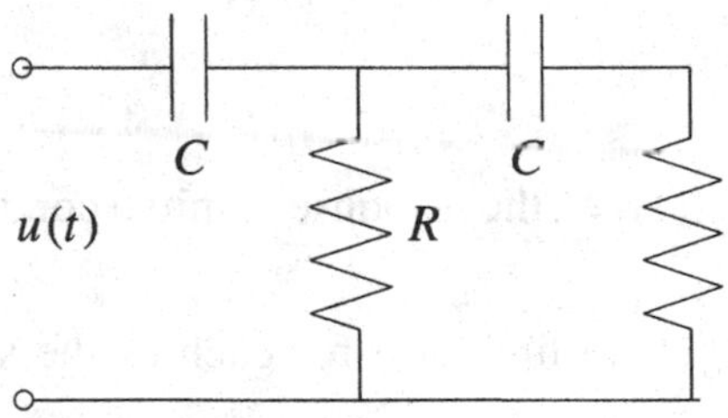

a) Find the A matrix.
b) When $RC = 1$, find $\phi(t)$ by Methods 2 and 3.

4.9 a) Repeat Problem 4.5 using Methods 2 and 3.
b) Find $h(t)$.

4.10 a) Repeat Problem 4.6 using Methods 2 and 3.
b) Find $h(t)$ in each case.

4.11 For the system

$$\dot{x}_1(t) = x_2(t) + u_2(t)$$
$$\dot{x}_2(t) = -4.5x_1(t) - 4.5x_2(t) + u_1(t)$$
$$y(t) = x_1(t)$$

find $\phi(t)$ and $h(t)$.

4.12 For the system

$$\dot{x}_1(t) = -x_1(t) + x_2(t) + u(t)$$
$$\dot{x}_2(t) = x_3(t)$$
$$\dot{x}_3(t) = -2x_2(t) - 3x_3(t) + u(t)$$
$$y_1(t) = x_1(t) + u(t)$$
$$y_2(t) = x_3(t)$$

find the transition matrix $\phi(t)$ and the impulse response matrix $H(t)$.

4.13 Find $\phi(t) = e^{At}$ for the matrix

$$A = \begin{bmatrix} 2 & 0 & 0 \\ -\frac{1}{2} & 2 & 0 \\ \frac{1}{2} & 0 & 2 \end{bmatrix}.$$

4.14 Prove that for the time-varying system $\dot{\boldsymbol{x}}(t) = A(t)\boldsymbol{x}(t)$ the transition matrix is given by

$$\phi(t, t_0) = e^{F(t,t_0)} \quad \text{where} \quad F(t, t_0) \triangleq \int_{t_0}^{t} A(\tau)\, d\tau$$

if and only if $A(t)F(t, t_0) = F(t, t_0)A(t)$.

4.15 For the system below with $f(t) = 1/(t + 1)$ find the transition matrix

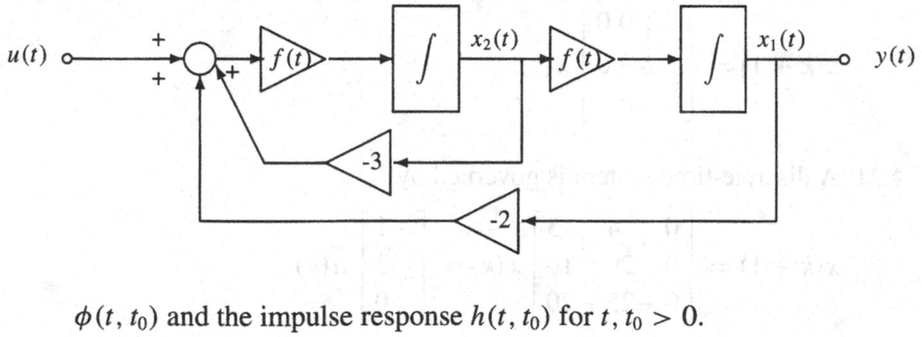

$\phi(t, t_0)$ and the impulse response $h(t, t_0)$ for $t, t_0 > 0$.

4.16 Find the transition matrix $\phi(t, t_0)$ for each of the system matrices below.

a) $A_a(t) = \begin{bmatrix} -2 & -t \\ 0 & 0 \end{bmatrix}$ b) $A_b(t) = \begin{bmatrix} -2t & -2t \\ 0 & 0 \end{bmatrix}$

c) $A_c(t) = \begin{bmatrix} -2t & 1 \\ 0 & -2t \end{bmatrix}$ d) $A_d(t) = \begin{bmatrix} -1 & \sin(t) \\ 0 & 0 \end{bmatrix}.$

4.17 Find the transition matrix $\phi(t, t_0)$ when $A(t) = \begin{bmatrix} -1-2t & 1 \\ -1 & 1-2t \end{bmatrix}$.

4.18 We know that the transition matrix $\phi(t, t_0)$ for a system matrix $A(t)$ satisfies $\dot{\phi}(t, t_0) = A(t)\phi(t, t_0)$ and $\phi(t_0, t_0) = I$.

a) Verify that the transition matrix $\phi(t, t_0)$ that was found in Example 4.12 is correct by showing that these conditions are satisfied.

b) Do the same for the transition matrix $\phi(t, t_0)$ that was found in Example 4.13.

4.19 Find $\phi(k, k_0)$ and $h(k, k_0)$ for the discrete system shown.

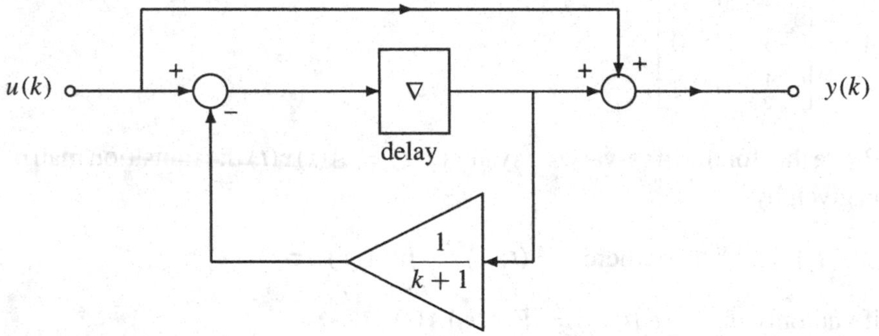

4.20 Find the transition matrix $\phi(k - k_0)$ for the discrete-time system

$$x(k+1) = \begin{bmatrix} \frac{1}{2} & 0 & 0 \\ -\frac{1}{2} & \frac{1}{2} & 0 \\ -\frac{1}{2} & 0 & \frac{1}{2} \end{bmatrix} x(k).$$

4.21 A discrete-time system is governed by

$$x(k+1) = \begin{bmatrix} 0 & 4 & 3 \\ 0 & 20 & 16 \\ 0 & -25 & -20 \end{bmatrix} x(k) + \begin{bmatrix} -1 \\ 3 \\ 0 \end{bmatrix} u(k)$$

$$y(k) = \begin{bmatrix} -1 & 3 & 0 \end{bmatrix} x(k) + 4u(k).$$

a) Find $\phi(k - k_0)$.

b) Find $h(k - k_0)$.

c) When $u(k) = 1$ for all $k \geq 0$ and $x(0) = \begin{bmatrix} 1 \\ 0 \\ 0 \end{bmatrix}$ find $y(k)$ for $k \geq 0$.

4.22 For the system

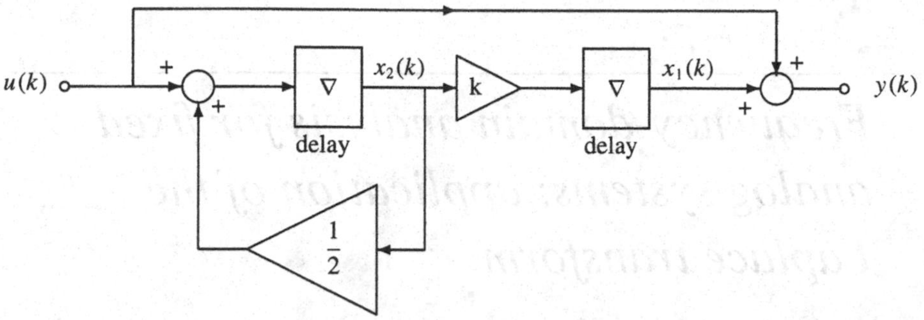

a) Find the transition matrix $\phi(k, j)$.
b) Find the delta response $h(k, j)$.

4.23 For the time-varying discrete system described by the state equations

$$q_1(k+1) = a^{-k}q_1(k) + \tfrac{1}{2}a^{-k}q_2(k)$$
$$q_2(k+1) = 2a^{-k}q_2(k) + kx(k)$$
$$y(k) = 2q_1(k) + q_2(k)$$

where a^{-k} is a time-varying gain:
a) Find the transition matrix $\phi(k, k_0)$.
b) Find the delta response $h(k, k_0)$ (i.e., the zero-state response to the input $x(k) = \delta(k - k_0)$).

5

Frequency-domain analysis for fixed analog systems: application of the Laplace transform

5.1 Basic principles

We have mentioned that in the indirect approach to analyzing linear systems, an arbitrary input $\{u(t)\}$ can be represented as a linear combination of elementary time functions. For time-domain analysis of analog systems, the input signals are considered to be resolved into impulses or step functions, and the response to a typical impulse or step is found. The response to the input $\underset{\sim}{u}$ is then just a linear combination of the responses to the elementary functions. Frequency-domain analysis of analog systems is based on resolution of arbitrary inputs into complex exponentials $\{e^{st}\}$. These elementary functions are used because of the following fundamental property.

Fundamental property of fixed linear analog systems

Theorem 5.1 *If S is a fixed linear analog system then the zero-state response*

$$S\{e^{st}\} = H(s)\{e^{st}\}. \tag{5.1}$$

Proof: Let $\underset{\sim}{y} = \{f(t, s)\} = S\{e^{st}\}$ (i.e., the output could depend on t and the complex number s) so

$$S\{e^{st}\} = \frac{\{f(t, s)\}}{\{e^{st}\}} \cdot \{e^{st}\} = \{H(t, s)\}\{e^{st}\}$$

but since the system is fixed, we have for any τ

$$S\{e^{s(t-\tau)}\} = \{H(t - \tau, s)\}\{e^{s(t-\tau)}\}. \tag{5.2}$$

Since the system is linear, by homogeneity

$$S\{e^{-s\tau}e^{st}\} = e^{-s\tau}S\{e^{st}\} = e^{-s\tau}\{H(t, s)\}\{e^{st}\} = \{H(t, s)\}\{e^{s(t-\tau)}\}. \tag{5.3}$$

This shows that $\{H(t - \tau, s)\} = \{H(t, s)\}$ for any τ.

In order for the two functions to be equal they must be independent of time, which gives

$$\{H(t, s)\} = H(s).$$

QED.

Thus the response of a fixed linear system to $\{e^{st}\}$ has exactly the same shape as the input. Only the (complex) amplitude is changed. $H(s)$ is called the transfer function of the system.

5.2 Application

If $\{u(t)\}$ is a periodic input with period T, which over one period satisfies the Dirichlet conditions,

1 $\underset{\sim}{u}$ has only a finite number of maxima and minima,

2 $\underset{\sim}{u}$ has only a finite number of discontinuities, and

3 $\int_{-T/2}^{T/2} |u(t)|\,dt < \infty$

then $\{u(t)\}$ can be resolved into a Fourier series

$$u(t) = \sum_{n=-\infty}^{\infty} U_n e^{jn\omega_0 t} \quad \text{where } \omega_0 = \frac{2\pi}{T} \tag{5.4}$$

for all t at which $\underset{\sim}{u}$ is continuous. The complex amplitudes U_n are given by

$$U_n = \frac{1}{T}\int_{-T/2}^{T/2} u(t) e^{-jn\omega_0 t}\,dt. \tag{5.5}$$

Applying the input (5.4) to a fixed linear system gives

$$y(t) = S\{u(t)\} = \sum_{n=-\infty}^{\infty} U_n S\{e^{jn\omega_0 t}\} = \sum_{n=-\infty}^{\infty} U_n H(jn\omega_0) e^{jn\omega_0 t}$$

so the complex amplitudes Y_n of the output are given by

$$Y_n = H(jn\omega_0) U_n \tag{5.6}$$

(i.e., Y_n is U_n multiplied by $H(s)$ evaluated at the complex frequency $s = jn\omega_0$).

Nonperiodic inputs $\underset{\sim}{u}$ can be resolved into complex exponentials $\{e^{st}\}$ via the inverse Laplace transform

$$u(t) = \frac{1}{2\pi j}\int_{c-j\infty}^{c+j\infty} U(s) e^{st}\,ds \tag{5.7}$$

where the complex amplitude $U(s)$ is given by the Laplace transform

$$U(s) = \int_0^\infty u(t) e^{-st}\,dt.$$

Note: $\underset{\sim}{u}$ must satisfy Dirichlet conditions on $[0, \infty]$ where $\sigma = \Re e\ s$ must make

$$\int_0^\infty |u(t)|e^{-\sigma t}\,dt < \infty \quad \text{and} \quad c < \sigma.$$

Equation (5.7) can be viewed as an expression which shows how the time function $\{u(t)\}$ is composed of a linear combination of the functions $\{e^{st}\}$, where s is a continuous indexing variable, $U(s)$ shows how much of each elementary function $\{e^{st}\}$ is present in $\{u(t)\}$, and the integral represents a summation which in the limit has a continuous index.

Therefore, when a scalar function $u(t)$ is applied to a linear analog system, from (5.7) and Assertion 2 of Section 1.4,

$$y(t) = S\{u(t)\} = \frac{1}{2\pi j}\int_{c-j\infty}^{c+j\infty} U(s)S\{e^{st}\}\,ds. \tag{5.8}$$

When the system is also fixed, we see from (5.1) that

$$y(t) = \frac{1}{2\pi j}\int_{c-j\infty}^{c+j\infty} U(s)H(s)e^{st}\,ds \tag{5.9}$$

so the complex amplitude of $y(t)$ is $Y(s) = H(s)U(s)$.

Therefore, in studying fixed, linear, analog systems using the frequency-domain approach, we can concentrate on finding $Y(s)$. At the very end $y(t)$ is obtained using (5.9). Thus we see that the use of $\mathcal{L}$-transforms is restricted to *fixed, linear*, analog systems.

Note: If the input $\{u(t)\} = \{\delta(t)\}$ then $U(s) = 1$ (this assumes that δ occurs at $0+$ or that the integral in the definition of the $\mathcal{L}$-transform is from $0-$), so $Y(s) = H(s)$. In other words, the transfer function is just the $\mathcal{L}$-transform of the impulse response $h(t)$.

For a multiple-input, multiple-output system if we consider the ith output when only the kth input is applied (i.e., all other inputs are zero) then

$$y_i(t) = \frac{1}{2\pi j}\int_{c-j\infty}^{c+j\infty} H_{ik}(s)U_k(s)e^{st}\,ds. \tag{5.10}$$

When all inputs are applied at the same time, then by superposition

$$y_i(t) = \frac{1}{2\pi j}\int_{c-j\infty}^{c+j\infty} \sum_{k=1}^{r} H_{ik}(s)U_k(s)e^{st}\,ds \tag{5.11}$$

and combining all the equations for $i = 1, \ldots, m$ into a vector expression gives

$$\mathbf{y}(t) = \frac{1}{2\pi j}\int_{c-j\infty}^{c+j\infty} \mathrm{H}(s)\mathbf{U}(s)e^{st}\,ds \tag{5.12}$$

where H(s) is the $m \times r$ matrix whose elements are the transfer functions $H_{ik}(s)$. This matrix is called the transfer function matrix, and it relates the vector of complex amplitudes $\boldsymbol{Y}(s)$ (i.e., the $\mathscr{L}$-transform of $\boldsymbol{y}(t)$) to $\boldsymbol{U}(s)$ by $\boldsymbol{Y}(s) = \mathrm{H}(s)\boldsymbol{U}(s)$.

5.3 Solution of the state differential equations using $\mathscr{L}$-transforms

We are familiar with the application of Laplace transforms to simplify the analysis of fixed linear analog systems. The state differential equations of a fixed, linear, analog system can also be solved using Laplace transforms. For

$$\begin{aligned} \dot{\boldsymbol{x}}(t) &= A\boldsymbol{x}(t) + B\boldsymbol{u}(t) \\ \boldsymbol{y}(t) &= C\boldsymbol{x}(t) + D\boldsymbol{u}(t) \end{aligned} \tag{5.13}$$

we can take the $\mathscr{L}$-transform of both sides to give

$$\begin{aligned} \mathscr{L}\{\dot{\boldsymbol{x}}(t)\} &= A\mathscr{L}\{\boldsymbol{x}(t)\} + B\mathscr{L}\{\boldsymbol{u}(t)\} \\ \mathscr{L}\{\boldsymbol{y}(t)\} &= C\mathscr{L}\{x(t)\} + D\mathscr{L}\{u(t)\}. \end{aligned} \tag{5.14}$$

Let $\mathscr{L}\{\boldsymbol{u}(t)\} = \boldsymbol{U}(s)$ (a vector of $\mathscr{L}$-transforms), $\mathscr{L}\{\boldsymbol{y}(t)\} = \boldsymbol{Y}(s)$, and $\mathscr{L}\{\boldsymbol{x}(t)\} = \boldsymbol{X}(s)$.

From the differentiation theorem we have

$$\mathscr{L}\{\dot{\boldsymbol{x}}(t)\} = s\boldsymbol{X}(s) - \boldsymbol{x}(0).$$

Therefore the transformed equations become

$$\begin{aligned} s\boldsymbol{X}(s) - \boldsymbol{x}(0) &= A\boldsymbol{X}(s) + B\boldsymbol{U}(s) \\ \boldsymbol{Y}(s) &= C\boldsymbol{X}(s) + D\boldsymbol{U}(s) \end{aligned} \tag{5.15}$$

so

$$\begin{aligned} \boldsymbol{X}(s) &= [sI - A]^{-1}\boldsymbol{x}(0) + [sI - A]^{-1}B\boldsymbol{U}(s) \\ \boldsymbol{Y}(s) &= C[sI - A]^{-1}\boldsymbol{x}(0) + [C[sI - A]^{-1}B + D]\boldsymbol{U}(s). \end{aligned} \tag{5.16}$$

Recall that from the direct solution of the state equations we found that for fixed analog systems

$$\boldsymbol{x}(t) = \phi(t - t_0)\boldsymbol{x}(t_0) + \int_{t_0}^{t} \phi(t - \tau)B\boldsymbol{u}(\tau)\,d\tau$$

and

$$\boldsymbol{y}(t) = C\phi(t - t_0)\boldsymbol{x}(t_0) + \int_{t_0}^{t} C\phi(t - \tau)B\boldsymbol{u}(\tau)\,d\tau + D\boldsymbol{u}(t).$$

If we take the initial time $t_0 = 0$ and then $\mathcal{L}$-transform each of these equations we obtain

$$\begin{aligned} \boldsymbol{X}(s) &= \mathcal{L}\{\phi(t)\}\boldsymbol{x}(0) + \mathcal{L}\left\{\int_0^t \phi(t-\tau)B\boldsymbol{u}(\tau)\,d\tau\right\} \\ &= \mathcal{L}\{\phi(t)\}\boldsymbol{x}(0) + \mathcal{L}\{\phi(t)\}B\boldsymbol{U}(s) \end{aligned} \tag{5.17}$$

and

$$\boldsymbol{Y}(s) = C\mathcal{L}\{\phi(t)\}\boldsymbol{x}(0) + [C\mathcal{L}\{\phi(t)\}B + D]\boldsymbol{U}(s). \tag{5.18}$$

Comparing (5.17) and (5.18) with (5.16) we see that

$$\mathcal{L}\{\phi(t)\} = [sI - A]^{-1}. \tag{5.19}$$

This is an alternative way (not necessarily the best way) to find the transition matrix when A is constant (i.e., A does *not* depend on t). Also note that when $\boldsymbol{x}(0) = \boldsymbol{0}$ (i.e., zero initial conditions) the transformed zero-state response $\boldsymbol{Y}_{zs}(s)$ is related to the transformed input $\boldsymbol{U}(s)$ by the matrix

$$[C(sI - A)^{-1}B + D] = \mathrm{H}(s) \tag{5.20}$$

which is the transfer function matrix.

Also, when the system is initially in the zero state at $t_0 = -\infty$, for a fixed system we obtain, from (3.39),

$$\boldsymbol{y}(t) = \int_{-\infty}^{\infty} H(t-\tau)\boldsymbol{u}(\tau)d\tau. \tag{5.21}$$

Now if $u_i(\tau) = 0$ for all $i = 1, \ldots, r$ and all $\tau < 0$ (so the system is in the zero state at time 0) then (5.21) becomes

$$\boldsymbol{y}(t) = \int_0^{\infty} H(t-\tau)\boldsymbol{u}(\tau)\,d\tau. \tag{5.22}$$

Taking the $\mathcal{L}$-transform of this convolution integral gives

$$\boldsymbol{Y}(s) = \mathcal{L}\{H(t)\}\boldsymbol{U}(s). \tag{5.23}$$

We found from (5.12) that $\boldsymbol{Y}(s) = \mathrm{H}(s)\boldsymbol{U}(s)$, so $\mathcal{L}\{H(t)\} = \mathrm{H}(s)$; that is, the Laplace transform of the impulse response matrix is the transfer function matrix, just as for the case of single-input, single-output systems.

Example 5.1

$$\begin{aligned} \dot{x}_1 &= x_2 \\ \dot{x}_2 &= -x_2 + u \\ y &= x_1 \end{aligned}$$

so

$$A = \begin{bmatrix} 0 & 1 \\ 0 & -1 \end{bmatrix}; \qquad B = \begin{bmatrix} 0 \\ 1 \end{bmatrix}; \qquad C = \begin{bmatrix} 1 & 0 \end{bmatrix}; \qquad D = 0.$$

This gives

$$[sI - A]^{-1} = \begin{bmatrix} s & -1 \\ 0 & s+1 \end{bmatrix}^{-1} = \frac{1}{s(s+1)} \begin{bmatrix} s+1 & 1 \\ 0 & s \end{bmatrix} = \begin{bmatrix} \frac{1}{s} & \frac{1}{s(s+1)} \\ 0 & \frac{1}{s+1} \end{bmatrix}.$$

Therefore

$$\phi(t) = e^{At} = \begin{bmatrix} 1 & 1 - e^{-t} \\ 0 & e^{-t} \end{bmatrix}$$

or

$$H(s) = C[sI - A]^{-1}B = \frac{1}{s(s+1)} \quad \text{so} \quad h(t) = 1 - e^{-t} \quad \text{for} \quad t \geq 0.$$

Using the transfer function matrix we can obtain the transform $\boldsymbol{Y}_{zs}(s)$, of the zero-state response $\boldsymbol{y}_{zs}(t)$ for any arbitrary input. The zero-state response $\boldsymbol{y}_{zs}(t)$ is found by simply taking the inverse Laplace transform. Of course, when nonzero initial conditions are present, the contribution $C\phi(t)\boldsymbol{x}(0)$ must be added to get the entire output $\boldsymbol{y}(t)$.

5.4 Solutions of systems of differential equations

Suppose we are given a system of differential equations with constant coefficients which are not in the form of state differential equations. Chapter 2 (Section 2.7) showed how to put such equations into the form of state differential equations, and we saw how we could obtain the transfer function matrix from these state equations. It is not necessary, however, to use such a procedure if only the transfer function matrix is desired, since the matrix $\mathcal{H}(s)$ can be obtained directly. In fact, the equations can be solved directly after taking the $\mathcal{L}$-transform.

Example 5.2

For the system

$$\ddot{y}_1 + \dot{y}_1 + \dot{y}_2 + y_2 = u$$

$$\dot{y}_1 + y_1 - \ddot{y}_2 - \dot{y}_2 = \ddot{u} - \dot{u} + u$$

when we take the $\mathcal{L}$-transform of both sides, since $\mathcal{L}\{\dot{y}\} = sY(s) - y(0)$ and $\mathcal{L}\{\ddot{y}\} = s^2Y(s) - sy(0) - \dot{y}(0)$, we find that

$$(s^2 + s)Y_1(s) - sy_1(0) - \dot{y}_1(0) - y_1(0) + (s+1)y_2(s) - y_2(0) = U(s)$$

and

$$(s+1)y_1(s) - y_1(0) - (s^2+s)y_2(s) + sy_2(0) + \dot{y}_2(0) + y_2(0) = (s^2-s+1)U(s).$$

(*Note:* We consider that $u(t)$ starts at $t = 0$ so $u^{(i)}(0-) = 0$.) Thus

$$\underbrace{\begin{bmatrix} (s^2+s) & (s+1) \\ (s+1) & -(s^2+s) \end{bmatrix}}_{J(s)} \underbrace{\begin{bmatrix} Y_1(s) \\ Y_2(s) \end{bmatrix}}_{Y(s)} = \underbrace{\begin{bmatrix} 1 \\ (s^2-s+1) \end{bmatrix}}_{K(s)} U(s) + \underbrace{\begin{bmatrix} I_1(s) \\ I_2(s) \end{bmatrix}}_{I(s)}$$

where

$$I_1(s) = (s+1)y_1(0) + \dot{y}_1(0) + y_2(0)$$

$$I_2(s) = -(s+1)y_2(0) - \dot{y}_2(0) - y_1(0).$$

We see that $\boldsymbol{Y}(s)$ has the form

$$\boldsymbol{Y}(s) = \boldsymbol{J}^{-1}(s)\boldsymbol{K}(s)\boldsymbol{U}(s) + \boldsymbol{J}^{-1}(s)\boldsymbol{I}(s).$$

Since all the initial condition terms are contained in $\boldsymbol{I}(s)$ we see that for a system with zero initial conditions (i.e., in the zero state so $\boldsymbol{I}(s) = \boldsymbol{0}$) that

$$\boldsymbol{Y}(s) = \boldsymbol{J}^{-1}(s)\boldsymbol{K}(s)\boldsymbol{U}(s) = \mathrm{H}(s)\boldsymbol{U}(s)$$

so the transfer function matrix is just

$$\mathrm{H}(s) = \boldsymbol{J}^{-1}(s)\boldsymbol{K}(s) = -\frac{1}{(s+1)(s^2+1)}\begin{bmatrix} -s & -1 \\ -1 & s \end{bmatrix}\begin{bmatrix} 1 \\ (s^2-s+1) \end{bmatrix}$$

$$= -\frac{1}{(s+1)(s^2+1)}\begin{bmatrix} -s-s^2+s-1 \\ s^3-s^2+s-1 \end{bmatrix}$$

$$= -\frac{1}{(s+1)(s^2+1)}\begin{bmatrix} -(s^2+1) \\ (s-1)(s^2+1) \end{bmatrix}.$$

When the common factors are canceled, this becomes

$$\mathrm{H}(s) = \frac{1}{(s+1)}\begin{bmatrix} 1 \\ 1-s \end{bmatrix}.$$

In Chapter 2 (Example 2.10) we found that this system could be simulated by

$$\dot{x} = -x + u; \qquad \begin{bmatrix} y_1 \\ y_2 \end{bmatrix} = \begin{bmatrix} 0 \\ 1 \end{bmatrix} x + \begin{bmatrix} 0 \\ -1 \end{bmatrix} u.$$

Using the formula $\mathrm{H}(s) = C[sI - A]^{-1}B + D$ we get

$$\mathrm{H}(s) = \begin{bmatrix} 1 \\ 2 \end{bmatrix}\frac{1}{(s+1)} + \begin{bmatrix} 0 \\ -1 \end{bmatrix} = \frac{1}{(s+1)}\begin{bmatrix} 1 \\ 1-s \end{bmatrix}$$

which is the same as above.

This method can be used whenever a system is characterized by a set of differential equations which are linear and time-invariant. The $\mathcal{L}$-transform approach is directly applicable to finding the solutions.

Let $J_{ij}(D)$ and $K_{ij}(D)$ represent the linear differential operators of the form

$$[D^p + a_{p-1}D^{p-1} + \cdots + a_1 D + a_0]y \triangleq J(D)y.$$

For a system described by

$$\begin{aligned} J_{11}(D)\underset{\sim}{y}_1 + \cdots + J_{1m}(D)\underset{\sim}{y}_m &= K_{11}(D)\underset{\sim}{u}_1 + \cdots + K_{1n}(D)\underset{\sim}{u}_h \\ &\ \vdots \\ J_{m1}(D)\underset{\sim}{y}_1 + \cdots + J_{mm}(D)\underset{\sim}{y}_m &= K_{m1}(D)\underset{\sim}{u}_1 + \cdots + K_{mn}(D)\underset{\sim}{u}_n \end{aligned} \tag{5.24}$$

we can write this in matrix form as

$$J(D)\boldsymbol{y}(t) = K(D)\boldsymbol{u}(t) \tag{5.25}$$

where $J(D)$ is an $m \times m$ matrix of operators and $K(D)$ is an $m \times n$ matrix of operators.

Taking the $\mathcal{L}$-transform gives

$$J(s)\boldsymbol{Y}(s) = K(s)\boldsymbol{U}(s) + \boldsymbol{I}(s) \tag{5.26}$$

where the term $\boldsymbol{I}(s)$ contains all initial condition terms (since $\mathcal{L}\left\{\frac{d^k f}{dt^k}\right\} = s^k F(s)$ minus initial condition terms). Thus we find that

$$\boldsymbol{Y}(s) = \boldsymbol{J}^{-1}(s)\boldsymbol{K}(s)\boldsymbol{U}(s) + \boldsymbol{J}^{-1}(s)\boldsymbol{I}(s) \tag{5.27}$$

where $J^{-1}(s)$ denotes the matrix inverse of $J(s)$. In the absence of initial conditions (5.27) becomes

$$\boldsymbol{Y}_{zs}(s) = J^{-1}(s)K(s)\boldsymbol{U}(s) = \mathrm{H}(s)\boldsymbol{U}(s). \tag{5.28}$$

Thus the transfer function matrix can be obtained directly from the differential equations.

Always remember that the $\mathcal{L}$-transform method is valid *only* for fixed, linear, analog systems.

Example 5.3

Consider the mechanical system of Example (2.3). To get the transfer function we start by taking the $\mathcal{L}$-transform of the equations of motion (2.15) and (2.16). This gives

$$M_1 s V_1(s) = K[Z_2(s) - Z_1(s)] + D_1[V_2(s) - V_1(s)] \tag{5.29}$$

$$M_2 s V_2(s) = F(s) - K[Z_2(s) - Z_1(s)] - D_2 V_2(s) - D_1[V_2(s) - V_1(s)] \tag{5.30}$$

when $v_1(0) = v_2(0) = 0$ and $z_2(0) - z_1(0) = 0$. Rearranging and using $z_1(0) = 0$ so $V_i(s) = sZ_i(s)$, we find that (5.30) gives

$$(M_2 s + D_2 + D_1)sZ_2(s) + KZ_2(s) = F(s) + KZ_1(s) + D_1 sZ_1(s) \tag{5.31}$$

while (5.29) gives

$$(M_1 s + D_2 + D_1)sZ_1(s) + KZ_1(s) = (K + sD_1)Z_2(s). \tag{5.32}$$

Substitute $Z_2(s)$ from (5.31) into (5.32) and solve for $Z_1(s)$ to get

$$Z_1(s) = \frac{D_1 s + K}{\Delta(s)} F(s)$$

where the denominator polynomial $\Delta(s) = M_1 M_2 s^4 + (D_1(M_1 + M_2) + D_2 M_1)s^3 + (K(M_1 + M_2) + D_1 D_2)s^2 + D_2 K s$. We thus see that the transfer function is

$$\mathrm{H}(s) = \frac{Z_1(s)}{F(s)} = \frac{D_1 s + K}{\Delta(s)}.$$

The reader should verify that using $\mathrm{H}(s) = C[sI - A]^{-1}B + D$ gives the same result.

5.5 Leverrier's method

As we have seen, when using the $\mathcal{L}$-transform method to solve for the transition matrix, it is necessary to invert the matrix $(sI - A)$. This is not too difficult to do directly for $n = 2$ or even 3, but for larger values of n a direct inversion via the adjoint matrix is impractical. We will discuss an algorithm for calculating $(sI - A)^{-1}$ iteratively which is also suitable for machine computation. The derivation of the method is given in Sections A.17 and A.18 of the Appendix.

Let $p(s) = \det(sI - A) = s^n + a_{n-1}s^{n-1} + \cdots + a_1 s + a_0$ be the characteristic polynomial of A and

$$\Gamma(s) = \text{ the adjoint matrix of } (sI - A)$$

(i.e., $\Gamma_{ij}(s) =$ the cofactor of the jith element of $(sI - A)$). From matrix theory (see Appendix, Section A.9) we know that the inverse is given by

$$(sI - A)^{-1} = \frac{\Gamma(s)}{p(s)} \tag{5.33}$$

so the matrix $\Gamma(s)$ and the polynomial $p(s)$ must be found. Note that since each element of $\Gamma(s)$ is obtained from the determinant of an $(n - 1) \times (n - 1)$ matrix, each element $\Gamma_{ij}(s)$ is a polynomial in s of degree no greater than s^{n-1}.

Thus we can write for $\Gamma(s)$

$$\Gamma(s) = K_{n-1}s^{n-1} + K_{n-2}s^{n-2} + \cdots + K_1 s + K_0 \tag{5.34}$$

where K_i are $(n \times n)$ constant matrices.
The algorithm generates these matrices K_i by iterating the equation

$$K_{n-(j+1)} = K_{n-j}A + a_{n-j}I \quad \text{for } j = 1, 2, \ldots, n-1 \quad \text{with } K_{n-1} = I. \tag{5.35}$$

Thus we have a procedure for iteratively generating the matrices K_i when the coefficients a_i are known. *Note:* The equation with $j = n$ can be considered as a check on the roundoff error in computation since we must have $K_{-1} = \Theta$ (the zero matrix).

The coefficients a_i can also be calculated iteratively using

$$a_{n-j} = -\frac{1}{j}\text{tr}\,(K_{n-j}A) \quad \text{for } j = 1, \ldots, n \tag{5.36}$$

where tr denotes the trace, defined as the sum of the diagonal elements of a square matrix.

Let us summarize the iterative procedure.

1 Start with $K_{n-1} = I$ (we know that $a_n = 1$). Then for $j = 1$,
2 calculate $a_{n-j} = -\frac{1}{j}\text{tr}\,(K_{n-j}A)$,
3 calculate $K_{n-j-1} = K_{n-j}A + a_{n-j}I$,
4 increase j by 1 and return to (2) if $j \leq n$. (This also gives the check on $K_{-1} = \Theta$.)

This procedure has also been credited to Souriau (1948), Frame (1949), and Faddeev (1949) who found it independently in the late 1940s. In his book, Gantmacher (1959:87) says that Leverrier (1840) presented a method for finding only the coefficients a_i of the characteristic equation, using a different approach, and attributes the procedure for finding both the a_i and K_i to Faddeev (1949). Householder (1964:166–8, 172) credits Leverrier (1840) with the first iterative method for finding the coefficients a_i and describes the algorithm given here, calling it somewhat of an improvement. It is called Leverrier's method here for purely historical reasons and to avoid a designation such as LSFF.

Example 5.4

$$A = \begin{bmatrix} 3 & 0 & 3 \\ 2 & 4 & 1 \\ 1 & 1 & 2 \end{bmatrix}$$

so $n = 3,\ K_2 = I$, and $a_3 = 1$. We get

$$a_2 = -\text{tr } A = -9$$

$$K_1 = A - 9I = \begin{bmatrix} -6 & 0 & 3 \\ 2 & -5 & 1 \\ 1 & 1 & -7 \end{bmatrix}$$

$$a_1 = -\frac{1}{2}\text{tr } K_1 A = -\frac{1}{2}\text{tr}\begin{bmatrix} -15 & 3 & -12 \\ -3 & -19 & 3 \\ -2 & -3 & -10 \end{bmatrix} = 22$$

$$K_0 = K_1 A + a_1 I = \begin{bmatrix} 7 & 3 & -12 \\ -3 & 3 & 3 \\ -2 & -3 & 12 \end{bmatrix}$$

$$a_0 = -\frac{1}{3}\text{tr } K_0 A = -\frac{1}{3}[15 + 15 + 15] = -15.$$

We check the result by finding

$$K_0 A + a_0 I = \begin{bmatrix} 15 & 0 & 0 \\ 0 & 15 & 0 \\ 0 & 0 & 15 \end{bmatrix} - 15I = \Theta$$

which is correct. This gives

$$(sI - A)^{-1} = \frac{1}{s^3 - 9s^2 + 22s - 15}\begin{bmatrix} s^2 - 6s + 7 & 3 & 3s - 12 \\ 2s - 3 & s^2 - 5s + 3 & s + 3 \\ s - 2 & s - 3 & s^2 - 7s + 12 \end{bmatrix}.$$

Problems

5.1 Find the transition matrix $\phi(t)$ using $[sI - A]^{-1}$ and the transfer function H(s) for the system of Problem 4.5.

5.2 Calculate $[sI - A]^{-1}$, the transfer function H(s), and then $\phi(t)$ and $h(t)$ for each of the systems of Problem 4.6.

5.3 Find the transition matrix $\phi(t)$ using $[sI - A]^{-1}$ and the transfer function matrix H(s) for the system of Problem 4.12.

5.4 Calculate the transfer function matrix of the network shown in Fig. P5.4
   a) Using the transformed loop or node equations.
   b) By first writing the state differential equations in standard form.

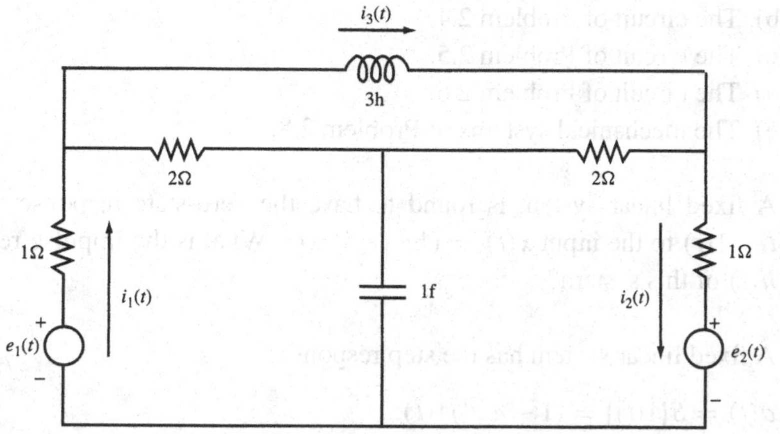

Fig. P5.4.

The voltage sources e_1 and e_2 are the inputs, and the currents i_1, i_2, and i_3 are the outputs.

5.5 Find the transition matrix $\phi(t)$ of the network in Problem 5.4.

5.6 For the system below find $H(s)$ and $\phi(t)$.

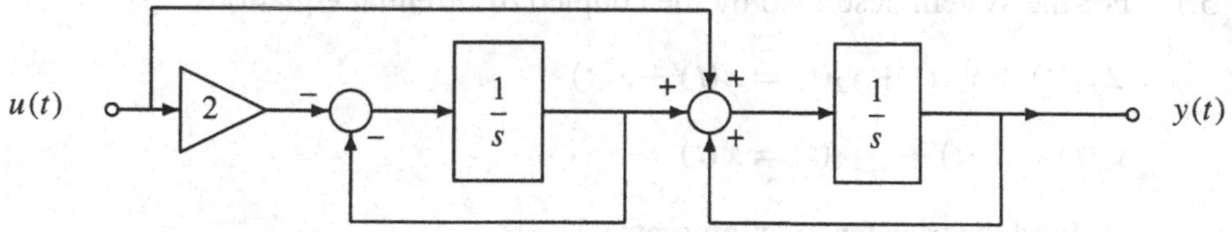

5.7 Using $\mathcal{L}$-transform methods find $\phi(t)$ for
a) The circuit of Problem 4.8.
b) The A matrix of Problem 4.13.

5.8 Find the transfer function matrix for the system described by the system of differential equations

$$\ddot{y}_1(t) + \dot{y}_1(t) + \ddot{y}_2(t) + y_2(t) + \dot{y}_3(t) = \dot{u}_1(t) - u_2(t)$$

$$\dot{y}_1(t) + \ddot{y}_2(t) + y_2(t) + \dot{y}_3(t) + y_3(t) = u_1(t)$$

$$\ddot{y}_1(t) + y_1(t) + y_2(t) + \ddot{y}_3(t) + \dot{y}_3(t) = u_2(t).$$

5.9 Find the transfer function matrix for
a) The circuit of Problem 2.3.

b) The circuit of Problem 2.4.
c) The circuit of Problem 2.5.
d) The circuit of Problem 2.6.
e) The mechanical systems of Problem 2.8.

5.10 A fixed linear system is found to have the zero-state response $y(t) = te^{-t}1(t)$ to the input $x(t) = (1 - e^{-t})1(t)$. What is the impulse response $h(t)$ of this system?

5.11 A fixed linear system has the step response

$$a(t) = S\{1(t)\} = (1 - e^{-2t})1(t).$$

Find the differential equation governing the dynamic behavior of this system.

5.12 A fixed, linear system is found to have the zero-state response $y(t) = t\ 1(t)$ to the input $x(t) = e^{-t}\ 1(t)$.
a) Find the transfer function $H(s)$ of this system.
b) Find the differential equation of a system which has this $H(s)$.

5.13 For the system described by the coupled differential equations

$$2y_1(t) + \dot{y}_2(t) + y_2(t) = \dot{x}(t) + x(t)$$

$$\dot{y}_1(t) + y_1(t) + \frac{1}{2}y_2(t) = x(t)$$

a) Find the transfer function matrix H(s).
b) Find $y_1(t)$ for $t \geq 0$ when $y_1(0) = 0,\ y_2(0) = 0$, and $x(t) = 1(t)$.

6

The $\mathcal{Z}$-transformation of discrete-time signals

6.1 Introduction

When studying continuous-time systems we found that for systems which are linear and fixed, it is extremely useful to employ the Laplace transform in the analysis. For linear, fixed, discrete-time systems there is a transformation method called the $\mathcal{Z}$-transformation, which is useful in the same way. We shall see that many of the applications (as well as the limitations) of the transform approach are the same for discrete systems as for analog systems.

6.2 Definition of the $\mathcal{Z}$-transform

The $\mathcal{Z}$-transformation of a discrete-time signal $\{f(k)\}$ is defined as the infinite power series in z^{-1} whose coefficients are the values of the signal at the discrete times $k = 0, 1, 2, \ldots$. That is,

$$Z\{f(k)\} \triangleq \sum_{k=0}^{\infty} f(k)z^{-k} \tag{6.1}$$

where z is a complex variable. This is also called the *one-sided $\mathcal{Z}$-transform* to distinguish it from the two-sided $\mathcal{Z}$-transform in which the sum is taken from $-\infty$ to $+\infty$.

The first question we consider is for what class of discrete functions does the $\mathcal{Z}$-transform exist?

The convergence theorem for $\mathcal{Z}$-transforms

Theorem 6.1 *If $f(k)$ is a discrete function satisfying*

1 $f(k)$ is bounded for all $k < \infty$ (i.e., $|f(k)| \leq M < \infty$), for $0 \leq k < \infty$,

2 $f(k)$ is of exponential order as $k \to \infty$ (i.e., there exist positive constants $N, r, K < \infty$ such that $|f(k)| \leq Kr^k$ for all $k \geq N$),

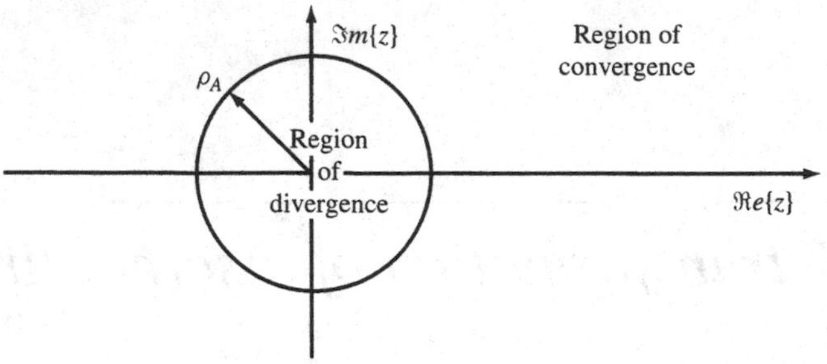

Fig. 6.1. The circle of convergence

then

$$\mathcal{Z}\{f(k)\} = \sum_{k=0}^{\infty} f(k)z^{-k}$$

converges absolutely for all complex numbers z with $|z| > r$.

Proof: For the finite integer N,

$$|\mathcal{Z}\{f(k)\}| \leq \sum_{k=0}^{N-1} |f(k)||z^{-k}| + \sum_{k=N}^{\infty} |f(k)||z^{-k}|$$

$$\leq M \sum_{k=0}^{N-1} |z|^{-k} + K \sum_{k=N}^{\infty} \left(\frac{r}{|z|}\right)^k . \tag{6.2}$$

Since $N < \infty$, the first term on the right-hand side of (6.2) is finite for all $|z| \neq 0$, and the second term is finite for all $|z| > r$. QED.

Definition 6.1 *The radius of absolute convergence ρ_A is the smallest r such that the defining series (6.1) converges for all $|z| > \rho_A$.*

Since z can be a complex number in general, we see that the region of convergence of the defining series is the outside of a circle centered at the origin and having radius ρ_A. This circle is called the circle of convergence (see Figure 6.1). Now if we can find a function $\mathcal{F}(z)$ of the variable z, which can be expanded into an infinite power series in z^{-1}, and this power series is identical to the defining series (6.1) for $\mathcal{Z}\{f(k)\}$ in the region outside the circle of convergence, then by a theorem from complex variable theory $\mathcal{F}(z)$ is a unique representation of $\mathcal{Z}\{f(k)\}$ in this region. $\mathcal{F}(z)$, however, may also be defined for values of z inside the circle. In this way we extend our definition of $\mathcal{Z}\{f(k)\}$ to be equal to this $\mathcal{F}(z)$ (this is what is called analytic continuation) for all points for which $\mathcal{F}(z)$ is defined (i.e., the points at which $\mathcal{F}(z)$ is analytic or equivalently points at which $\mathcal{F}(z)$ has no singularities).

Example 6.1
For $f(k) = \delta(k - k_0)$ we see that

$$\mathcal{Z}\{\delta(k - k_0)\} = \sum_{k=0}^{\infty} \delta(k - k_0) z^{-k} = \begin{cases} 0 & \text{if } k_0 < 0 \\ z^{-k_0} & \text{if } k_0 \geq 0. \end{cases}$$

Example 6.2
When $f(k) = 1$ for all $k \geq 0$ then $\{f(k)\} = \{1(k)\}$ = the discrete unit-step function:

$$\mathcal{Z}\{1(k)\} = \sum_{k=0}^{\infty} z^{-k} = \frac{1}{1 - z^{-1}} = \frac{z}{z - 1}$$

for $|z^{-1}| < 1$ which is $|z| > 1$. Here $\rho_A = 1$, and we say that $\mathcal{Z}\{1(k)\} = z/(z-1)$. Note this function of z is defined everywhere except at $z = 1$ which is a first-order pole of $\mathcal{F}(z)$.

Example 6.3
The $\mathcal{Z}$-transform of the function $f(k) = w^k$ is

$$\mathcal{Z}\{w^k\} = \sum_{k=0}^{\infty} w^k z^{-k} = \sum_{k=0}^{\infty} \left(wz^{-1}\right)^k = \frac{1}{1 - wz^{-1}} = \frac{z}{z - w}$$

for $|wz^{-1}| < 1$ or $|z| > |w|$. Here $\mathcal{F}(z) = z/(z - w)$ is defined for all z except at $z = w$ which is a first-order pole of $\mathcal{F}(z)$.

Example 6.4
Find

$$\mathcal{Z}\{k\} = \sum_{k=0}^{\infty} k z^{-k}.$$

It is difficult to see the closed form directly, but note that since

$$\sum_{k=0}^{\infty} z^{-k} = \frac{1}{1 - z^{-1}} \quad \text{for all } |z| \geq 1,$$

if we take d/dz of both sides (we are permitted to take the derivative of the sum, term by term since the sum is absolutely convergent) then

$$-\sum_{k=0}^{\infty} k z^{-(k+1)} = -\frac{z^{-2}}{(1 - z^{-1})^2}$$

which gives

$$\sum_{k=0}^{\infty} kz^{-k} = \frac{z^{-1}}{(1-z^{-1})^2} = \frac{z}{(z-1)^2} \quad \text{for all } |z| > 1.$$

Thus $\rho_A = 1$ and $\mathcal{F}(z)$ has a second-order pole at $z = 1$.

6.3 The inverse transformation

As we mentioned already, the infinite series which defines $\mathcal{Z}\{f(k)\}$ is equal to the power series expansion in z^{-1} for $\mathcal{F}(z)$ which converges in the region outside the circle of convergence. (This is the Laurent series expansion of complex variable theory.) Since such a power series representation of a function $\mathcal{F}(z)$ is unique, any method which gives this infinite series representation for $\mathcal{F}(z)$ will be acceptable. Note that once we have the power series $\sum_{k=0}^{\infty} f(k)z^{-k}$ then the values of the time function $f(k)$ are merely the coefficients of the powers z^{-k}.

When $\mathcal{F}(z)$ is a finite degree polynomial in z^{-1} such as

$$\mathcal{F}(z) = \sum_{i=0}^{n} a_i z^{-i} \tag{6.3}$$

then using the relation $\mathcal{Z}\{\delta(k-i)\} = z^{-i}$ found in Example 6.1, we see that

$$f(k) = \sum_{i=0}^{n} a_i \delta(k-i) \quad \text{for} \quad k \geq 0. \tag{6.4}$$

Since $\delta(k-i) = 0$ for $k \neq i$, this means that $f(k) = a_k$ for $0 \leq k \leq n$ and $f(k) = 0$ otherwise.

More generally, when $\mathcal{F}(z)$ is a rational function

$$\mathcal{F}(z) = \frac{\mathcal{N}(z)}{\mathcal{D}(z)} \tag{6.5}$$

where $\mathcal{N}(z)$ and $\mathcal{D}(z)$ are finite degree polynomials in z, then we must find the coefficients of the power series expansion. Note that for the one-sided $\mathcal{Z}$-transform the $\lim_{z\to\infty} \mathcal{F}(z)$ must be finite. Thus the degree of the numerator polynomial $\mathcal{N}(z)$ must be less than or equal to the degree of the denominator $\mathcal{D}(z)$.

Method 1: Long division

This method is useful if we want *only a few terms* of $\{f(k)\}$. Basically we arrange the denominator and numerator in decreasing powers of z and divide.

Example 6.5

For $\mathcal{F}(z) = 3z/(z^2 + z - 2)$ find $f(k)$ for $k \geq 0$.

$$
\begin{array}{r l}
 & \underline{3z^{-1} - 3z^{-2} + 9z^{-3} - 15z^{-4} \ldots} \\
z^2 + z - 2 \,\big) & 3z \\
 & \underline{3z + 3 - 6z^{-1}} \\
 & \quad -3 + 6z^{-1} \\
 & \underline{\quad -3 - 3z^{-1} + 6z^{-2}} \\
 & \qquad 9z^{-1} - 6z^{-2} \\
 & \underline{\qquad 9z^{-1} + 9z^{-2} - 18z^{-3}} \\
 & \qquad\quad -15z^{-2} + 18z^{-3}
\end{array}
$$

and we see that $f(0) = 0,\ f(1) = 3,\ f(2) = -3,\ f(3) = 9,\ f(4) = -15$. *Note*: It is not obvious how a general term $f(k)$ behaves as a function of k.

Method 2: Partial fraction expansion

This method is analogous to a similar method for getting the inverse $\mathcal{L}$-transform. Here we express $\mathcal{F}(z)/z$ as a sum of terms having only simple poles. Then from our table of known $\mathcal{Z}$-transforms we can find the inverse transform of each part.

Example 6.6

From

$$\mathcal{F}(z) = \frac{3z}{z^2 + z - 2} = \frac{3z}{(z+2)(z-1)} = \frac{Az}{(z+2)} + \frac{Bz}{(z-1)}$$

we see that

$$A = (z+2)\left.\frac{\mathcal{F}(z)}{z}\right|_{z=-2} = \frac{3}{-3} = -1$$

$$B = (z-1)\left.\frac{\mathcal{F}(z)}{z}\right|_{z=1} = \frac{3}{3} = 1.$$

Thus $\mathcal{F}(z) = -z/(z+2) + z/(z-1)$, so $f(k) = [1 - (-2)^k]$ for $k \geq 0$, since we recognize that $z/(z-1)$ is the transform of $f_b(k) = 1$ and $z/(z+2)$ is the transform of $f_a(k) = (-2)^k$.

Note that $f(0) = 0,\ f(1) = 3,\ f(2) = -3,\ f(3) = 9,\ f(4) = -15$ which checks with the values we found using Method 1. However, here we have a general form for all values of $k \geq 0$.

Method 3: Inversion formula

An inversion formula can be obtained for the coefficients of the power series expansion in z^{-1} for $\mathcal{F}(z)$ by applying a result from complex variables known as the Laurent series expansion theorem. When applied to our problem we obtain the result that

$$f(k) = \sum \text{residues of} \quad \mathcal{F}(z)z^{(k-1)}$$

(evaluated at poles lying inside the circle of convergence).

However, since $\mathcal{F}(z)$ has no singularities outside the circle of convergence we can say alternately that

$$f(k) = \sum \text{residues of} \quad \frac{\mathcal{F}(z)}{z} z^k \quad \text{for} \quad k \geq 0. \tag{6.6}$$

Thus for

$$\frac{\mathcal{F}(z)}{z} = \frac{\mathcal{N}(z)}{(z-\alpha_1)^{n_1}(z-\alpha_2)^{n_2}\ldots(z-\alpha_s)^{n_s}} \tag{6.7}$$

where $\mathcal{N}(z)$ is a polynomial in z, $\mathcal{F}(z)/z$ has poles at $z = \alpha_i$ of order n_i for $i = 1, \ldots, s$. Applying the residue formula to (6.6) gives

$$f(k) = \sum_{i=1}^{s} \frac{1}{(n_i-1)!} \frac{d^{n_i-1}}{dz^{n_i-1}} \left[(z-\alpha_i)^{n_i} \frac{\mathcal{F}(z)z^k}{z} \right]_{z=\alpha_i} \tag{6.8}$$

as our inversion formula.

Note: In some cases there can be poles of $\mathcal{F}(z)z^k/z$ at $z = 0$ for small values of k. It is easiest to use either long division to find $f(k)$ for these values of k or the time-translation theorem (Theorem 6.4).

Example 6.7

For $\quad \mathcal{F}(z) = \dfrac{3z}{z^2+z-2} = \dfrac{3z}{(z+2)(z-1)}$

there are two simple poles at $z = -2$ and $z = 1$. Thus

$$f(k) = \left.\frac{3z^k}{z+2}\right|_{z=1} + \left.\frac{3z^k}{z-1}\right|_{z=-2} = 1-(-2)^k \quad \text{for } k \geq 0.$$

Example 6.8

For $\mathcal{F}(z) = \dfrac{1}{(z+1)(z-1/2)}$

we see that

$$\frac{\mathcal{F}(z)}{z} = \frac{1}{z(z+1)(z-1/2)}$$

has poles at $z = 0, \ -1, \ 1/2$.

Since $\mathcal{F}(z)/z$ has a pole at $z = 0$ we can let

$$\mathcal{F}(z) = z^{-1}\mathcal{F}_1(z) \quad \text{so} \quad \mathcal{F}_1(z) = z\mathcal{F}(z) = \frac{z}{(z+1)(z-1/2)}.$$

Then $\mathcal{F}_1(z)/z$ has no poles at $z = 0$, and we can find $f_1(k)$ from

$$f_1(k) = \left.\frac{z^k}{z+1}\right|_{z=1/2} + \left.\frac{z^k}{z-1/2}\right|_{z=-1} = \frac{2}{3}\left(\frac{1}{2}\right)^k - \frac{2}{3}(-1)^k \quad \text{for} \quad k \geq 0.$$

How is $f(k)$ related to $f_1(k)$? The time-translation theorem (to be proved in Section 6.4) says that when $\mathcal{F}(z) = z^{-n}\mathcal{F}_1(z)$ for $n \geq 0$ then $f(k) = f_1(k-n)\,1(k-n)$. When applied to our $f_1(k)$ we get

$$f(k) = \frac{2}{3}\left[\left(\frac{1}{2}\right)^{k-1} - (-1)^{k-1}\right] 1(k-1) \quad \text{for} \quad k \geq 0$$

or we could have seen that since

$$\frac{\mathcal{F}(z)}{z}z^k = \frac{z^{k-1}}{(z+1)(z-1/2)}$$

has no pole at $z = 0$ for $k \geq 1$ then

$$f(k) = \left.\frac{z^{k-1}}{z+1}\right|_{z=1/2} + \left.\frac{z^{k-1}}{z-1/2}\right|_{z=-1} = \frac{2}{3}\left[\left(\frac{1}{2}\right)^{k-1} - (-1)^{k-1}\right] \quad \text{for} \quad k \geq 1.$$

This does not give the correct answer for $k = 0$. To get $f(0) = 0$ just use long division.

Example 6.9

For $\mathcal{F}(z) = (z^4 + 2z + 1)/[z^3(z - 1/2)]$ we see that

$$\frac{\mathcal{F}(z)}{z} = \frac{z^4 + 2z + 1}{z^4(z-1/2)} = z^{-4}\frac{z^4 + 2z + 1}{(z-1/2)}.$$

But note that if we try to define

$$\frac{\mathcal{F}_1(z)}{z} = \frac{z^4 + 2z + 1}{(z-1/2)} \quad \text{then} \quad \mathcal{F}_1(z) = \frac{z^5 + 2z^2 + z}{(z-1/2)}$$

and this function cannot be the one-sided $\mathcal{Z}$-transform of any $f_1(k)$. Therefore we must be careful to define $\mathcal{F}_1(z)$ so that its denominator degree is greater than or equal to its numerator degree. In this example we could express $\mathcal{F}(z)$ as

$$\mathcal{F}(z) = \frac{z}{z-1/2} + 2z^{-3}\frac{z}{z-1/2} + z^{-4}\frac{z}{z-1/2}.$$

Thus letting $\mathcal{F}_1(z) = z/(z - 1/2)$ means that $f_1(k) = \left(\frac{1}{2}\right)^k$. From the time-translation theorem we get

$$f(k) = f_1(k) + 2f_1(k-3)\,1(k-3) + f_1(k-4)\,1(k-4)$$
$$= \left(\frac{1}{2}\right)^k + \left(\frac{1}{2}\right)^{k-3} 1(k-3) + \left(\frac{1}{2}\right)^{k-4} 1(k-4) \quad \text{for} \quad k \geq 0.$$

6.4 Properties of $\mathcal{Z}$-transforms

Theorem 6.2 *Linearity*

$$\mathcal{Z}\{\alpha_1 f_1(k) + \alpha_2 f_2(k)\} = \alpha_1 \mathcal{Z}\{f_1(k)\} + \alpha_2 \mathcal{Z}\{f_2(k)\}. \tag{6.9}$$

Proof: This follows directly from the definition, since

$$\mathcal{Z}\{\alpha_1 f_1(k) + \alpha_2 f_2(k)\} = \sum_{k=0}^{\infty} [\alpha_1 f_1(k) + \alpha_2 f_2(k)] z^{-k}$$
$$= \alpha_1 \sum_{k=0}^{\infty} f_1(k) z^{-k} + \alpha_2 \sum_{k=0}^{\infty} f_2(k) z^{-k}$$

and both sums converge for $|z| \geq \max[\rho_{a1}, \rho_{a2}]$. Therefore

$$\mathcal{Z}\{\alpha_1 f_1(k) + \alpha_2 f_2(k)\} = \alpha_1 \mathcal{Z}\{f_1(k)\} + \alpha_2 \mathcal{Z}\{f_2(k)\}.$$

QED.

Theorem 6.3 *Convolution*

If $f(k) = \sum_{j=0}^{k} f_1(k-j)\, f_2(j)$ *then*

$$\mathcal{Z}\{f(k)\} = \mathcal{Z}\{f_1(k)\}\mathcal{Z}\{f_2(k)\} \quad \textit{for} \quad |z| \geq \max[\rho_{a1}, \rho_{a2}]. \tag{6.10}$$

Proof:

$$\mathcal{Z}\{f(k)\} \triangleq \sum_{k=0}^{\infty} \left[\sum_{j=0}^{k} f_1(k-j)\, f_2(j) \right] z^{-k}$$
$$= \sum_{k=0}^{\infty} \left[\sum_{j=0}^{k} f_1(k-j)\, 1(k-j)\, f_2(j) \right] z^{-k}$$
$$= \sum_{j=0}^{\infty} f_2(j) \sum_{k=0}^{\infty} f_1(k-j)\, 1(k-j) z^{-k}. \tag{6.11}$$

Now let $\ell = k - j$ so

$$\mathcal{Z}\{f(k)\} = \sum_{j=0}^{\infty} f_2(j) \sum_{\ell=-j}^{\infty} f_1(\ell)\, 1(\ell) z^{-(l+j)}. \tag{6.12}$$

The lower limit for ℓ in (6.12) can be changed to zero because of the step function $1(\ell)$, so

$$\mathcal{Z}\{f(k)\} = \left[\sum_{j=0}^{\infty} f_2(j)z^{-j}\right]\left[\sum_{\ell=0}^{\infty} f_1(\ell)z^{-\ell}\right] = \mathcal{Z}\{f_1(k)\}\mathcal{Z}\{f_2(k)\}$$

provided that $|z| \geq \max[\rho_{a1}, \rho_{a2}]$.

The same procedure applies to matrices; that is, for

$$A(k) = \sum_{j=0}^{\infty} A_1(k-j)\, A_2(j)$$

where A_1 and A_2 are matrices conformable for multiplication then

$$\mathcal{Z}\{A(k)\} = \mathcal{Z}\{A_1(k)\}\mathcal{Z}\{A_2(k)\}. \tag{6.13}$$

This follows since the im element of $A(k)$ has

$$\begin{aligned}\mathcal{Z}\{A(k)_{im}\} &= \mathcal{Z}\left\{\sum_{j=0}^{k}\sum_{l=1}^{s}[A_1(k)]_{il}[A_2(k-j)]_{lm}\right\}\\ &= \sum_{l=1}^{s}\mathcal{Z}\left\{\sum_{j=0}^{k}[A_1(k)]_{il}[A_2(k-j)]_{lm}\right\}\\ &= \sum_{l=1}^{s}\mathcal{Z}\{[A_1(k)]_{il}\}\mathcal{Z}\{[A_2(k)]_{lm}\}.\end{aligned}$$

QED.

Theorem 6.4 *Time translation*

1 *For functions which are advanced in time (shifted to the left)*

$$\mathcal{Z}\{f(k+m)\} = z^m\left[\mathcal{Z}\{f(k)\} - \sum_{i=0}^{m-1} f(i)z^{-i}\right] \quad \text{for} \quad m \geq 0. \tag{6.14}$$

2 *For functions which are delayed in time (shifted to the right)*

$$\mathcal{Z}\{f(k-m)\} = z^{-m}\left[\mathcal{Z}\{f(k)\} + \sum_{i=1}^{m} f(-i)z^{i}\right] \quad \text{for} \quad m \geq 0. \tag{6.15}$$

Proof: First consider case (1) with $m = 1$. If we advance the function $\{f(k)\}$ by one, we get a new function $\{g(k)\} = \{f(k+1)\}$. Since the $\mathcal{Z}$-transform of $\{g(k)\}$ only includes values of $g(k)$ for $k \geq 0$, we expect that the value $g(-1) = f(0)$ which is included in $\mathcal{F}(z)$ must be removed somehow from $\mathcal{G}(z)$.

Formally we get from the definition

$$\mathcal{Z}\{f(k+1)\} = \sum_{k=0}^{\infty} f(k+1)z^{-k} = z\sum_{k=0}^{\infty} f(k+1)z^{-(k+1)}$$

$$= z\left[\sum_{j=1}^{\infty} f(j)z^{-j} + f(0) - f(0)\right] = z[\mathcal{F}(z) - f(0)].$$

For the general case

$$\mathcal{Z}\{f(k+m)\} = \sum_{k=0}^{\infty} f(k+m)z^{-k} = z^m \sum_{k=0}^{\infty} f(k+m)z^{-(k+m)}$$

$$= z^m\left[\sum_{j=m}^{\infty} f(j)z^{-j} + \sum_{j=0}^{m-1} f(j)z^{-j} - \sum_{j=0}^{m-1} f(j)z^{-j}\right]$$

$$= z^m\left[\mathcal{F}(z) - \sum_{j=0}^{m-1} f(j)z^{-j}\right].$$

For a delayed function, we expect the opposite effect. That means that values of $f(k)$ for negative k now must be included in the transform of the shifted function. These values are not present in $\mathcal{F}(z)$, so they must be inserted. Again starting with the definition

$$\mathcal{Z}\{f(k-m)\} = \sum_{k=0}^{\infty} f(k-m)z^{-k} = z^{-m}\sum_{k=0}^{\infty} f(k-m)z^{-(k-m)}$$

$$= z^{-m}\left[\sum_{j=-m}^{\infty} f(j)z^{-j}\right] = z^{-m}\left[\mathcal{F}(z) + \sum_{j=-m}^{-1} f(j)z^{-j}\right]$$

$$= z^{-m}\left[\mathcal{F}(z) + \sum_{i=1}^{m} f(-i)z^{i}\right].$$

QED.

Note: If $f(k) = 0$ for all $k < 0$ the summation is zero, so

$$\mathcal{Z}\{f(k-m)1(k-m)\} = z^{-m}\mathcal{F}(z). \tag{6.16}$$

Theorem 6.5 *Initial value*

$$f(0) = \lim_{z\to\infty} \mathcal{F}(z) = \lim_{z\to\infty} (1 - z^{-1})\mathcal{F}(z). \tag{6.17}$$

The proof of this follows directly from the definition.

Theorem 6.6 *Final value*

$$\lim_{N\to\infty} f(N) = \lim_{z\to 1}(1 - z^{-1})\mathcal{F}(z) \tag{6.18}$$

provided $(1 - z^{-1})\mathcal{F}(z)$ *has no poles for* $|z| \geq 1$.

Proof:

$$\mathcal{F}(z) = \lim_{N\to\infty} \sum_{k=0}^{N} f(k)z^{-k} \triangleq \lim_{N\to\infty} \mathcal{F}_N(z) \tag{6.19}$$

$$z^{-1}\mathcal{F}(z) = \lim_{N\to\infty} \sum_{k=0}^{N-1} f(k)z^{-(k+1)} \triangleq \lim_{N\to\infty} z^{-1}\mathcal{F}_{N-1}(z). \tag{6.20}$$

Subtracting (6.20) from (6.19),

$$(1-z^{-1})\mathcal{F}(z) = \lim_{N\to\infty} [\mathcal{F}_N(z) - z^{-1}\mathcal{F}_{N-1}(z)] \triangleq \lim_{N\to\infty} \mathcal{G}_N(z) \tag{6.21}$$

and

$$f(N) = \lim_{z\to 1} [\mathcal{F}_N(z) - z^{-1}\mathcal{F}_{N-1}(z)] \triangleq \lim_{z\to 1} \mathcal{G}_N(z). \tag{6.22}$$

It turns out that when $(1-z^{-1})\mathcal{F}(z)$ has no poles for all $|z| \geq 1$ then $\mathcal{G}_N(z)$ converges uniformly to the limit $(1-z^{-1})\mathcal{F}(z)$ as $N \to \infty$ for every $|z| \geq 1$. Thus the taking of limits can be interchanged and

$$\lim_{N\to\infty} \lim_{z\to 1} \mathcal{G}_N(z) = \lim_{z\to 1} \lim_{N\to\infty} \mathcal{G}_N(z). \tag{6.23}$$

Inserting (6.21) and (6.22) in (6.23) completes the proof. QED.

Theorem 6.7 *Scale change*
Given a discrete-time function $\{f(k)\}$, consider a new function $\{g(k)\}$ constructed by spreading out $\{f(k)\}$ using the rule $g(kK) = f(k)$ but otherwise $g(k) = 0$, where K is a constant integer. Then

$$\mathcal{Z}\{g(k)\} = \mathcal{G}(z) = \mathcal{F}(z^K). \tag{6.24}$$

Proof: From the definition

$$\mathcal{Z}\{g(k)\} = \sum_{k=0}^{\infty} g(k)z^{-k}$$

and since only the values of $g(k)$ for $k = iK,\ i = 0, 1, 2, \ldots$ are nonzero we see that

$$\mathcal{G}(z) = \sum_{i=0}^{\infty} g(iK)z^{-iK} = \sum_{i=0}^{\infty} f(i)\left[z^K\right]^{-i} = \mathcal{F}(z^K).$$

QED.

Some other useful properties are listed below. The proofs are left as exercises (see Problem 6.10).

Theorem 6.8 *Summation For* $g(k) \triangleq \sum_{j=-\infty}^{k} f(j)$

$$\mathcal{G}(z) = \frac{\mathcal{F}(z)}{1-z^{-1}} + \frac{g(-1)}{1-z^{-1}}. \tag{6.25}$$

Theorem 6.9 *Multiplication by a function* a^k

$$\mathcal{Z}\{a^k f(k)\} = \mathcal{F}\left(\frac{z}{a}\right). \tag{6.26}$$

Theorem 6.10 *Time multiplication*

$$\mathcal{Z}\{k f(k)\} = -z\frac{d\mathcal{F}(z)}{dz}. \tag{6.27}$$

More generally, if $U = -z(d/dz)$ *denotes the operation of taking the derivative with respect to* z *and then multiplying by* $(-z)$*, then*

$$\mathcal{Z}\{k^n f(k)\} = U^n\{\mathcal{F}(z)\} \tag{6.28}$$

where U^n *means repeat the operation* U n *times.*

Problems

6.1 For each of the functions below, determine whether the $\mathcal{Z}$-transform does or does not exist. Justify your answer.

a) $f(k) = \dfrac{1}{k-1}$ b) $f(k) = \dfrac{1}{k+1}$

c) $f(k) = e^{-k}$ d) $f(k) = e^{k}$

e) $f(k) = e^{-k^2}$ f) $f(k) = e^{k^2}$

g) $f(k) = k!$ h) $f(k) = \dfrac{1}{k!}$

6.2 Find the $\mathcal{Z}$-transform of each function below.

a) $f(k) = (k)^2 e^{-ak}$

b) $f(k) = a^k$

6.3 Find the $\mathcal{Z}$-transform of each function below.

a) $f(k) = \cos \beta k$

b) $f(k) = \sin^2 \beta k$

6.4 Find the $\mathcal{Z}$-transform of the functions below.

a) $f(k) = k(k-1)(k-2)$

b) $f(k)$ is the sequence $0, 1\times 2, 2\times 3, 3\times 4$, etc., for $k = 0, 1, 2, 3, \ldots$.

6.5 Find the $\mathcal{Z}$-transform of $f(k) = [3k2^{k-1} - 2^k + 1]1(k)$.

6.6 Find the $\mathcal{Z}$-transform of the function $f(k) = 1/k!$.

6.7 Find the inverse $\mathcal{Z}$-transform of the following functions for all $k \geq 0$.

a) $\mathcal{F}(z) = \dfrac{1}{(1 - z^{-1})(1 - 0.5z^{-1})}$ b) $\mathcal{F}(z) = \dfrac{z}{(z - 0.5)^2}$

c) $\mathcal{F}(z) = \dfrac{z^{-1} + z^{-2}}{(1 - z^{-1})^2}$ d) $\mathcal{F}(z) = \dfrac{z + 1}{z^3(3z + 1)}$

e) $\mathcal{F}(z) = \dfrac{z^2 + 2z + 1}{z^2 + 2z + 2}$

6.8 Find the inverse $\mathcal{Z}$-transform of the following functions:

a) $\mathcal{F}(z) = \dfrac{z(z + 1)}{(2z - 1)(z - 0.5)}$ b) $\mathcal{F}(z) = \dfrac{z^{-1}}{(1 - z^{-1})^2(2 + z^{-1})^2}$

c) $\mathcal{F}(z) = \dfrac{1}{z(z^2 - 1)(z + 1)}$ d) $\mathcal{F}(z) = \dfrac{z^{-1}}{(1 - z^{-2})^2}$

6.9 Show that the inverse $\mathcal{Z}$-transform of $\mathcal{F}(z) = z/(z - a)^{p+1}$ is

$$f(k) = \frac{k!}{p!\,(k - p)!} a^{k-p}\, 1(k - p).$$

6.10 Derive the following $\mathcal{Z}$-transform properties:

a) $\mathcal{Z}\left\{ \sum_{j=-\infty}^{k} f(j) \right\} = \dfrac{\mathcal{F}(z)}{1 - z^{-1}} + \dfrac{1}{1 - z^{-1}} \sum_{j=-\infty}^{-1} f(j)$

b) $\mathcal{Z}\{a^k f(k)\} = \mathcal{F}(a^{-1}z)$

c) $\mathcal{Z}\{k^n f(k)\} = [-z(d/dz)]^n \mathcal{F}(z)$ (where the notation on the right side of (c) means repeat the operation $[-z(d/dz)]$ n times)

6.11 Use the $\mathcal{Z}$-transform to get closed form expressions for the sums

a) $S_a(k) = \sum_{i=0}^{k} i$

b) $S_b(k) = \sum_{i=0}^{k} i^2$

6.12 Find the $\mathcal{Z}$-transform of

$$f(k) = \frac{1}{1 + k}.$$

6.13 Find the inverse $\mathcal{Z}$-transform of

$$\mathcal{F}(z) = -\log(1 - z^{-1}).$$

Hint: For the last three problems use some of the $\mathcal{Z}$-transform properties rather than proceeding directly.

7

Frequency-domain analysis of discrete systems and application of $\mathcal{Z}$-transforms

7.1 Introduction

Just as in the case of analog systems, we can develop a transform method of analyzing fixed discrete systems. This is based on resolving arbitrary discrete-time inputs into the elementary functions $\{z^k\}$ where z is a complex number. These functions are used because of the following fundamental theorem.

Fundamental property of fixed linear discrete systems

Theorem 7.1 *If S is a fixed linear discrete-time system, then the zero-state response*

$$S\{z^k\} = \mathcal{H}(z)z^k. \tag{7.1}$$

Proof: (Similar to the analog case.) Let $y(k) = S\{z^k\} = g(z,k)$ (i.e., the output depends on the time k and the value of z). Thus we can write

$$y(k) = \frac{g(z,k)}{z^k} z^k \triangleq \mathcal{G}(z,k)z^k. \tag{7.2}$$

But, since the system is fixed, we have for any k_0

$$S\{z^{(k-k_0)}\} = \mathcal{G}(z,k-k_0)z^{(k-k_0)} \tag{7.3}$$

and by linearity

$$S\{z^{-k_0}z^k\} = z^{-k_0}S\{z^k\} = z^{-k_0}\mathcal{G}(z,k)z^k = \mathcal{G}(z,k)z^{(k-k_0)}. \tag{7.4}$$

Thus equating (7.3) and (7.4), we see that $\mathcal{G}(z,k-k_0) = \mathcal{G}(z,k)$ for any k_0 and all k. In order for this to be true, $\mathcal{G}$ cannot depend on the time k, so $\mathcal{G}(z,k) = \mathcal{H}(z)$ which is called the discrete transfer function.

Thus we have shown that the zero-state response of a fixed, linear, discrete system to the input $\{z^k\}$ has the same shape as the input; only the complex amplitude is changed. QED.

7.2 Application to the indirect analysis approach

From the inversion integral for $\mathcal{Z}$-transforms, we find that a function $\mathcal{F}(z)$ with no poles outside a circle in the z-plane gives a bounded function of exponential order

$$f(k) = \frac{1}{2\pi} \oint_C \mathcal{F}(z) z^{k-1}\, dz. \tag{7.5}$$

The expression (7.5) can be considered to be a formula that shows the resolution of $\{f(k)\}$ into the elementary functions $\{z^{k-1}\}$, where the complex amplitudes are $\mathcal{F}(z)$. Thus, if we want to find the zero-state response to an arbitrary input $\{u(k)\}$, then for a linear system

$$y(k) = S\{u(k)\} = \frac{1}{2\pi} \oint_C \mathcal{U}(z) S\{z^{k-1}\}\, dz. \tag{7.6}$$

If, in addition, the system is fixed, then using (7.1)

$$y(k) = \frac{1}{2\pi} \oint_C \mathcal{U}(z)\mathcal{H}(z) z^{k-1}\, dz.$$

Thus the complex amplitude of $y(k)$ is $\mathcal{Y}(z) = \mathcal{H}(z)\mathcal{U}(z)$.

In studying linear, fixed, discrete systems, we can concentrate on finding $\mathcal{Y}(z)$ (the $\mathcal{Z}$-transform of $y(k)$), which is just the product of the transfer function and the $\mathcal{Z}$-transform of $u(k)$. *Note:*

$$\text{If} \quad u(k) = \begin{cases} 1 & \text{when } k = 0 \\ 0 & \text{when } k \neq 0 \end{cases}$$

then $u(k)$ is the discrete delta function and $\mathcal{U}(z) = 1$, so $\mathcal{Y}(z) = \mathcal{H}(z)$. But since $y(k)$ is also the delta response $h(k)$, we see that $\mathcal{Y}(z) = \mathcal{Z}\{h(k)\}$, and that the $\mathcal{Z}$-transform of the delta response $h(k)$ is the discrete transfer function $\mathcal{H}(z)$. For multiple-input multiple-output systems the preceding approach can be used (just as for analog systems) to show that

$$\boldsymbol{\mathcal{Y}}(z) = \boldsymbol{\mathcal{H}}(z)\boldsymbol{\mathcal{U}}(z) \tag{7.7}$$

where $\boldsymbol{\mathcal{H}}(z)$ is the discrete transfer function matrix. The element $\mathcal{H}_{ij}(z)$ (located at the ith row and jth column of $\boldsymbol{\mathcal{H}}(z)$) is the discrete transfer function from the jth input to the ith output.

7.3 Solution of the state equations using $\mathcal{Z}$-transforms

Just as in the continuous-time case, if we have a fixed, linear, discrete-time system, the transform approach can be used:

$$\begin{aligned} \boldsymbol{x}(k+1) &= \boldsymbol{A}\boldsymbol{x}(k) + \boldsymbol{B}\boldsymbol{u}(k) \\ \boldsymbol{y}(k) &= \boldsymbol{C}\boldsymbol{x}(k) + \boldsymbol{D}\boldsymbol{u}(k) \end{aligned} \tag{7.8}$$

where A, B, C, and D are constant matrices. Taking the $\mathcal{Z}$-transform of both sides gives

$$z[\mathcal{Z}\{\boldsymbol{x}(k)\} - \boldsymbol{x}(0)] = A\mathcal{Z}\{\boldsymbol{x}(k)\} + B\mathcal{Z}\{u(k)\}$$
$$\mathcal{Z}\{\boldsymbol{y}(k)\} = C\mathcal{Z}\{\boldsymbol{u}(k)\} + D\mathcal{Z}\{\boldsymbol{u}(k)\}. \tag{7.9}$$

Letting $\mathcal{U}(z) = \mathcal{Z}\{\boldsymbol{u}(k)\}$, $\mathcal{Y}(z) = \mathcal{Z}\{\boldsymbol{y}(k)\}$, $\mathcal{X}(z) = \mathcal{Z}\{\boldsymbol{x}(k)\}$ gives

$$(zI - A)\mathcal{X}(z) = z\boldsymbol{x}(0) + B\mathcal{U}(z)$$
$$\text{so} \quad \mathcal{X}(z) = z(zI - A)^{-1}\boldsymbol{x}(0) + (zI - A)^{-1}B\mathcal{U}(z)$$

and therefore

$$\mathcal{Y}(z) = Cz(zI - A)^{-1}\boldsymbol{x}(0) + [C(zI - A)^{-1}B + D]\mathcal{U}(z). \tag{7.10}$$

Also note that, when the initial state is zero (i.e., $\boldsymbol{x}(0) = \mathbf{0}$), then the transform of the output and the transform of the input are related by

$$\mathcal{Y}(z) = [C(zI - A)^{-1}B + D]\,\mathcal{U}(z). \tag{7.11}$$

Comparing (7.11) and (7.7), we see that the discrete transfer function matrix is given by

$$\mathcal{H}(z) = C(zI - A)^{-1}B + D. \tag{7.12}$$

Also recall that we found for $\mathbf{x}(k_0) = \mathbf{0}$ and $k_0 = -\infty$ that

$$\boldsymbol{y}(k) = \sum_{j=-\infty}^{k} H(k - j)\boldsymbol{u}(j)$$

where $H(k - j)$ is the delta response matrix. When $\boldsymbol{u}(j) = \mathbf{0}$ for all $j < 0$, then the lower limit is 0. Taking the $\mathcal{Z}$-transform gives

$$\mathcal{Y}(z) = \mathcal{Z}\{H(k)\}\mathcal{U}(z)$$

so we see that

$$\mathcal{Z}\{H(k)\} = \mathcal{H}(z).$$

The $\mathcal{Z}$-transform of the delta response matrix is the discrete transfer function matrix. From the direct solution of the state equations with $k_0 = 0$, we found (see (4.56))

$$\boldsymbol{y}(k) = C\phi(k)\boldsymbol{x}(0) + \sum_{i=0}^{k-1} C\phi(k - i - 1)B\boldsymbol{u}(i) + D\boldsymbol{u}(k). \tag{7.13}$$

Taking the $\mathcal{Z}$-transform of (7.13) gives

$$\mathcal{Y}(z) = C\mathcal{Z}\{\phi(k)\}\boldsymbol{x}(0) + \mathcal{Z}\left\{\sum_{i=0}^{k-1} C\phi(k - i - 1)B\boldsymbol{u}(i)\right\} + D\mathcal{U}(z). \tag{7.14}$$

Let

$$f(k-1) \triangleq \sum_{i=0}^{k-1} \phi(k-1-i)Bu(i)$$

and note that

$$\mathcal{Z}\{f(k)\} = \mathcal{Z}\{\phi(k)\}B\mathcal{U}(z)$$

$$\text{since} \qquad \mathcal{Z}\{f(k-1)\} = z^{-1}[\mathcal{Z}\{f(k)\} + f(-1)z]$$

and $f(-1) = \mathbf{0}$, since we take 0 as the time at which the input starts. We get $\mathcal{Z}\{f(k-1)\} = z^{-1}\mathcal{Z}\{\phi(k)\}B\mathcal{U}(z)$, so

$$\mathcal{Y}(z) = C\mathcal{Z}\{\phi(k)\}x(0) + Cz^{-1}\mathcal{Z}\{\phi(k)\}B\mathcal{U}(z) + D\mathcal{U}(z). \tag{7.15}$$

Comparing with (7.10), we see that

$$\mathcal{Z}\{\phi(k)\} = z(zI - A)^{-1} \tag{7.16}$$

which can be used to find $\phi(k)$ for a discrete-time system with A constant. Note that Leverrier's algorithm can be used to find $(zI - A)^{-1}$.

Example 7.1

Let us solve Example 4.14 using $\mathcal{Z}$-transforms. Since

$$\mathcal{Z}\{\phi(k\} = z[zI - A]^{-1} = z\begin{bmatrix} z+\frac{3}{2} & 1 \\ -1 & z-1 \end{bmatrix}^{-1} = \frac{\begin{bmatrix} z(z-1) & -z \\ z & z\left(z+\frac{3}{2}\right) \end{bmatrix}}{(z+1)\left(z-\frac{1}{2}\right)}$$

$$\text{let } \mathcal{F}_1(z) = \frac{z}{(z+1)(z-1/2)} \qquad \text{so}$$

$$f_1(k) = \sum \text{res.} \frac{z^k}{(z+1)(z-1/2)} = -\frac{2}{3}(-1)^k + \frac{2}{3}(1/2)^k.$$

$$\text{Let } \mathcal{F}_2(z) = \frac{z^2}{(z+1)(z-1/2)} \qquad \text{so}$$

$$\begin{aligned} f_2(k) &= \sum \text{res.} \frac{z^{(k+1)}}{(z+1)(z-1/2)} = -\frac{2}{3}(-1)^{(k+1)} + \frac{2}{3}(1/2)^{(k+1)} \\ &= \frac{2}{3}(-1)^k + \frac{1}{3}(1/2)^k. \end{aligned}$$

We see that

$$\phi_{11}(k) = f_2(k) - f_1(k) = \frac{4}{3}(-1)^k - \frac{1}{3}(1/2)^k$$

$$\phi_{21}(k) = -\phi_{12}(k) = f_1(k) = \frac{2}{3}(1/2)^k - \frac{2}{3}(-1)^k$$

$$\phi_{22}(k) = f_2(k) + \frac{3}{2}f_1(k) = -\frac{1}{3}(-1)^k + \frac{4}{3}(1/2)^k$$

which is the same as we found earlier.

To get the response to $u(k) = 1$ for $k \geq 0$ with $\boldsymbol{x}(0) = \mathbf{0}$, we find

$$\mathcal{H}(z) = C[zI - A]^{-1}B + D = \frac{z^2 + 3/2z}{(z+1)(z-1/2)}$$

$$\mathcal{U}(z) = \frac{z}{z-1}$$

$$\text{so} \quad \mathcal{Y}(z) = \frac{(z^2 + 3/2\,z)z}{(z-1)(z+1)(z-1/2)}$$

which gives

$$\begin{aligned} y(k) &= \sum \text{res.} \frac{(z^2+1.5z)z^k}{(z-1)(z+1)(z-0.5)} \\ &= \frac{2.5}{2 \cdot 0.5} - \frac{0.5(-1)^k}{-2(-1.5)} + \frac{(0.5)^k}{-0.5(1.5)} \quad \text{for } k \geq 0 \\ &= 2.5 - \frac{1}{6}(-1)^k - \frac{4}{3}(0.5)^k \quad \text{for } k \geq 0. \end{aligned}$$

It can easily be verified that this is the same as the result found earlier for all $k \geq 0$.

7.4 Solution of systems of difference equations

Just as in the case of continuous-time systems, the $\mathcal{Z}$-transform can be taken directly of both sides of the difference equations, and the discrete transfer function matrix can be obtained.

Example 7.2

Consider a system governed by

$$y(k+3) - 1.5y(k+2) + 0.5y(k) = u(k+2) - u(k+1) - 2u(k).$$

If we took the $\mathcal{Z}$-transform of this equation directly, we would need the values $y(0)$, $y(1)$, $y(2)$, $u(0)$, and $u(1)$ to get $\mathcal{Y}(z)$. It is usually more likely that the values before zero are known. That is, usually $u(k) = 0$ for $k < 0$ and $y(k)$ for $k < 0$ are known. (If the system is in the zero state, then $y(k) = 0$ for $k < 0$.) It is therefore preferable to rewrite the equation as

$$y(k) - 1.5y(k-1) + 0.5y(k-3) = u(k-1) - u(k-2) - 2u(k-3)$$

(i.e., just shift the time argument by three time instants). Applying the $\mathcal{Z}$-transform to both sides gives

$$\begin{aligned} &\mathcal{Y}(z) - 1.5z^{-1}[\mathcal{Y}(z) + y(-1)z] + 0.5z^{-3}[\mathcal{Y}(z) + y(-1)z + y(-2)z^2y(-3)z^3] \\ &= z^{-1}\mathcal{U}(z) - z^{-2}\mathcal{U}(z) - 2z^{-3}\mathcal{U}(z) \end{aligned}$$

since the input is first applied for $k \geq 0$. Note that the values of $u(k)$ for $k < 0$ are zero. Rearranging gives

$$\mathcal{Y}(z) = \frac{z^{-1} - z^{-2} - 2z^{-3}}{1 - 1.5z^{-1} + 0.5z^{-3}}\mathcal{U}(z) + \mathcal{I}(z)$$

where $\mathcal{I}(z)$ denotes the initial condition terms. If the system is initially at rest, then $y(-1)$, $y(-2)$, $y(-3)$ are zero and $\mathcal{I}(z) = 0$. If $y(-1)$, $y(-2)$, $y(-3)$ are known, then they go right into the equation.

In general, if we are given a system of m difference equations, we proceed almost exactly as in the continuous-time case, except for the shifting of the time argument to remove all advance operations. Here let $J_{ij}(E^{-1})$ and $K_{ij}(E^{-1})$ be operators of the form

$$[1 + b_1E^{-1} + \cdots + b_pE^{-p}]y \triangleq J(E^{-1})y. \tag{7.17}$$

Therefore, this system is described by

$$\begin{aligned}
J_{11}(E^{-1})y_1(k) + \cdots + J_{1m}(E^{-1})y_m(k) &= K_{11}(E^{-1})u_1(k) + \cdots + K_{1r}(E^{-1})u_r(k)\\
&\vdots\\
J_{m1}(E^{-1})y_1(k) + \cdots + J_{mm}(E^{-1})y_m(k) &= K_{m1}(E^{-1})u_1(k) + \cdots + K_{mr}(E^{-1})u_r(k).
\end{aligned}$$

In matrix form, this is

$$J(E^{-1})\boldsymbol{y}(k) = K(E^{-1})\boldsymbol{u}(k). \tag{7.18}$$

Taking the $\mathcal{Z}$-transform gives the form

$$\mathcal{J}(z)\boldsymbol{\mathcal{Y}}(z) = \mathcal{K}(z)\boldsymbol{\mathcal{U}}(z) + \boldsymbol{\mathcal{I}}(z) \tag{7.19}$$

where $\mathcal{J}(z)$ is an $m \times m$ matrix of $\mathcal{Z}$-transforms and $\mathcal{K}(z)$ is an $m \times r$ matrix of $\mathcal{Z}$-transforms. So

$$\boldsymbol{\mathcal{Y}}(z) = \mathcal{J}^{-1}(z)\mathcal{K}(z)\boldsymbol{\mathcal{U}}(z) + \mathcal{J}^{-1}(z)\boldsymbol{\mathcal{I}}(z) \tag{7.20}$$

so we see that $\mathcal{H}(z) = \mathcal{J}^{-1}(z)\mathcal{K}(z)$ in this case. *Note:* $\mathcal{J}^{-1}(z)$ means the matrix inverse of $\mathcal{J}(z)$.

Example 7.3

A system is governed by the set of equations

$$\begin{aligned}
y_1(k) - y_2(k-1) + y_3(k) &= 0\\
y_1(k) - y_2(k+1) &= 0\\
y_1(k) - 2y_2(k+1) &= -u(k).
\end{aligned}$$

Find $y_3(k)$ for all $k \geq 0$ when $u(k) = 1(k)$ and the system is initially in the zero state.

We rewrite the equations as

$$y_1(k) - y_2(k-1) + y_3(k) = 0$$
$$y_1(k-1) - y_2(k) = 0$$
$$y_1(k-1) - 2y_2(k) = -u(k-1).$$

Taking the $\mathcal{Z}$-transform gives

$$\mathcal{Y}_1(z) - z^{-1}\mathcal{Y}_2(z) + \mathcal{Y}_3(z) = 0$$
$$z^{-1}\mathcal{Y}_1(z) - \mathcal{Y}_2(z) = 0$$
$$z^{-1}\mathcal{Y}_1(z) - 2\mathcal{Y}_2(z) = -z^{-1}\mathcal{U}(z)$$

or
$$\underbrace{\begin{bmatrix} 1 & -z^{-1} & 1 \\ z^{-1} & -1 & 0 \\ z^{-1} & -2 & 0 \end{bmatrix}}_{\mathcal{J}(z)} \begin{bmatrix} \mathcal{Y}_1(z) \\ \mathcal{Y}_2(z) \\ \mathcal{Y}_3(z) \end{bmatrix} = \underbrace{\begin{bmatrix} 0 \\ 0 \\ -z^{-1} \end{bmatrix}}_{\mathcal{K}(z)} \mathcal{U}(z)$$

$$\mathcal{J}^{-1}(z) = \begin{bmatrix} 0 & 2z & -z \\ 0 & 1 & -1 \\ 1 & z^{-1} - 2z & z - z^{-1} \end{bmatrix}$$

so
$$\mathcal{Y}(z) = \begin{bmatrix} \mathcal{Y}_1(z) \\ \mathcal{Y}_2(z) \\ \mathcal{Y}_3(z) \end{bmatrix} = \mathcal{J}^{-1}(z)\mathcal{K}(z)\mathcal{U}(z) = \begin{bmatrix} 1 \\ z^{-1} \\ z^{-2} - 1 \end{bmatrix} \mathcal{U}(z).$$

Since $u(k) = 1(k)$, we find $\mathcal{U}(z) = \frac{z}{z-1}$, thus

$$\mathcal{Y}_3(z) = \frac{z^{-2} - 1}{z - 1} z = -\frac{(z^2 - 1)}{z(z-1)} = -\frac{(z+1)}{z} = -1 - z^{-1}.$$

Taking the inverse $\mathcal{Z}$-transform, we find

$$y(k) = -\delta(k) - \delta(k-1)$$

or
$$y(k) = \begin{cases} -1 & \text{for } k = 0, 1 \\ 0 & \text{for all other } k. \end{cases}$$

7.5 Sampled signals and sampled data systems

We have shown that by using $\mathcal{Z}$-transforms, we can characterize discrete, time-invariant systems by their transfer functions in exactly the same way as for analog systems (see Figure 7.1).

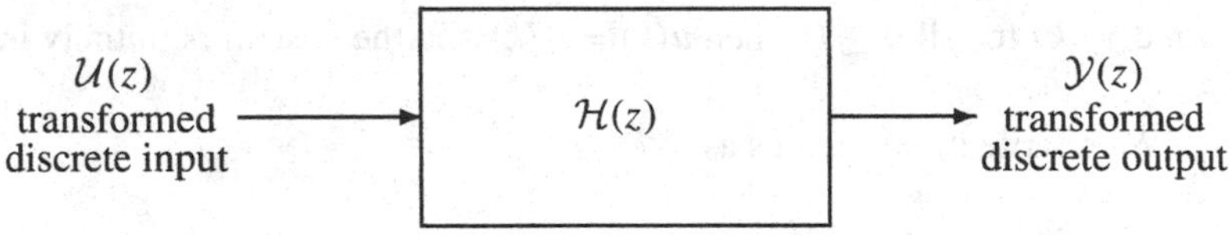

Fig. 7.1. A discrete system

It is appropriate to mention at this point that there are few physical linear systems which are completely discrete. A digital computer is indeed a discrete system, but it is usually not linear, and most linear systems (governed by linear differential equations) may have discrete portions but are essentially analog.

For the most part, our linear discrete analysis methods are useful in dealing with "hybrid" systems (sometimes called sampled-data systems), that is, systems that are essentially analog but have some discrete portions. This occurs for example when a digital computer is inserted into a system (such as a control system) to control a portion of the system linearly as shown in Figure 7.2.

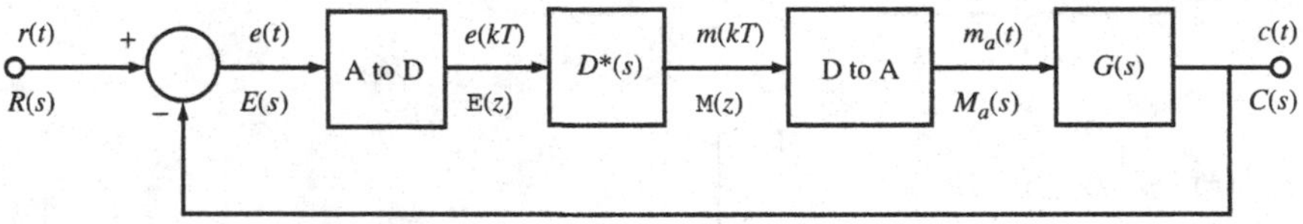

Fig. 7.2. A hybrid system

The A to D converter, as shown in Figure 7.3, changes an analog signal into a discrete signal.

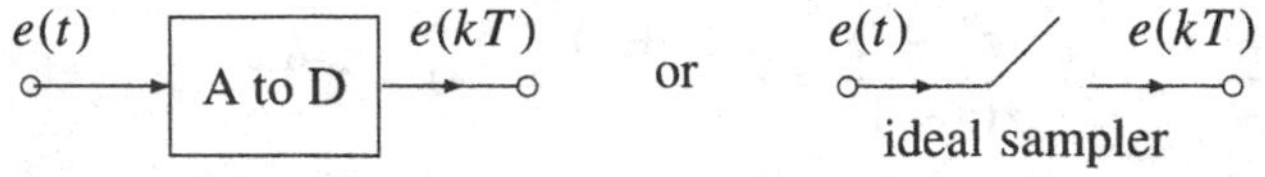

Fig. 7.3. Analog to discrete signal conversion using ideal sampling

The D to A converter, as shown in Figure 7.4, changes a discrete signal into an analog signal.

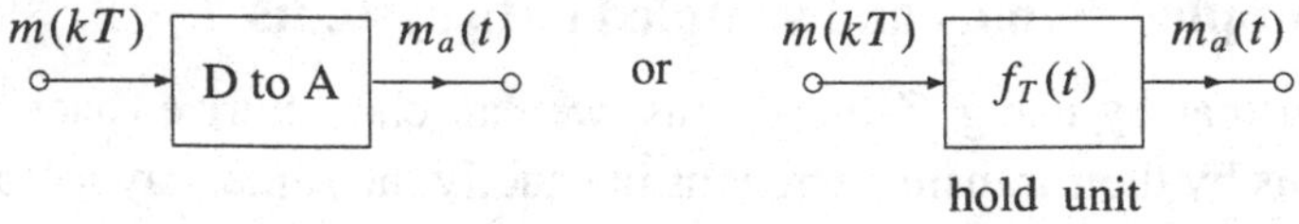

Fig. 7.4. Discrete to analog conversion requires a hold unit

Note: The reconstructed analog signal $m_a(t)$ depends entirely on its values at the discrete time instants $k = 0, 1, \ldots$. The reconstruction is usually done by just holding the discrete value constant until the next value is obtained at the next discrete time as shown in Figure 7.5.

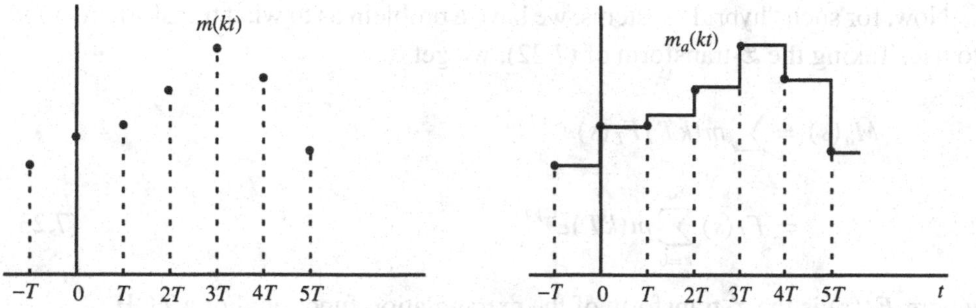

Fig. 7.5. Piecewise constant reconstruction of a discrete signal

This device is called the "zero-order hold," abbreviated as ZOH. In other words, the successive operations A to D followed by D to A gives $m_a(t)$ (see Figure 7.6), which is not the same as $m(t)$ for all t. However, at the sample times $t = 0, T, 2T, \ldots$, they are equal.

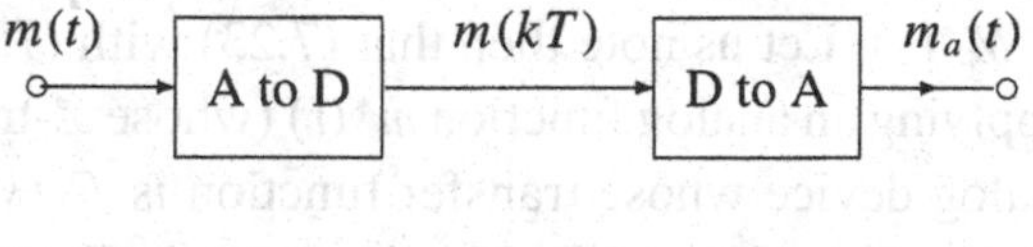

Fig. 7.6.

The "zero-order hold" is characterized by the extrapolation function

$$f_T(t) = \begin{cases} 1 & \text{for } 0 \leq t < T \\ 0 & \text{for all other } t \end{cases} \tag{7.21}$$

which has the form shown in Figure 7.7.

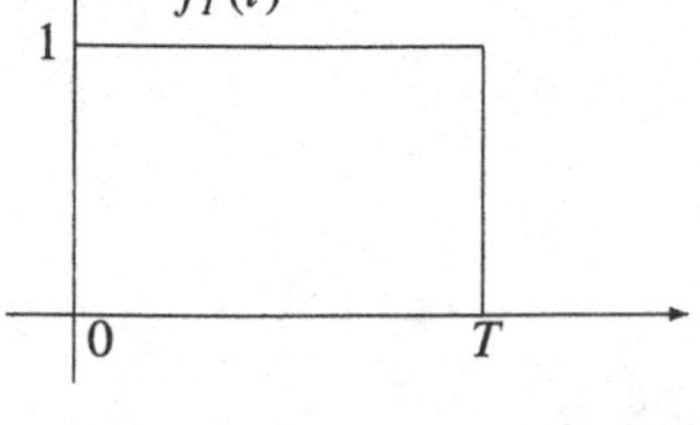

Fig. 7.7. Extrapolation function for a zero-order hold

Thus we can express

$$m_a(t) = \sum_{k=-\infty}^{\infty} m(kT) f_T(t - kT) \tag{7.22}$$

(or if $m(kT) = 0$ for $k < 0$, the sum starts at 0).

Now, for such "hybrid" systems, we have a problem as to what transform method to use. Taking the $\mathcal{L}$-transform of (7.22), we get

$$\begin{aligned} M_a(s) &= \sum_{k=0}^{\infty} m(kT) F_T(s) e^{-kTs} \\ &= F_T(s) \sum_{k=0}^{\infty} m(kT) e^{-kTs} \end{aligned} \tag{7.23}$$

where $F_T(s)$ is the $\mathcal{L}$-transform of the extrapolation function. For a ZOH

$$F_T(s) = \frac{1 - e^{-sT}}{s}. \tag{7.24}$$

We should recognize that the expression

$$\sum_{k=0}^{\infty} m(kT)(e^{sT})^{-k} = \mathcal{M}(e^{sT}) \triangleq M^*(s). \tag{7.25}$$

That is, the sum in (7.25) is the same as $\mathcal{Z}\{m(kT)\} = \mathcal{M}(z)$ with z replaced by e^{sT}. We denote such a function as $M^*(s)$. Let us note then that (7.23) with (7.25) says that $m_a(t)$ can be formed by applying an analog function $m^*(t)$ (whose $\mathcal{L}$-transform is $M^*(s)$) to a linear fixed analog device whose transfer function is $F_T(s)$. When the device is a ZOH, then $F_T(s)$ is given by (7.25). This is shown in Figure 7.8.

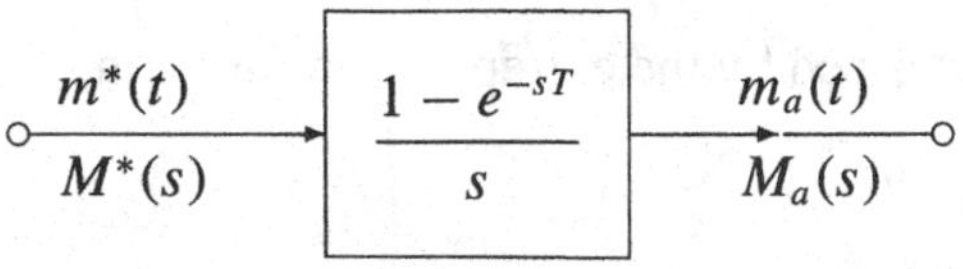

Fig. 7.8. Filter transfer function for a zero-order hold

What kind of function is $m^*(t)$? To find out, let us take the inverse Laplace transform of $M^*(s)$:

$$\begin{aligned} m^*(t) &= \mathcal{L}^{-1}\{M^*(s)\} \\ &= \sum_{k=0}^{\infty} m(kT) \mathcal{L}^{-1}\{e^{kTs}\} \\ &= \sum_{k=0}^{\infty} m(kT) \delta(t - kT) \quad \text{for } t \geq 0. \end{aligned} \tag{7.26}$$

We see from (7.26) that $m^*(t)$ is a sequence of impulse functions occurring every T. The area of each impulse is $m(kT)$. This kind of function is called an "impulse-modulated" function because it can be generated by multiplying $m(t)$ by an infinite sequence of evenly spaced unit impulses. That is, for

$$\delta_T(t) \triangleq \sum_{k=-\infty}^{\infty} \delta(t - kT) \tag{7.27}$$

($\delta_T(t)$ is called an "impulse train"), $m^*(t)$ can be generated by

$$m(t) \cdot \delta_T(t) = \sum_{k=-\infty}^{\infty} m(t)\delta(t - kT) = \sum_{k=-\infty}^{\infty} m(kT)\delta(t - kT) = m^*(t). \tag{7.28}$$

Note that the area of each impulse in (7.28) is the value $m(kT)$. Thus we see that the successive physical operations shown in Figure 7.9 can be replaced by the mathematical equivalent shown in Figure 7.10 where the impulse modulator

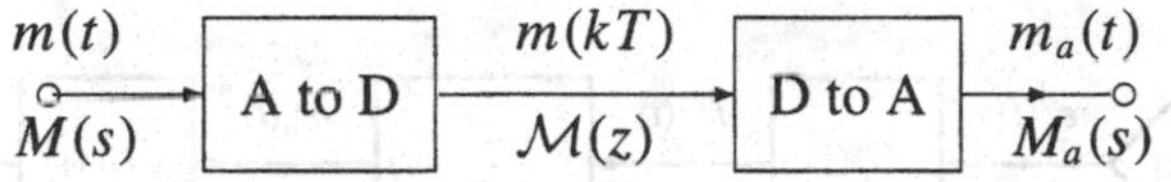

Fig. 7.9. Sampling and holding

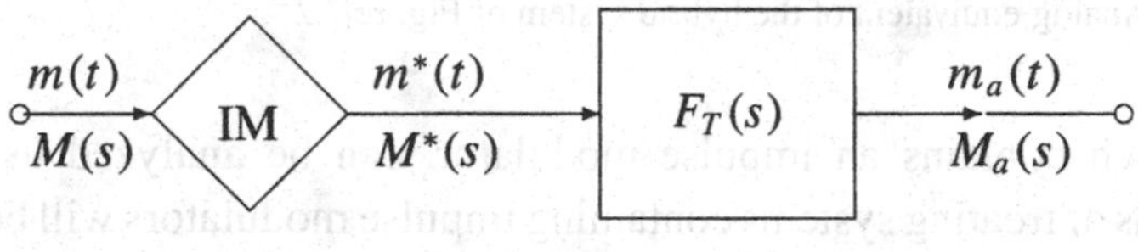

Fig. 7.10. Impulse modulating and filtering

IM converts $m(t)$ into $m^*(t)$. In other words, we have shown that the operation of impulse modulating followed by filtering is identical to the combined operation of sampling and holding when the "hold" function is identical to the impulse response of the filter. Also, we have seen that there is a relation between $M^*(s)$, the $\mathcal{L}$-transform of an impulse-modulated signal, and $\mathcal{M}(z)$, the $\mathcal{Z}$-transform of a discrete signal, both of which are derived from the same continuous-time signal $m(t)$. This same reasoning shows that the physical operations shown in Figure 7.11 can be replaced by the equivalent operations shown in Figure 7.12.

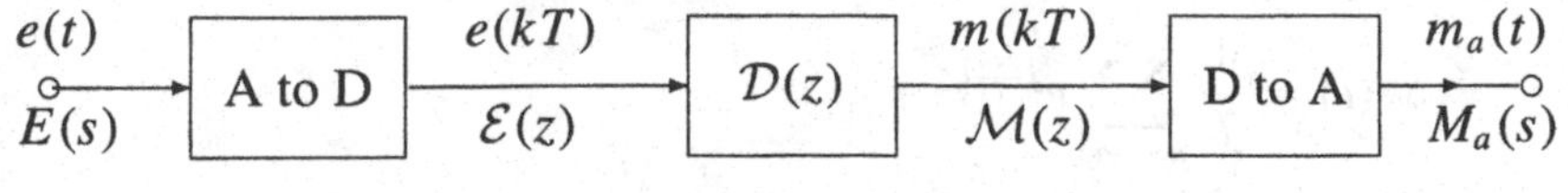

Fig. 7.11. Sampling, $\mathcal{Z}$-transforming, and holding

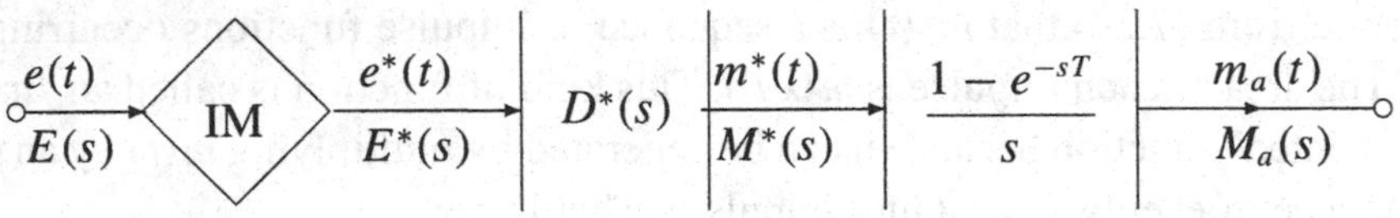

Fig. 7.12. Impulse modulating, transforming, and filtering

Also,

$$\text{since} \quad M_a(s) = \frac{1-e^{-sT}}{s}\mathcal{M}(z)\Big|_{z=e^{sT}} \qquad \text{and} \qquad \mathcal{M}(z) = \mathcal{D}(z)\mathcal{E}(z)$$

$$\text{then} \quad M_a(s) = \frac{1-e^{-sT}}{s}\mathcal{D}(e^{sT})\mathcal{E}(e^{sT}) = \frac{1-e^{-sT}}{s}D^*(s)E^*(s). \qquad (7.29)$$

Thus the hybrid system shown in Figure 7.2 can be replaced by the analog system of Figure 7.13.

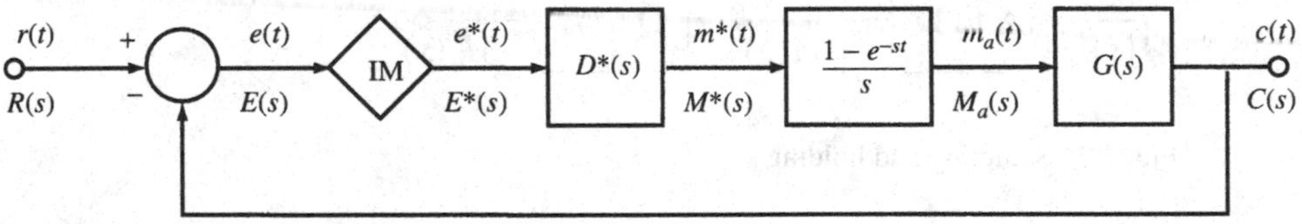

Fig. 7.13. Analog equivalent of the hybrid system of Figure 7.2

This system, which contains an impulse modulator, can be analyzed using $\mathcal{L}$-transforms. Details of treating systems containing impulse modulators will be given in Section 7.7.

A useful expression for $F^*(s)$ can be obtained directly from the Fourier series expansion of the periodic function $\delta_T(t)$ of (7.27). Since $\delta_T(t)$ is periodic, we know that it can be expanded in the Fourier series

$$\delta_T(t) = \sum_{n=-\infty}^{\infty} C_n e^{j\frac{2\pi n}{T}t} \qquad (7.30)$$

$$\text{where} \qquad C_n = \frac{1}{T}\int_{-T/2}^{T/2} \delta_T(t)e^{-j\frac{2\pi n}{T}t}\,dt = \frac{1}{T}.$$

(*Note:* $\delta_T(t) = \delta(t)$ in the interval $(-T/2, T/2)$.) Thus

$$\delta_T(t) = \frac{1}{T}\sum_{n=-\infty}^{\infty} e^{j\frac{2\pi n}{T}t} \qquad (7.31)$$

$$\text{and} \qquad f^*(t) = f(t)\delta_T(t) = \frac{1}{T}\sum_{n=\infty}^{\infty} f(t)e^{j\frac{2\pi n}{T}t}. \qquad (7.32)$$

Taking the $\mathcal{L}$-transform gives

$$F^*(s) = \frac{1}{T}\sum_{n=-\infty}^{\infty} \mathcal{L}\{f(t)e^{j\frac{2\pi n}{T}t}\} = \frac{1}{T}\sum_{n=-\infty}^{\infty} F\left(s + j\frac{2\pi}{T}n\right). \tag{7.33}$$

This expression tells us that $F^*(s)$ is periodic in the frequency domain (i.e., in the $j\omega$ direction of the s-plane), which means that

$$F^*\left(s + j\frac{2\pi}{T}k\right) = F^*(s) \quad \text{for any integer } k. \tag{7.34}$$

This will be a useful property in analyzing systems with impulse-modulated signals. To illustrate this periodicity, consider a function $f(t)$ with $|F(j\omega)|$ (i.e., amplitude spectrum) as shown in Figure 7.14. The amplitude spectrum of the impulse-modulated function $F^*(j\omega)$ is the sum of many such spectra shifted in frequency as shown in Figure 7.15. We can see this better if $f(t)$ is band lim-

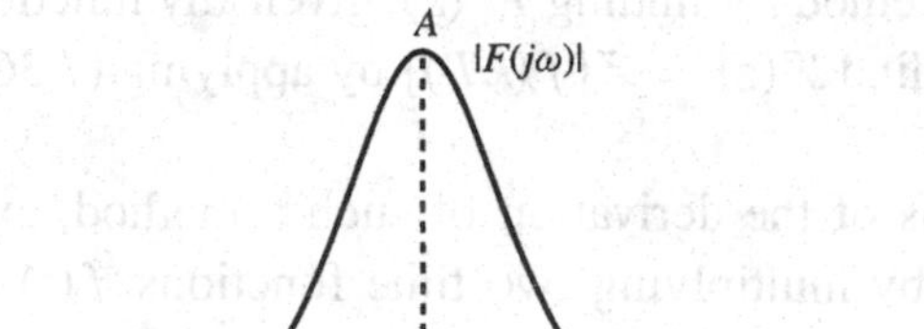

Fig. 7.14. Amplitude spectrum of an ordinary analog signal

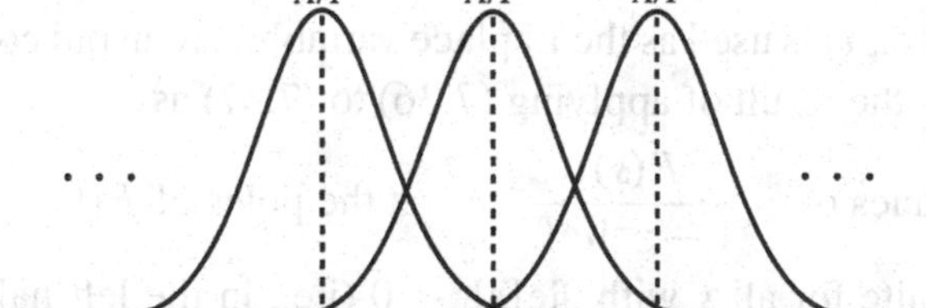

Fig. 7.15. Amplitude spectrum of an impulse-modulated signal

ited (i.e., if $|F(j\omega)| \equiv 0$ for all $|\omega| > \omega_0$). When $\omega_0 < \pi/T$, so that the highest frequency contained in the continuous-time signal is less than half the sampling frequency, then the spectrum of $F^*(j\omega)$ does not overlap as shown in Figure 7.16. If this signal is passed through an ideal low pass filter with pass band $-\pi/T$ to π/T, the exact signal $f(t)$ can be recovered.

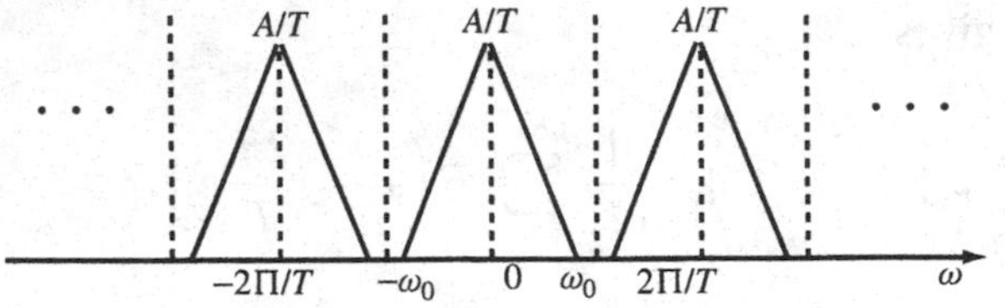

Fig. 7.16. Amplitude spectrum of an impulse-modulated signal

7.6 Method for obtaining the $\mathcal{Z}$-transform from the $\mathcal{L}$-transform

In the preceding section, we saw that the $\mathcal{Z}$-transform of the discretized form of a function $f(t)$ is directly related to the $\mathcal{L}$-transform of the impulse-modulated form of $f(t)$. That is, (7.25) says that

$$F^*(s) = \mathcal{F}(e^{sT}) = \mathcal{F}(z)|_{z=e^{sT}} \tag{7.35}$$

so, conversely,

$$\mathcal{F}(z) = F^*(s)|_{e^{sT}=z} \,. \tag{7.36}$$

Therefore if we develop a good method for finding $F^*(s)$, given any function $f(t)$, we can easily use that method to find $\mathcal{F}(z) = \mathcal{Z}\{f(kT)\}$ by applying (7.36) to the result.

We will not go into the details of the derivation of such a method, except to note that since $f^*(t)$ is formed by multiplying two time functions $f(t)$ and $\underset{\sim}{\delta}_T$ (the impulse train) then the $\mathcal{L}$-transform of the product function is given by a convolution in the complex s-plane, which must be evaluated by a contour integral in the complex plane.

The net result of this operation is the expression

$$F^*(s) = \sum \text{ residues of } \frac{F(w)}{1 - e^{-sT}e^{wT}} \text{ at the poles of } F(w) \tag{7.37}$$

where $F(w)$ is $\mathcal{L}\{f(t)\}$ (i.e., w is used as the Laplace variable). To avoid confusion between s and w, express the result of applying (7.36) to (7.37) as

$$\mathcal{F}(z) = \sum \text{ residues of } \frac{F(s)}{1 - z^{-1}e^{sT}} \text{ at the poles of } F(s) \tag{7.38}$$

provided $\lim_{s\to\infty} sF(s)$ is finite for all s with $\Re e\{s\} < 0$ (i.e., in the left half of the s-plane).

Example 7.4

Find $\mathcal{Z}\{(kT)^3\}$.

We see that if we sample the function $\{f(t)\} = \{t^3\}$, we obtain the required discrete function $f(kT) = (kT)^3$. Thus using $\mathcal{L}\{t^3\} = 3!/s^4$ we have

$$\mathcal{Z}\{(kT)^3\} = 3! \times \text{res. } \frac{1}{s^4(1 - z^{-1}e^{sT})} \quad \text{at } s = 0.$$

Since the function $F(s)$ has a fourth-order pole at $s = 0$,

$$\begin{aligned}
\mathcal{F}(z) &= 3! \times \frac{1}{3!}\frac{d^3}{ds^3}\left[\frac{1}{1-z^{-1}e^{sT}}\right]_{s=0} = \frac{d^2}{ds^2}\left[\frac{Tz^{-1}e^{sT}}{(1-z^{-1}e^{sT})^2}\right]_{s=0} \\
&= T\frac{d}{ds}\left[\frac{(1-z^{-1}e^{sT})Tz^{-1}e^{sT} + 2z^{-1}e^{sT}z^{-1}e^{sT}T}{(1-z^{-1}e^{sT})^3}\right]_{s=0} \\
&= T^2\frac{d}{ds}\left[\frac{z^{-1}e^{sT}(1+z^{-1}e^{sT})}{(1-z^{-1}e^{sT})^3}\right]_{s=0} \\
&= T^2\left.\frac{[Tz^{-1}e^{sT} + 2Tz^{-2}e^{2sT}]}{(1-z^{-1}e^{sT})^3} - \frac{z^{-1}e^{sT}(1+z^{-1}e^{sT})3(-Tz^{-1}e^{sT})}{(1-z^{-1}e^{sT})^4}\right|_{s=0} \\
\mathcal{F}(z) &= T^3\frac{z^{-1}(1+4z^{-1}+z^{-2})}{(1-z^{-1})^4}.
\end{aligned}$$

Example 7.5

Find $\mathcal{Z}\{1(kT - T)\}$.

From $\mathcal{L}\{1(t - T)\} = e^{-sT}/s$, we see that the $\lim_{s\to\infty} e^{-sT}$ is not finite for $\Re e\{s\} < 0$, thus (7.38) might not give the correct answer. From the time-translation theorem of $\mathcal{Z}$-transforms, we know that $\mathcal{Z}\{1(kT - T)\} = z^{-1}\ z/(z-1) = 1/(z-1)$, but from (7.38), we get

$$\mathcal{F}(z) = \left.\frac{e^{-sT}}{1-z^{-1}e^{sT}}\right|_{s=0} = \frac{z}{z-1} \qquad \text{which is wrong.}$$

7.7 Analysis of systems containing impulse-modulated signals

In Section 7.5 we saw that hybrid systems could be analyzed via $\mathcal{L}$-transforms by modeling them as systems that contain impulse-modulated signals. In this section we give some details on performing such analyses. Consider the simple system of Figure 7.17.

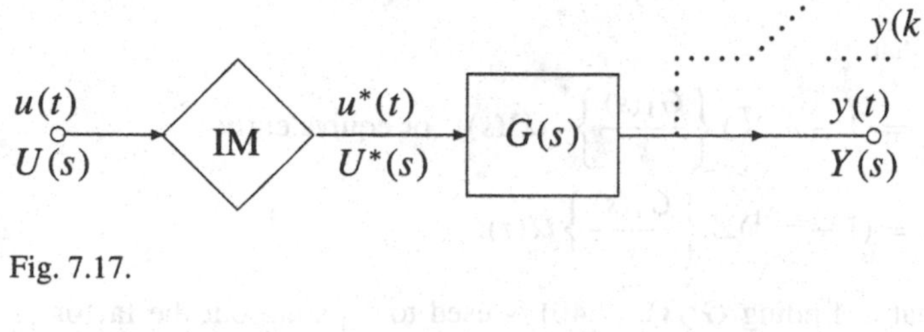

Fig. 7.17.

Applying Laplace transforms, we know that $Y(s) = G(s)U^*(s)$.

Note: $G(s)$ will usually be a rational function (ratio of polynomials) in s, whereas $U^*(s)$ will be a rational function in e^{sT}. It is thus quite difficult to find $y(t)$ for all $t \geq 0$. However, if we only wish to find $y(kT)$ as indicated by the dotted sampling switch in Figure 7.17, then we can first get

$$
\begin{aligned}
Y^*(s) &= \frac{1}{T}\sum_{n=-\infty}^{\infty} Y(s + j\frac{2\pi}{T}n) \\
&= \frac{1}{T}\sum_{n=-\infty}^{\infty} G(s + j\frac{2\pi}{T}n)U^*(s + j\frac{2\pi}{T}n). \qquad (7.39)
\end{aligned}
$$

Since $U^*(s)$ is periodic in $2\pi j/T$ (i.e., $U^*(s) = U^*(s + j(2\pi/T)n)$ for all n), therefore

$$
Y^*(s) = \frac{1}{T}\sum_{n=-\infty}^{\infty} G(s + j\frac{2\pi}{T}n)U^*(s) = G^*(s)U^*(s) \qquad (7.40)
$$

or in terms of $\mathcal{Z}$-transforms

$$
\mathcal{Y}(z) = \mathcal{G}(z)\mathcal{U}(z). \qquad (7.41)
$$

Thus $y(kT)$ can be readily found. Denoting the impulse-modulation operation by $\{\}^*$, this result can be written as

$$
\{G(s)U^*(s)\}^* = G^*(s)U^*(s) \qquad (7.42)
$$

or equivalently as

$$
\mathcal{Z}\{G(s)U^*(s)\} = \mathcal{G}(z)\mathcal{U}(z). \qquad (7.43)
$$

Note: Equation (7.43) really means that the $\mathcal{Z}$-transform is taken of the discrete-time signal obtained from sampling the continuous-time signal $\mathcal{L}^{-1}\{G(s)U^*(s)\}$.

Example 7.6

If $G(s) = [(1 - e^{-sT})/s]G_1(s)$ (i.e., the filter for a zero-order hold followed by the transfer function $G_1(s)$) then $Y^*(s) = G^*(s)U^*(s)$ or $\mathcal{Y}(z) = \mathcal{G}(z)\mathcal{U}(z)$, and since

$$
G^*(s) = \left\{\frac{1 - e^{-sT}}{s}G_1(s)\right\}^* = (1 - e^{-sT})\left\{\frac{G_1(s)}{s}\right\}^*
$$

we see that

$$
Y^*(s) = (1 - e^{-sT})\left\{\frac{G_1(s)}{s}\right\}^* U^*(s) \quad \text{or equivalently}
$$

$$
\mathcal{Y}(z) = (1 - z^{-1})\mathcal{Z}\left\{\frac{G_1(s)}{s}\right\}\mathcal{U}(z).
$$

Note that in finding $G^*(s)$, (7.40) is used to separate out the factor $(1 - e^{-sT})$ before applying (7.37) to find $\{G_1(s)/s\}^*$. This is because $\lim_{s\to\infty}(1 - e^{-sT})$ is not finite for s with $\Re e\{s\} < 0$, so (7.37) cannot be applied directly.

Next, consider the system shown in Figure 7.18:

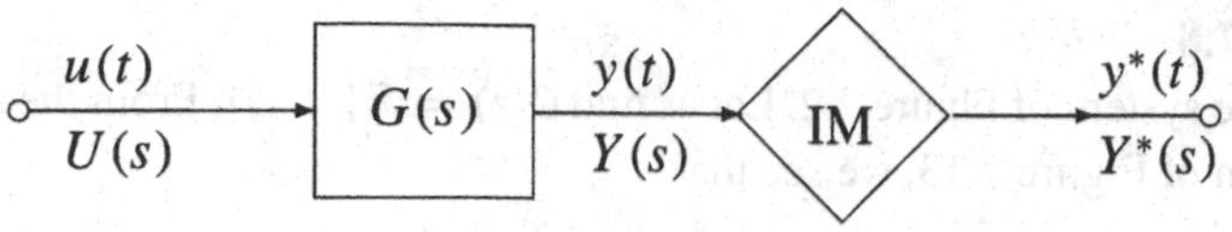

Fig. 7.18.

$$Y(s) = G(s)U(s)$$
$$Y^*(s) = \{G(s)U(s)\}^* = (GU)^*(s) \tag{7.44}$$
$$\text{so} \quad \mathcal{Y}(z) = \mathcal{Z}\{G(s)U(s)\} = \mathcal{GU}(z). \tag{7.45}$$

Note: This is not the same as $\mathcal{G}(z)\mathcal{U}(z)$, so we use the notation $(GU)^*(s)$ and $\mathcal{GU}(z)$ to indicate that G and U cannot be separated in performing the operation $\{\}^*$ or $\mathcal{Z}\{\}$. Using these two basic results, the response at sample instants of more complicated structures can be found.

Example 7.7
Consider the system of Figure 7.19.

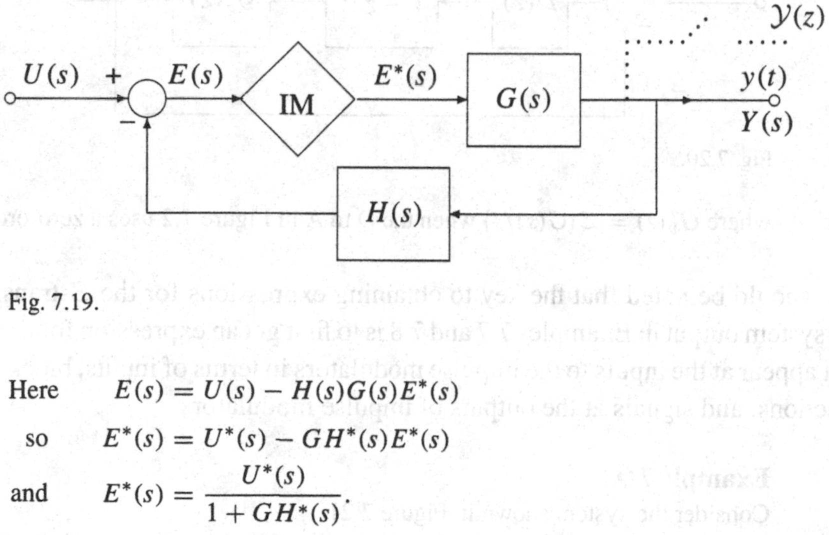

Fig. 7.19.

Here $\quad E(s) = U(s) - H(s)G(s)E^*(s)$

so $\quad E^*(s) = U^*(s) - GH^*(s)E^*(s)$

and $\quad E^*(s) = \dfrac{U^*(s)}{1 + GH^*(s)}.$

Now $\quad Y(s) = G(s)E^*(s)$

so $\quad Y^*(s) = G^*(s)E^*(s) = \dfrac{G^*(s)U^*(s)}{1 + GH^*(s)}$

or $\quad \mathcal{Y}(z) = \dfrac{\mathcal{G}(z)\mathcal{U}(z)}{1 + \mathcal{GH}(z)}.$

Thus, when $U(s)$, $G(s)$, and $H(s)$ are known, we can easily get $\mathcal{Y}(z)$ and then $y(kT)$.

Example 7.8

Consider the system of Figure 7.2. Let us find $\mathcal{C}(z) = \mathcal{Z}\{C(s)\}$. From the replacement system of Figure 7.13, we see that

$$E(s) = R(s) - G(s)M_a(s) = R(s) - G(s)\frac{1-e^{-sT}}{s}D^*(s)E^*(s).$$

$$\text{So} \quad E^*(s) = R^*(s) - (1-e^{-sT})\left\{\frac{G(s)}{s}\right\}^* D^*(s)E^*(s)$$

$$\text{or} \quad E^*(s) = \frac{R^*(s)}{1+(1-e^{-sT})\left\{\frac{G(s)}{s}\right\}^* D^*(s)}.$$

Since $C(s) = G(s)((1-e^{-sT})/s)D^*(s)E^*(s)$, we get the result

$$\mathcal{C}(z) = (1-z^{-1})\mathcal{Z}\left\{\frac{G(s)}{s}\right\}\mathcal{D}(z)\mathcal{E}(z) = \frac{(1-z^{-1})\mathcal{Z}\left\{\frac{G(s)}{s}\right\}\mathcal{D}(z)\mathcal{R}(z)}{1+(1-z^{-1})\mathcal{Z}\left\{\frac{G(s)}{s}\right\}\mathcal{D}(z)}.$$

This example shows that at discrete-time instants, the hybrid system of Figure 7.2 can be modeled by the completely discrete system shown in Figure 7.20,

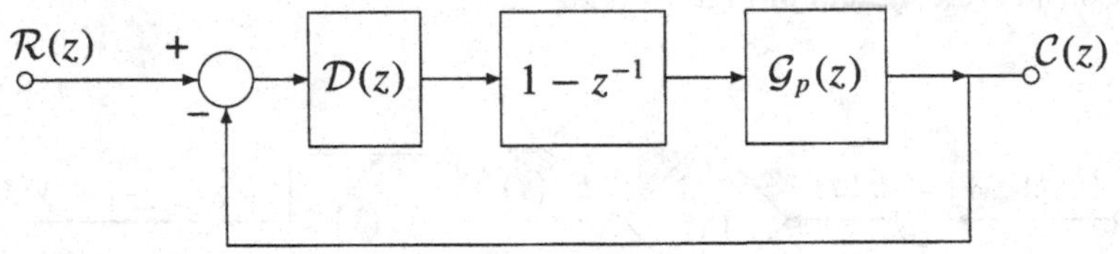

Fig. 7.20.

where $G_p(z) = \mathcal{Z}\{G(s)/s\}$ when the D to A in Figure 7.2 uses a zero-order hold.

It should be noted that the key to obtaining expressions for the $\mathcal{Z}$-transform of the system output in Examples 7.7 and 7.8 is to first get an expression for the signals that appear at the inputs to the impulse modulators in terms of inputs, block transfer functions, and signals at the outputs of impulse modulators.

Example 7.9

Consider the system shown in Figure 7.21:

This system is complicated by the existence of a feedback loop with strictly analog signals:

$$E(s) = R(s) - H(s)G_2(s)[G_1(s)E(s) + G_3(s)E^*(s)].$$

Note: Starring this expression would lock $E(s)$ to G_1G_2 and H. We thus solve for $E(s)$, giving

$$E(s) = \frac{R(s) - G_2(s)G_3(s)H(s)E^*(s)}{1+G_1(s)G_2(s)H(s)}.$$

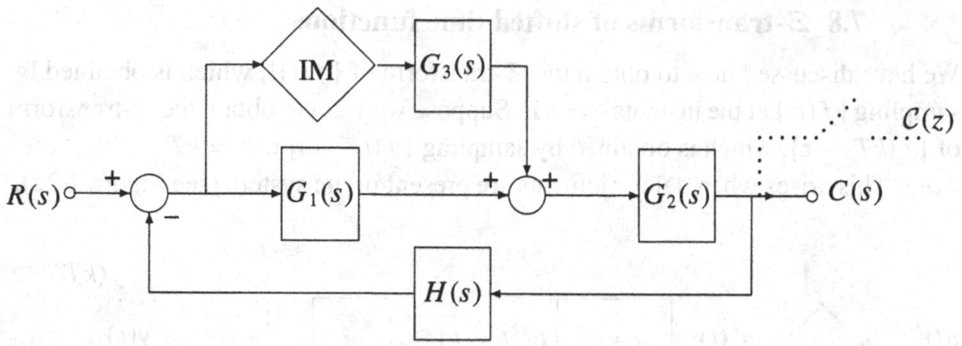

Fig. 7.21.

Now let us denote

$$A(s) \triangleq \frac{1}{1 + G_1(s)G_2(s)H(s)}$$

so $E(s) = A(s)R(s) - A(s)G_2(s)G_3(s)H(s)E^*(s)$

which becomes

$$\mathcal{E}(z) = \mathcal{Z}\{A(s)R(s)\} - \mathcal{Z}\{A(s)G_2(s)G_3(s)H(s)\}\mathcal{E}(z)$$

or $$\mathcal{E}(z) = \frac{\mathcal{AR}(z)}{1 + \mathcal{AG}_2\mathcal{G}_3\mathcal{H}(z)}.$$

Since

$$\begin{aligned} C(s) &= G_2(s)[G_3(s)E^*(s) + G_1(s)E(s)] \\ &= A(s)G_1(s)G_2(s)R(s) + [1 - A(s)G_1(s)G_2(s)H(s)]G_2(s)G_3(s)E^*(s) \\ &= A(s)G_1(s)G_2(s)R(s) + A(s)G_2(s)G_3(s)E^*(s) \end{aligned}$$

we have

$$\mathcal{C}(z) = \mathcal{AG}_1\mathcal{G}_2\mathcal{R}(z) + \mathcal{AG}_2\mathcal{G}_3(z)\mathcal{E}(z).$$

Note: The notation emphasizes that $\mathcal{AG}_1\mathcal{G}_2\mathcal{R}(z) = \mathcal{Z}\{A(s)G_1(s)G_2(s)R(s)\}$. Thus, for example, for $T = 1$ with $G_1(s) = H(s) = 1$ and $G_2(s) = G_3(s) = 1/s$, $R(s) = 1/s$, we have

$$A(s) = \frac{1}{1 + (1/s)} = \frac{s}{s+1}$$

$$\mathcal{E}(z) = \frac{\mathcal{Z}\left\{\frac{1}{s+1}\right\}}{1 + \mathcal{Z}\left\{\frac{1}{s(s+1)}\right\}}$$

$$\mathcal{C}(z) = \mathcal{Z}\left\{\frac{1}{s(s+1)}\right\}[1 + \mathcal{E}(z)].$$

7.8 $\mathcal{Z}$-transforms of shifted time functions

We have discussed how to obtain the $\mathcal{Z}$-transform of $\{f(t)\}$, which is obtained by sampling $\{f(t)\}$ at the instants $t = kT$. Suppose we want to obtain the $\mathcal{Z}$-transform of $\{f(kT - \tau\}$, which is obtained by sampling $\{f(t-\tau)\}$ at $t = kT$.
Note: This arises when delay elements are present in the system (see Figure 7.22).

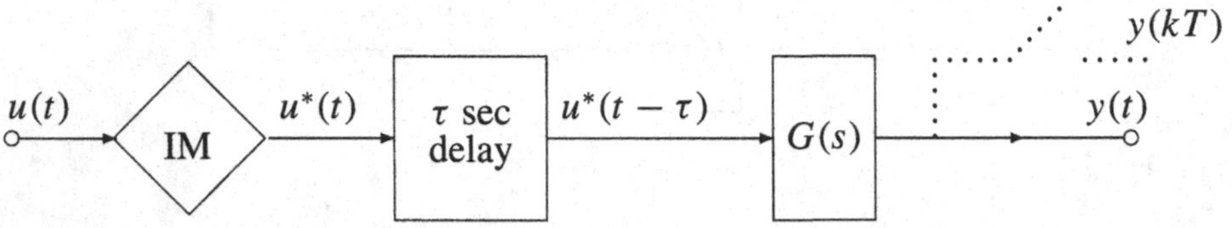

Fig. 7.22.

When an input is applied at $t = 0$, no response can occur for τ time units. If τ is not an integer multiple of T, the response will start between sample instants. Therefore, when $f(t) = 0$, for $t < 0$ we want $\mathcal{Z}\{f(kT-\tau)\}$. We found for $\tau = \ell T$ (with ℓ an integer) that

$$\mathcal{Z}\{f(kT-\tau)\} = z^{-\ell}\mathcal{Z}\{f(kT)\}.$$

When τ is not an integer multiple of T, let

$$\tau = \ell T - \Delta T \quad \text{where} \quad 0 \le \Delta < 1$$

and $\ell =$ the integer that is the least upper bound on τ/T. (This is known as the ceiling of τ/T.) Thus $\mathcal{Z}\{f(kT-\tau)\}$ is obtained from

$$\mathcal{L}\{f(t-\tau\} = \mathcal{L}\{f(t-\ell T + \Delta T\} = e^{-\ell Ts}e^{\Delta Ts}F(s) \qquad (7.46)$$

$$\text{so} \quad \mathcal{L}\{f^*(t-\tau)\} = \{e^{-\ell Ts}e^{\Delta Ts}F(s)\}^* = e^{-\ell Ts}\left\{e^{\Delta Ts}F(s)\right\}^* \qquad (7.47)$$

$$\text{thus} \quad \mathcal{Z}\{f(kT-\tau)\} = z^{-\ell}\mathcal{Z}\left\{e^{\Delta Ts}F(s)\right\} = z^{-\ell}\mathcal{F}(z,\Delta) \qquad (7.48)$$

$$\text{where} \quad \mathcal{F}(z,\Delta) \triangleq \mathcal{Z}\{f(kT+\Delta T)\} = \sum_{k=0}^{\infty} f(kT+\Delta T)z^{-k} \qquad (7.49)$$

is called the *modified* $\mathcal{Z}$-transform. We see therefore that

$$\mathcal{F}(z,\Delta) = \sum \text{res.} \quad \frac{e^{\Delta Ts}F(s)}{1-z^{-1}e^{sT}} \quad \text{at the poles of } F(s). \qquad (7.50)$$

Note: We must factor out $z^{-\ell}$, since taking the residue of $e^{-\tau s}F(s)/(1 - z^{-1} e^{sT})$ will not give correct results because $e^{-\tau s}F(s) \to \infty$ as $s \to \infty$ for $\Re\{s\} < 0$.

Example 7.10

1 $\mathcal{Z}\left\{e^{-\alpha(kT+\Delta T)}\right\}$. Since $\mathcal{L}\left\{e^{-\alpha(t+\Delta T)}\right\} = e^{\Delta Ts}/(s+\alpha)$, we find

$$\mathcal{F}(z,\Delta) = \text{res.}\ \frac{e^{\Delta Ts}/(s+\alpha)}{1-z^{-1}e^{sT}} = \left.\frac{e^{-\Delta Ts}}{1-z^{-1}e^{sT}}\right|_{s=-\alpha} = \frac{ze^{-\alpha\Delta T}}{z-e^{-\alpha T}}.$$

2 $\mathcal{Z}\left\{a^{(kT+\Delta T)}\right\}$. Since $\alpha = -\ln a \longleftrightarrow e^{-\alpha} = a$ we get $a^t = e^{-\alpha t}$, so $\mathcal{F}(z,\Delta) = za^{\Delta T}/(z-a^T)$.

3 $\mathcal{Z}\left\{(kT+\Delta T)e^{-\alpha(kT+\Delta T)}\right\}$. Since $\mathcal{L}\{f(t)\} = \mathcal{L}\{te^{-\alpha t}\} = F(s) = 1/(s+\alpha)^2$ we find

$$\begin{aligned}\mathcal{F}(z,\Delta) &= \text{res.}\ \left.\frac{e^{\Delta Ts}/(s+\alpha)^2}{(1-z^{-1}e^{sT})}\right|_{s=-\alpha} = \frac{d}{ds}\left[\frac{e^{\Delta Ts}}{(1-z^{-1}e^{sT})}\right]_{s=-\alpha}\\ &= \left.\frac{(1-z^{-1}e^{sT})\Delta Te^{\Delta Ts} - e^{\Delta Ts}(-Tz^{-1}e^{sT})}{(1-z^{-1}e^{sT})^2}\right|_{s=-\alpha}\\ &= e^{-\alpha\Delta T}T\frac{\Delta + (1-\Delta)e^{-\alpha T}z^{-1}}{(1-z^{-1}e^{-\alpha T})^2}.\end{aligned}$$

These modified $\mathcal{Z}$-transforms can be used to evaluate the response of hybrid systems at times between the sample instants. Thus if a hybrid system has been modeled as in Figure 7.23 we know that we can get $y(t)$ at the sample instants $t = kT$ from $\mathcal{Y}(z)$.

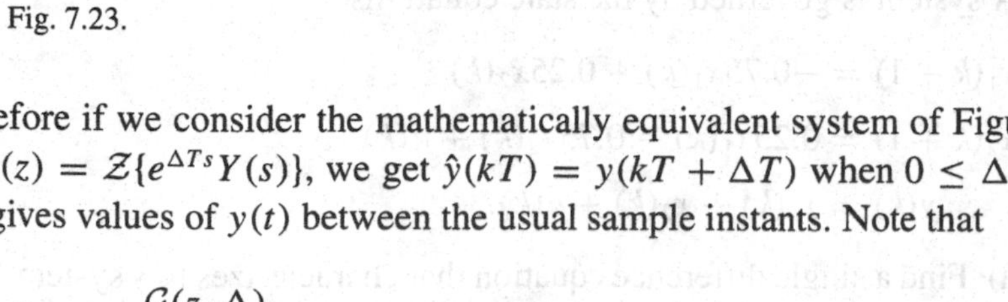

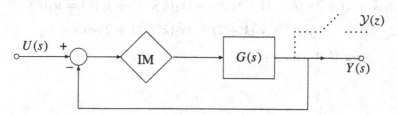

Fig. 7.23.

Therefore if we consider the mathematically equivalent system of Figure 7.24, since $\hat{\mathcal{Y}}(z) = \mathcal{Z}\{e^{\Delta Ts}Y(s)\}$, we get $\hat{y}(kT) = y(kT+\Delta T)$ when $0 \le \Delta < 1$, so $\hat{y}(kT)$ gives values of $y(t)$ between the usual sample instants. Note that

$$\hat{\mathcal{Y}}(z) = \frac{\mathcal{G}(z,\Delta)}{1+\mathcal{G}(z)}\mathcal{U}(z).$$

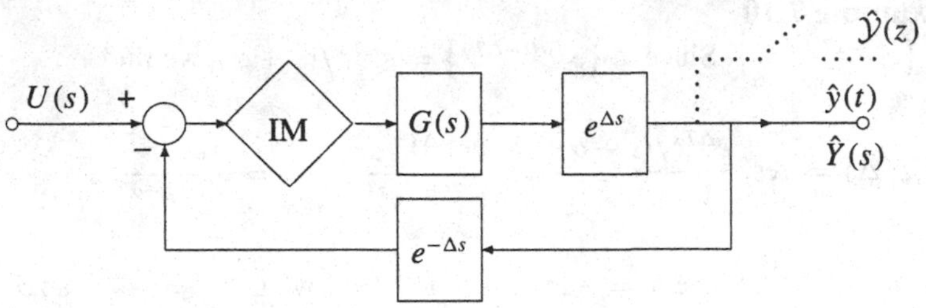

Fig. 7.24.

Problems

7.1 Find $\phi(k) = A^k$, using $\mathcal{Z}$-transforms for the matrices below.

a) $A = \begin{bmatrix} 0.5 & -0.5 \\ 0.5 & 0.5 \end{bmatrix}$ b) $A = \begin{bmatrix} 0 & 2 & 0.25 \\ 0 & 1 & 0 \\ -1 & 1 & 1 \end{bmatrix}$

7.2 Solve Problem 4.21, using $\mathcal{Z}$-transforms.

7.3 Solve Problem 3.19, using $\mathcal{Z}$-transforms.

7.4 Solve Problem 3.22, using $\mathcal{Z}$-transforms.

7.5 Find the discrete transfer function matrix of the system described by the difference equations

$$y_1(k+2) + 2y_1(k+1) + y_1(k) + 2y_2(k+1) + y_2(k) = u_1(k)$$
$$2y_1(k+1) + y_1(k) + y_2(k+2) + 2y_2(k+1) + 2y_2(k)$$
$$= u_1(k+1) + u_2(k).$$

7.6 A system is governed by the state equations

$$x_1(k+1) = -0.75x_1(k) + 0.25x_2(k)$$
$$x_2(k+1) = 0.25x_1(k) - 0.75x_2(k) + u(k)$$
$$y(k) = x_1(k) + x_2(k) + u(k).$$

a) Find a single difference equation that characterizes this system. That is, evaluate the α and β for

$$y(k+2) + \alpha_1 y(k+1) + \alpha_0 y(k) = \beta_2 u(k+2) + \beta_1 u(k+1) + \beta_0 u(k).$$

b) Find the transfer function of this system.
c) Calculate $y(k)$ for $k \geq 0$ when $u(k) = \partial(k) + \partial(k-1)$ and $\boldsymbol{x}(0) = \mathbf{0}$, using $\mathcal{Z}$-transforms.

7.7 A fixed, linear, discrete system has the zero-state delta response

$$h(k) = k1(k) - 2(k-2)1(k-2) + (k-4)1(k-4).$$

Find the zero-state response to $u(k) = k1(k)$.

7.8 A fixed, linear, discrete system is in the zero state. When the input $u(k) = (1/2)^k\ 1(k)$ is applied, the output is $y(k) = k\ 1(k)$.
a) Find the difference equation which governs the dynamic behavior of this system.
b) Find the delta response $h(k)$ of this system.

7.9 A fixed, linear, discrete system, which is in the zero state, is found to have the output $y(k) = k\ 1(k)$ when an input $u(k) = (-1)^k\ 1(k)$ is applied. Find the order of the system and the coefficients of the difference equation which governs the system.
In other words, for the difference equation

$$y(k+n) + \alpha_{n-1}y(k+n-1) + \cdots + \alpha_1 y(k+1) + \alpha_0 y(k)$$
$$= \beta_n u(k+n) + \beta_{n-1}u(k+n-1) + \cdots + \beta_1 u(k+1) + \beta_0 u(k)$$

determine the order n and the coefficients α_i and β_i.

7.10 Consider the discrete system described by the difference equations

$$y_1(k+1) + y_1(k) + y_2(k) = u(k+1)$$
$$y_1(k) + y_2(k+1) + y_2(k) = u(k)$$

with $y_1(-1) = 0,\ y_2(-1) = 1$, and $u(0) = 1, u(1) = -1$, and $u(k) = 0$ for all other k.
a) Find the discrete transfer function $\mathcal{H}(z)$ of this system.
b) Find $y_2(k)$ for all $k \geq 0$ when $y_1(-1) = 0,\ y_2(-1) = 1$, and $u(0) = 1, u(1) = -1$, and $u(k) = 0$ for all other k.

7.11 A fixed linear discrete system S has the transfer function

$$\mathcal{H}(z) = z^{-2}\frac{z+2}{z-1}.$$

a) Find the zero-state step response $a(k)$ for $k \geq 0$.

b) Draw a system which has this transfer function.

7.12 A fixed, linear, discrete system, which is in the zero state, is found to have the output $y(k) = 1(k)$ when the input $u(k)$ is applied where $u(0) = 1$, $u(1) = -1$, and $u(k) = 0$ for all other values of k.

a) Find the discrete transfer function $\mathcal{H}(z)$ of this system.

b) Suppose two systems with this transfer function are cascaded through a switch which samples only at even time instants (i.e., at $k = 0, 2, 4, 6, \ldots$) as shown below.

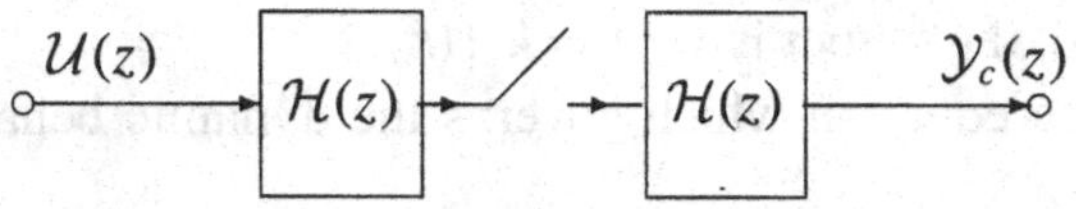

Find the transfer function $\mathcal{H}_c(z)$ of this cascaded system (i.e., find $\mathcal{Y}_c(z)$ of the combined system for the input $u(k) = \delta(k)$).

7.13 Find $\mathcal{F}(z)$ for the sampled function obtained from the following:

a) $F(s) = \dfrac{1}{(s+a)(s+b)}$

b) $F(s) = \dfrac{1}{(s+a)(s+b)^2}$

c) $F(s) = \dfrac{1}{(s+a)^3}$

d) $F(s) = \dfrac{1}{(s^2+\beta^2)^2}$

7.14 Find $\mathcal{Z}\{f(k)\}$ for the discrete functions given in Problems 6.2, 6.3, and 6.4 by first obtaining $F(s)$ for $f(t)$.

7.15 Find expressions for $C(s)$ and $\mathcal{C}(z)$ for each of the systems below.

a)

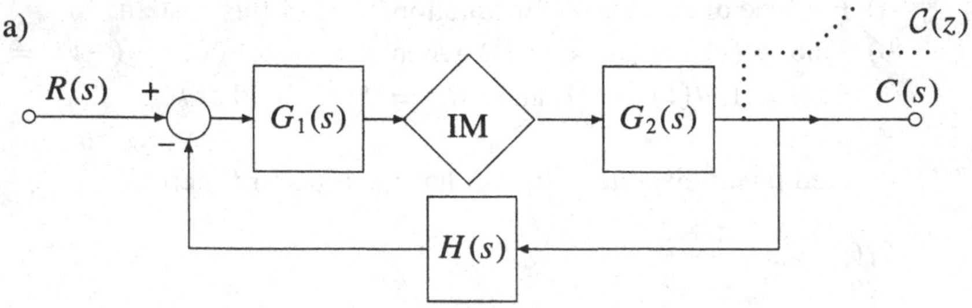

b)

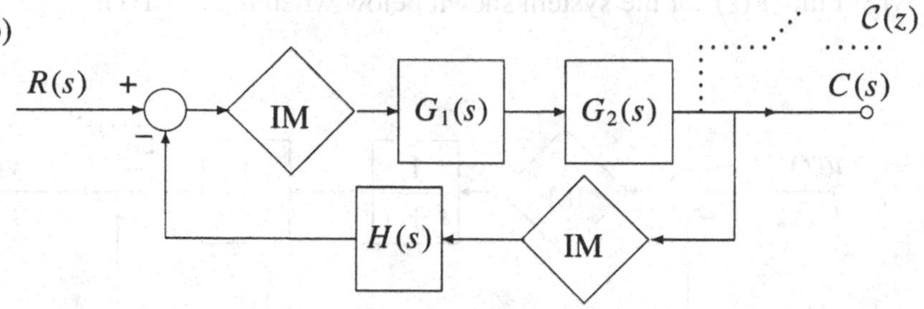

c)

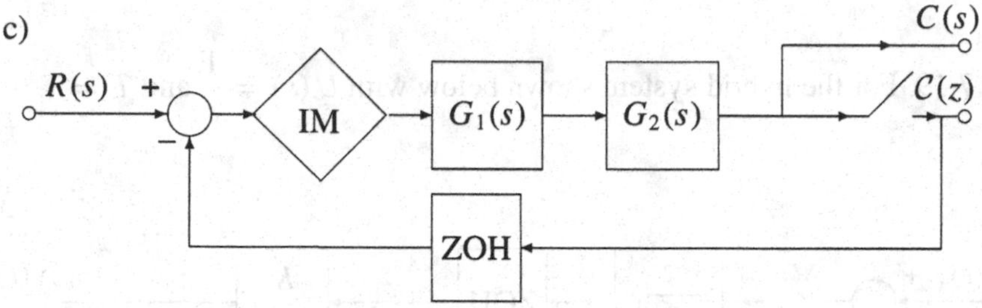

7.16 For the system shown below,

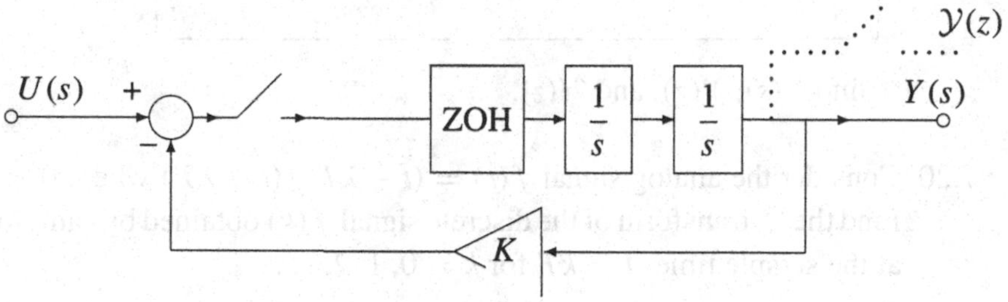

a) Find the discrete transfer function $\mathcal{H}(z)$ at the sample instants t = 0, 1, 2, ...

b) For $K = 1$, find $y(k)$ for $k \geq 0$ when $\mathcal{U}(z) = (z^{-1} + z^{-2})/(1 - z^{-1})^2$

7.17 For the system shown below with $U(s) = 1/s$,

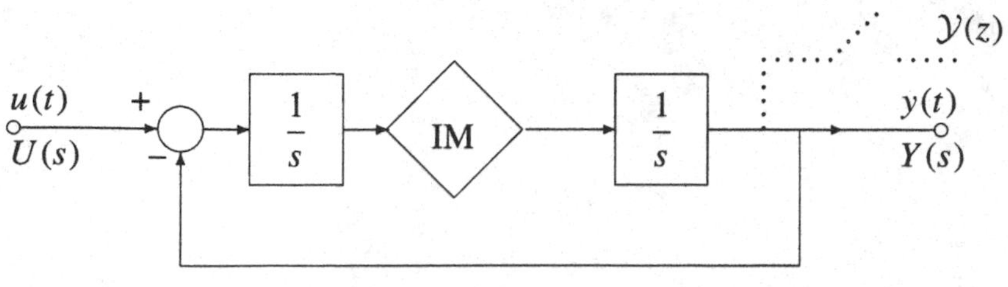

obtain expressions for $Y(s)$ and $\mathcal{Y}(z)$.

7.18 Find $\mathcal{Y}(z)$ for the system shown below when $u(t) = 1(t)$.

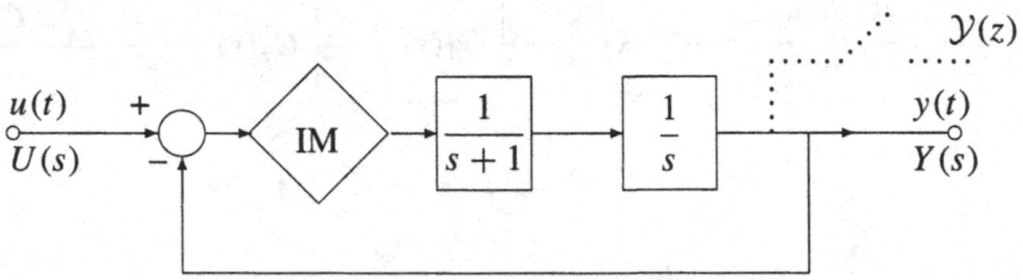

7.19 For the hybrid system shown below with $U(s) = \dfrac{1}{s}$ and $T = 1$

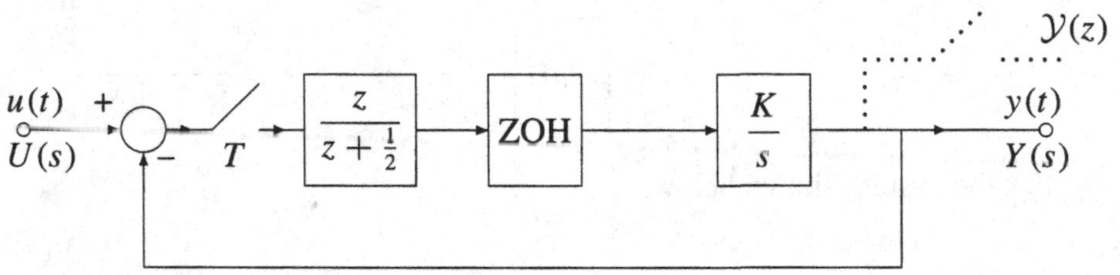

find $Y(s)$, $\mathcal{Y}(z)$, and $\mathcal{H}(z)$.

7.20 Consider the analog signal $f(t) = (t - \lambda T)1(t - \lambda T)$ where $0 < \lambda \leq 1$. Find the $\mathcal{Z}$-transform of the discrete signal $f(k)$ obtained by sampling $f(t)$ at the sample times $t = kT$ for $k = 0, 1, 2, \ldots$.

8

Controllability and observability of linear systems

8.1 Introduction and definition of controllability

The concepts of controllability and observability, which were first introduced by Kalman (1960), play an important role in modern system theory. We shall see later in this chapter how they are used to obtain minimal realizations for systems described by sets of differential equations, once any realization is obtained by the methods discussed in Chapter 2. These concepts also appear as necessary conditions for the existence of solutions to many optimal control problems and when studying stability properties of systems. We begin with the concept of controllability.

Definition 8.1 *A system is called (completely) controllable if for any initial time t_0, any initial state $\boldsymbol{x}(t_0)$ can be transferred to any final state $\boldsymbol{x}^*$ (i.e., $\boldsymbol{x}(t) = \boldsymbol{x}^*$) using some input $\underset{\sim}{\boldsymbol{u}}[t_0, t]$ over a finite time interval (i.e., t is finite).*

Note: In the literature, "completely" is used to denote a system that is controllable for any initial time t_0 (as in our definition) as opposed to a system controllable over a specified time interval. There are numerous refinements of this concept, as, for example:

Total controllability if the system is completely controllable over every (or almost every) finite interval.

Strong controllability if the system is controllable from each input terminal.

Output controllability if the system output (rather than state) can be set arbitrarily at some finite time t, by using an appropriate input.

We will consider only the definition given above and will omit the qualifying term "completely." For fixed systems, the particular initial time t_0 is immaterial, so we will usually choose it to be zero.

Example 8.1

The system in Figure 8.1 is not controllable.

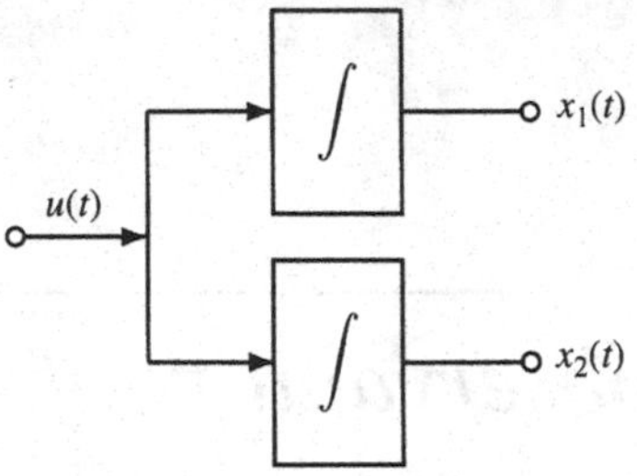

Fig. 8.1. A noncontrollable system

This can be seen from the expressions

$$x_1(t) = x_1(0) + \int_0^t u(\tau)\,d\tau$$

$$x_2(t) = x_2(0) + \int_0^t u(\tau)\,d\tau$$

so that $x_1(t) - x_1(0) = x_2(t) - x_2(0)$

or $x_1(t) = x_2(t) + x_1(0) - x_2(0)$.

Thus if $x_1(0) = x_2(0)$, then $x_1(t) = x_2(t)$ for all t, and we cannot reach a state for which $x_1 \neq x_2$.

For linear systems, various criteria can be developed for testing controllability of the system.

8.2 Controllability conditions for analog systems

For the system

$$\begin{aligned} \dot{\boldsymbol{x}}(t) &= A(t)\boldsymbol{x}(t) + B(t)\boldsymbol{u}(t) \\ \boldsymbol{y}(t) &= C(t)\boldsymbol{x}(t) + D(t)\boldsymbol{u}(t) \end{aligned} \tag{8.1}$$

with $A(t)$ bounded for all finite t, we know that

$$\boldsymbol{x}(t) = \phi(t, t_0)\boldsymbol{x}(t_0) + \int_{t_0}^t \phi(t, \tau)B(\tau)\boldsymbol{u}(\tau)\,d\tau \tag{8.2}$$

and $\phi(t, t_0)$ is nonsingular for all finite t. Let $\boldsymbol{x}^*$ be an arbitrary final state. We see that requiring $\boldsymbol{x}(t) = \boldsymbol{x}^*$ for some finite t is equivalent to requiring

$$\boldsymbol{x}^* - \phi(t, t_0)\boldsymbol{x}(t_0) \triangleq \boldsymbol{x}^d = \int_{t_0}^t \phi(t, \tau)B(\tau)\boldsymbol{u}(\tau)\,d\tau. \tag{8.3}$$

In other words, the ability to go from an arbitrary initial state $\boldsymbol{x}(t_0)$ to an arbitrary final state $\boldsymbol{x}^*$ is equivalent to the ability to go from the zero state to an arbitrary final state x^d. So the system (8.1) is controllable if any state can be reached from the origin in finite time. The following development is that described by Kreindler and Sarachik (1964).

In order to show under what conditions this is true, let us define the region $R(t_0, t)$ as the set of all points that can be reached from the origin, using a suitable input $\underset{\sim}{u}$ over the interval $[t_0, t]$. We assert that $R(t_0, t)$ is convex.

Definition 8.2 *A set R is called convex if for any points $\mathbf{r}_1$ and $\mathbf{r}_2$ belonging to the set R, the line segment joining these points also belongs to the set R (i.e., if $\mathbf{r}_3 = \alpha\mathbf{r}_1 + (1-\alpha)\mathbf{r}_2 \in R$ for all $0 \le \alpha \le 1$).*

To show that $R(t_0, t)$ is convex, let us pick any two states x_1 and x_2 in this set. Let $\underset{\sim}{u}_1$ and $\underset{\sim}{u}_2$ be the inputs that result in x_1 and x_2 respectively when $x(t_0) = \mathbf{0}$. We observe that the input $\underset{\sim}{u}_3 = \alpha\underset{\sim}{u}_1 + (1-\alpha)\underset{\sim}{u}_2$ gives

$$\int_{t_0}^{t} \phi(t,\tau)B(\tau)[\alpha u_1(\tau)+(1-\alpha)u_2(\tau)]\,d\tau = \alpha x_1+(1-\alpha)x_2 = x_3 \in R(t_0, t)$$

for any $0 \le \alpha \le 1$, so the set $R(t_0, t)$ is convex.

Now consider the projection of $x(t)$ on a nonzero vector $\boldsymbol{\lambda}$,

$$x(t)\cdot\boldsymbol{\lambda} = \boldsymbol{\lambda}^T x(t) = \sum_{i=1}^{n} \lambda_i x_i(t). \tag{8.4}$$

(Superscript T denotes transposition.) If for every $\boldsymbol{\lambda} \ne \mathbf{0}$ the projection on $\boldsymbol{\lambda}$ can be set to any arbitrary value, then we know that $x(t)$ can be made equal to any desired value x^d. (Since the projection of $x(t)$ can be set arbitrarily in any direction, its components in the unit directions can be set to any desired values.) Note that this will be true if for every $\boldsymbol{\lambda} \ne 0$, $\boldsymbol{\lambda}^T x(t) \ne 0$ for some value of $t > t_0$. On the other hand, if there were a direction $\boldsymbol{\lambda}_1$ for which $x(t)\cdot\lambda_1$ were zero for all $t \ge t_0$, this means that we could not make $x(t)$ equal to a x^d lying in the $\boldsymbol{\lambda}_1$ direction, so the system would not be controllable. Thus a necessary and sufficient condition for controllability is that there be no direction $\boldsymbol{\lambda}$ for which $\boldsymbol{\lambda}^T x(t) = 0$ for all $t \ge t_0$.

Consider our previous Example 8.1 where $x_1(0) = x_2(0)$ so $x_1(t) = x_2(t)$. Since $x(t)$ must always lie along line l with slope = 1 (as shown in Figure 8.2), the projection of $x(t)$ on $\boldsymbol{\lambda}$ perpendicular to l will always be zero. This system is not controllable.

From (8.2) with $x(t_0) = \mathbf{0}$, we get

$$\boldsymbol{\lambda}^T x(t) = \int_{t_0}^{t} \boldsymbol{\lambda}^T \phi(t,\tau)B(\tau)\underset{\sim}{u}(\tau)\,d\tau. \tag{8.5}$$

Let us define

$$v(\tau) \triangleq B^T(\tau)\phi^T(t,\tau)\boldsymbol{\lambda} \quad \text{for } t_0 \le \tau \le t \tag{8.6}$$

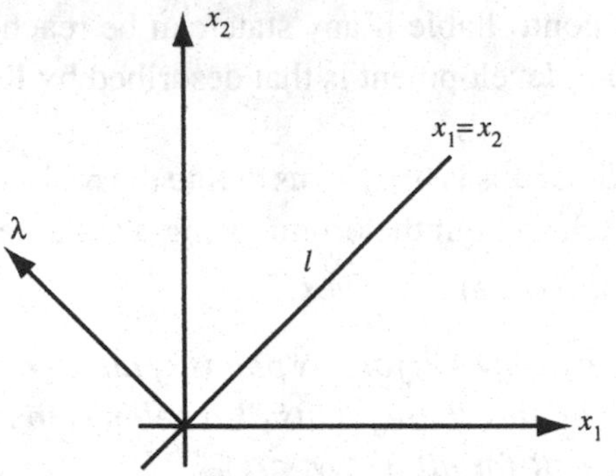

Fig. 8.2. The reachable states for Example 8.1

and choose the input $\boldsymbol{u}(\tau) = \alpha \boldsymbol{v}(\tau)$ so

$$\boldsymbol{\lambda}^T \boldsymbol{x}(t) = \alpha \int_{t_0}^{t} \boldsymbol{v}^T(\tau)\boldsymbol{v}(\tau)d\,\tau \tag{8.7}$$

$$= \alpha \boldsymbol{\lambda}^T \left[\int_{t_0}^{t} \phi(t,\tau)B(\tau)B^T(\tau)\phi^T(t,\tau)\,d\tau \right] \boldsymbol{\lambda} \tag{8.8}$$

Since α can be chosen arbitrarily, we see that $\boldsymbol{\lambda}^T \boldsymbol{x}(t)$ can be set to any value at t if and only if the integral in (8.7) is not zero. Observe that since the integral in (8.7) cannot be negative, the matrix inside the brackets of (8.8)

$$P(t_0, t) = \int_{t_0}^{t} \phi(t,\tau)B(\tau)B^T(\tau)\phi^T(t,\tau)\,d\tau \tag{8.9}$$

is positive semidefinite. The integral in (8.7) is not zero,and the system is controllable if and only if $\boldsymbol{\lambda}^T P(t_0, t)\boldsymbol{\lambda}$ is greater than zero for all $\boldsymbol{\lambda} \neq \mathbf{0}$. Thus we get the condition that $P(t_0, t)$ must be positive definite, but since it is already positive semidefinite, we need only require that it be nonsingular (i.e., det $P(t_0, t) \neq 0$).

Condition 1 *The system (8.1) is controllable if and only if given any t_0, $P(t_0, t)$ is nonsingular for some finite $t > t_0$.*

This condition will be convenient for obtaining some properties of controllable systems, but it is not easily applied in order to test whether a system is controllable. To obtain a better form for testing, we first note that we can express $\phi(t,\tau) = \phi(t, t_0)\phi(t_0, \tau)$, so (8.9) can be rewritten as

$$P(t_0, t) = \phi(t, t_0) \int_{t_0}^{t} \phi(t_0,\tau)B(\tau)B^T(\tau)\phi^T(t_0,\tau)\,d\tau\phi^T(t, t_0). \tag{8.10}$$

When $A(t)$ is bounded for all finite t, we know that $\phi(t, t_0)$ is nonsingular for all finite t, so $P(t_0, t)$ is nonsingular if and only if the matrix

$$\hat{S}(t_0, t) \triangleq \int_{t_0}^{t} \phi(t_0,\tau)B(\tau)B^T(\tau)\phi^T(t_0,\tau)\,d\tau \tag{8.11}$$

is nonsingular. This is equivalent to saying that for every constant vector $\boldsymbol{\mu} \neq \mathbf{0}$

$$\boldsymbol{\mu}^T \hat{S}(t_0, t)\boldsymbol{\mu} = \int_{t_0}^{t} \boldsymbol{\mu}^T \phi(t_0, \tau) B(\tau) B^T(\tau) \phi^T(t_0, \tau) \boldsymbol{\mu}\, d\tau > 0. \tag{8.12}$$

We see that the inequality in (8.12) will hold if the vector $\boldsymbol{z}(\tau) \triangleq B^T(\tau)\phi^T(t_0, \tau)\boldsymbol{\mu}$ is not identically zero for τ on $[t_0, t]$. On the other hand, if $\boldsymbol{z}(\tau)$ is identically zero over the entire interval $[t_0, t]$ then we see that the integral must be zero. Thus for a system to be controllable, it is necessary and sufficient that there exist a $t > t_0$ such that for every $\boldsymbol{\mu} \neq \mathbf{0}$, the resulting $\boldsymbol{z}(\tau)$ must be nonzero on some subinterval of $[t_0, t]$. This is stated as

Condition 2 *The continuous-time system (8.1) is controllable if and only if given any t_0 and for every $\boldsymbol{\mu} \neq \mathbf{0}$, the vector $\boldsymbol{z}(\tau) \triangleq B^T(\tau)\phi^T(t_0, \tau)\boldsymbol{\mu}$ is not identically zero for $\tau \geq t_0$.*

This condition is much easier to apply than Condition 1 for time-varying systems. For systems with constant A and B matrices, we will derive an even easier test in Section 8.4.

8.3 Controllability conditions for discrete systems

For the system

$$\begin{aligned} \boldsymbol{x}(k+1) &= A(k)\boldsymbol{x}(k) + B(k)\boldsymbol{u}(k) \\ \boldsymbol{y}(k) &= C(k)\boldsymbol{x}(k) + D(k)\boldsymbol{u}(k) \end{aligned} \tag{8.13}$$

we can use exactly the same reasoning as in Section 8.2 to show that the system is controllable if any state $\boldsymbol{x}^d$ can be reached from the origin. Here this requires that

$$\boldsymbol{x}(k) = \sum_{i=k_0}^{k-1} \phi(k, i+1) B(i) \boldsymbol{u}(i) \tag{8.14}$$

can be set equal to any arbitrary $\boldsymbol{x}^d$ for any k_0 and some finite k. Proceeding exactly as in the analog case, we get the corresponding conditions for discrete-time systems.

Condition 1′ *The system (8.13) is controllable if and only if given any k_0, there exists a finite $k > k_0$ such that the matrix*

$$P'(k_0, k) \triangleq \sum_{i=k_0}^{k-1} \phi(k, i+1) B(i) B^T(i) \phi^T(k, i+1) \tag{8.15}$$

is nonsingular.

Condition 2′ *The system (8.13) is controllable if and only if given any k_0 there exists a finite $k > k_0$ such that for every $\boldsymbol{\lambda} \neq \mathbf{0}$, the vector*

$v(i) = B^T(i)\phi^T(k, i+1)\boldsymbol{\lambda}$ is nonzero for some value of $i = k_0, \ldots, k-1$ (i.e., $\underset{\sim}{v} \neq \underset{\sim}{0}$ on $[k_0, k-1]$).

Condition 2′ is not quite equivalent to Condition 2, but if $A(k)$ is nonsingular for all $i \geq k_0$ then we can use

> **Condition 2″** *The system (8.13) is controllable if and only if given any k_0 and for every $\mu \neq 0$, the vector $z(i) = B^T(i)\phi^T(k_0, i+1)\mu$ is nonzero for some value of $i \geq k_0$.*

Here $\phi^T(k_0, i+1) \triangleq [A(i)A(i-1)\ldots A(k_0)]^{-1}$ for $i \geq k_0$.

8.4 Controllability test for fixed discrete systems

The difficulty with the conditions found so far is that we must be able to find the transition matrices. For time-varying systems, we might not be able to do this. For fixed systems simple tests for controllability can be found, which do not require finding the transition matrix.

For fixed discrete-time systems, the state difference equation becomes

$$x(k+1) = Ax(k) + B\underset{\sim}{u}(k) \tag{8.16}$$

and the transition matrix is

$$\phi(k, i+1) = \phi(k-i-1) = A^{k-i-1} \tag{8.17}$$

so

$$P'(0, k) = \sum_{i=0}^{k-1} A^{k-1-i} BB^T (A^{k-1-i})^T. \tag{8.18}$$

Note: By defining $\Gamma_k \triangleq [B \;\; AB \;\; A^2B \ldots A^{k-1}B]$, we can write (8.18) as $P'(0, k) = \Gamma_k \Gamma_k^T$, and it can be shown that $P'(0, k)$ is nonsingular if and only if the rank of $\Gamma_k = n$. So a fixed discrete system is controllable if and only if rank $\Gamma_k = n$ for some finite k. From the Cayley-Hamilton theorem of matrix theory (see Appendix, Section A.17), we know that powers of A higher than $n-1$ can be written as linear combinations of the lower powers. Therefore, increasing k beyond n can add no new linearly independent columns to Γ_n. We thus get the result

> **Condition 3** *A fixed system is controllable if and only if rank $\Gamma_n = n$ where n is the order of the system.*

The matrix Γ_n is called the controllability matrix, and when the discrete fixed system (8.16) is controllable, we say that (A, B) is a controllable pair.

8.5 Controllability test for fixed analog systems

We shall show that the same result holds for fixed continuous-time systems; that is, Condition 3 also holds for the fixed analog system

$$\dot{\boldsymbol{x}}(t) = A\boldsymbol{x}(t) + B\boldsymbol{u}(t). \tag{8.19}$$

By the Cayley-Hamilton theorem, the constant matrix A satisfies

$$A^n + a_{n-1}A^{n-1} + a_{n-2}A^{n-2} + \cdots + a_0 I = \Theta. \tag{8.20}$$

Premultiply (8.20) by $\boldsymbol{\mu}^T$ and postmultiply by $e^{A(t_0-\tau)}B$ to give

$$\boldsymbol{\mu}^T \left[A^n e^{A(t_0-\tau)} + a_{n-1}A^{n-1}e^{A(t_0-\tau)} + \cdots + a_0 e^{A(t_0-\tau)}\right] B = \boldsymbol{0}^T. \tag{8.21}$$

Now,

$$\frac{d^k}{d\tau^k}\phi(t_0-\tau) = \frac{d^k}{d\tau^k}e^{A(t_0-\tau)} = (-1)^k A^k e^{A(t_0-\tau)} \tag{8.22}$$

so (8.21) transposed becomes

$$\frac{d^n}{d\tau^n}B^T\phi^T(t_0-\tau)\boldsymbol{\mu} - a_{n-1}\frac{d^{n-1}}{d\tau^{n-1}}B^T\phi^T(t_0-\tau)\boldsymbol{\mu} + \cdots + (-1)^n a_0 B^T\phi^T(t_0-\tau)\boldsymbol{\mu} = \boldsymbol{0}. \tag{8.23}$$

We see therefore that the vector $\boldsymbol{z}(\tau) = B^T\phi^T(t_0-\tau)\boldsymbol{\mu}$ is governed by an nth-order unforced differential equation. We know that the solution can be identically zero (giving an uncontrollable system) if and only if all conditions

$$\left.\frac{d^k\boldsymbol{z}(\tau)}{d\tau^k}\right|_{\tau=t_0} = \boldsymbol{0} \quad \text{for } k = 0, 1, \ldots, n-1. \tag{8.24}$$

Therefore, if any one of these conditions is not zero, Condition 2 will hold, and the system will be controllable. Thus the fixed continuous-time system is controllable if and only if for every $\boldsymbol{\mu} \neq \boldsymbol{0}$ the vectors

$$\left[(-1)^k\frac{d^k\boldsymbol{z}(\tau)}{d\tau^k}\right]_{\tau=t_0} = \left[B^T(A^k)^T\phi^T(t_0-\tau)\boldsymbol{\mu}\right]_{\tau=t_0} = B^T(A^k)^T\boldsymbol{\mu} \neq \boldsymbol{0} \tag{8.25}$$

for at least one $k = 0, 1, 2, \ldots, n-1$. This is equivalent to requiring that for every $\boldsymbol{\mu} \neq \boldsymbol{0}$

$$\begin{bmatrix} B^T\boldsymbol{\mu} \\ B^T(A)^T\boldsymbol{\mu} \\ \vdots \\ B^T(A^{n-1})^T\boldsymbol{\mu} \end{bmatrix} \neq \begin{bmatrix} \boldsymbol{0} \\ \boldsymbol{0} \\ \vdots \\ \boldsymbol{0} \end{bmatrix}. \tag{8.26}$$

This can be written as $\Gamma_n^T\boldsymbol{\mu} \neq \boldsymbol{0}$, which is true for every $\boldsymbol{\mu} \neq \boldsymbol{0}$ if and only if the n columns of Γ_n^T are linearly independent, or equivalently if and only if the rank of the controllability matrix $\Gamma_n = n$. We see therefore, that Condition 3 is also a necessary and sufficient condition for controllability of the fixed analog system (8.19). Here, too, we say that (A, B) is a controllable pair if the fixed analog system (8.19) is controllable.

8.6 Observability: definition and conditions

Definition 8.3 *A system is called (completely) observable if for any initial time t_0, any initial state $\boldsymbol{x}(t_0)$ can be determined from observation of the output $\underset{\sim}{\boldsymbol{y}}[t_0, t]$ over a finite time interval (i.e., t is finite) when the input $\underset{\sim}{\boldsymbol{u}}[t_0, t]$ is known over the same interval.*

As in the case of controllability, the term "completely" is used to denote a system that is observable for any initial time t_0. Again, for fixed systems, such system properties do not depend on the initial time, and we will usually choose $t_0 = 0$.

Criteria for observability of linear systems
For the analog system (8.1), we know that

$$\boldsymbol{x}(t) = \phi(t, t_0)\boldsymbol{x}(t_0) + \int_{t_0}^{t} \phi(t, \tau)B(\tau)\boldsymbol{u}(\tau)\, d\tau \tag{8.27}$$

so

$$\boldsymbol{y}(t) = C(t)\phi(t, t_0)\boldsymbol{x}(t_0) + \int_{t_0}^{t} C(t)\phi(t, \tau)B(\tau)\boldsymbol{u}(\tau)\, d\tau + D(t)\boldsymbol{u}(t). \tag{8.28}$$

Rearranging gives

$$C(t)\phi(t, t_0)\boldsymbol{x}(t_0) = \boldsymbol{y}(t) - \int_{t_0}^{t} C(t)\phi(t, \tau)B(\tau)\boldsymbol{u}(\tau)\, d\tau - D(t)\boldsymbol{u}(t) \triangleq \hat{\boldsymbol{y}}(t). \tag{8.29}$$

Note that $\hat{\boldsymbol{y}}(t)$ is the zero-input response. Obviously, if $\boldsymbol{y}(\tau)$ and $\boldsymbol{u}(\tau)$ are known for all $\tau \in [t_0, t]$, then $\hat{\boldsymbol{y}}(\tau)$ is also known for $\tau \in [t_0, t]$. We thus seek to determine under what conditions $\boldsymbol{x}(t_0)$ can be determined when $\hat{\boldsymbol{y}}(\tau)$ is known over an interval $[t_0, t]$.

In (8.29), change t to τ, premultiply by $\phi^T(\tau, t_0)C^T(\tau)$, and integrate over $[t_0, t]$. This gives

$$\left[\int_{t_0}^{t} \phi^T(\tau, t_0)C^T(\tau)C(\tau)\phi(\tau, t_0)\, d\tau\right] \boldsymbol{x}(t_0) = \int_{t_0}^{t} \phi^T(\tau, t_0)C^T(\tau)\hat{\boldsymbol{y}}(\tau)\, d\tau. \tag{8.30}$$

A necessary and sufficient condition that this equation have a unique solution for $\boldsymbol{x}(t_0)$ is that the matrix

$$S(t_0, t) \triangleq \int_{t_0}^{t} \phi^T(\tau, t_0)C^T(\tau)C(\tau)\phi(\tau, t_0)\, d\tau \tag{8.31}$$

be nonsingular. We thus get

Condition 4 *The continuous-time system (8.1) is observable if and only if for any t_0, $S(t_0, t)$ is nonsingular for some finite $t > t_0$.*

Condition 4 is no easier to use to test for observability than was Condition 1 to test for controllability. Therefore by defining the vector $\boldsymbol{w}(\tau) = C(\tau)\phi(\tau, t_0)\boldsymbol{\mu}$ where $\boldsymbol{\mu}$ is a constant vector, and using exactly the same argument as we used to obtain Condition 2 from the nonsingularity of $\hat{S}(t_0, t)$, we get

Condition 5 *The analog system (8.1) is observable if and only if given any t_0, for every $\boldsymbol{\mu} \neq \mathbf{0}$, the vector $\boldsymbol{w}(\tau) \triangleq C(\tau)\phi(\tau, t_0)\boldsymbol{\mu}$ is not identically zero for all $\tau \geq t_0$.*

A similar result can be obtained for the discrete-time case. We would find that for the system (8.13)

$$C(k)\phi(k, k_0)\boldsymbol{x}(k_0) = \boldsymbol{y}(k) - \sum_{i=k_0}^{k-1} C(k)\phi(k, i+1)B(i)\boldsymbol{u}(i) - D(k)\boldsymbol{u}(k) \triangleq \hat{\boldsymbol{y}}(k). \tag{8.32}$$

The same reasoning as above leads to

Condition 4′ *The discrete system (8.13) is observable if and only if given any k_0 there exists a finite $k \geq k_0$ such that $S'(k_0, k)$ is nonsingular where*

$$S'(k_0, k) = \sum_{i=k_0}^{k-1} \phi^T(i, k_0)C^T(i)C(i)\phi(i, k_0). \tag{8.33}$$

For time-varying discrete systems it is easier to test the following condition:

Condition 5′ *The discrete-time system (8.13) is observable if and only if given any k_0, for every $\boldsymbol{\mu} \neq \mathbf{0}$ the vector $\boldsymbol{w}(i) \triangleq C(i)\phi(i, k_0)\boldsymbol{\mu}$ is nonzero for some finite value of $i \geq k_0$.*

For discrete-time systems with A and C constant we can set $k_0 = 0$, and we have $\phi(i, 0) = A^i$, so (8.33) gives

$$S'(0, k) = \sum_{i=k_0}^{k-1} (A^i)^T C^T C A^i = \Gamma_k^* (\Gamma_k^*)^T \tag{8.34}$$

where

$$\Gamma_k^* \triangleq \begin{bmatrix} C^T & A^T C^T & \dots & (A^{k-1})^T C^T \end{bmatrix}. \tag{8.35}$$

As before, since powers of A higher than $n - 1$ are linearly dependent on the lower powers, we need only check the rank for $k = n$. This leads to

Condition 6 *A fixed system is observable if and only if rank $\Gamma_n^* = n$.*

The matrix Γ_n^* is called the observability matrix, and when a fixed system is observable, we say that (A, C) is an observable pair.

8.7 Duality between observability and controllability

We could proceed from here as we did in the case of controllability to show that Condition 6 also holds for fixed analog systems. However, instead of doing this, let us observe that there is a very close relation between the conditions for observability and controllability. Suppose we have a system governed by

$$\dot{\boldsymbol{x}}^*(t) = A^*(t)\boldsymbol{x}^*(t) + B^*(t)\boldsymbol{u}^*(t)$$
$$\boldsymbol{y}^*(t) = C^*(t)\boldsymbol{x}^*(t) + D^*(t)\boldsymbol{u}^*(t). \tag{8.36}$$

A necessary and sufficient condition for controllability of this system is that the $\hat{S}^*(t_0, \tau)$ for this system

$$\hat{S}^*(t_0, t) = \int_{t_0}^{t} \phi^*(t_0, \tau)B^*(\tau)[B^*(\tau)]^T[\phi^*(t_0, \tau)]^T\, d\tau \tag{8.37}$$

be nonsingular.

If $A^*(t) = -A^T(t)$, we can get a relation between the transition matrix $\phi^*(t, \tau)$ of the system (8.36) and the transition matrix $\phi(t, \tau)$ of a system such as (8.1) with system matrix $A(t)$. We know that the transition matrix $\phi^*(t_0, \tau) = Q^*(t_0)[Q^*(\tau)]^{-1}$, and that $Q^*(t)$ must satisfy

$$\frac{dQ^*(t)}{dt} = -A^T(t)Q^*(t). \tag{8.38}$$

Recall that for system (8.1) we can express $\phi(t, \tau) = Q(t)Q^{-1}(\tau)$. We will now show that the matrix $Q^{-1}(\tau)$ satisfies the matrix differential equation

$$\frac{dQ^{-1}(\tau)}{d\tau} = -Q^{-1}(\tau)A(\tau). \tag{8.39}$$

To do this, note that $Q(\tau)Q^{-1}(\tau) = I =$ the identity matrix, so taking derivatives of both sides gives

$$\frac{dQ(\tau)}{d\tau}Q^{-1}(\tau) + Q(\tau)\frac{dQ^{-1}(\tau)}{d\tau} = \Theta \tag{8.40}$$

which is

$$\frac{dQ^{-1}(\tau)}{d\tau} = -Q^{-1}(\tau)\frac{dQ(\tau)}{d\tau}Q^{-1}(\tau). \tag{8.41}$$

Inserting

$$\frac{dQ(\tau)}{d\tau} = A(\tau)Q(\tau)$$

into (8.41) we get (8.39). Taking the transpose of (8.39) thus shows that

$$\frac{d[Q^{-1}(t)]^T}{dt} = -A^T(t)[Q^{-1}(t)]^T. \tag{8.42}$$

Comparing (8.38) and (8.42) we see that $Q^*(t) = [Q^{-1}(t)]^T = [Q^T(t)]^{-1}$, so

$$\phi^*(t_0, \tau) = [Q^{-1}(t_0)]^T Q^T(\tau) = [Q(\tau)Q^{-1}(t_0)]^T = \phi^T(\tau, t_0). \tag{8.43}$$

If also $B^*(t) = C^T(t)$, we see that $\hat{S}^*(t_0, t)$ becomes

$$\hat{S}^*(t_0, t) = \int_{t_0}^{t} \phi^T(\tau, t_0) C^T(\tau) C(\tau) \phi(\tau, t_0) d\tau. \tag{8.44}$$

Observe that this is $S(t_0, t)$ of (8.31).

Thus we see that the system of equation (8.36) with $A^*(t) = -A^T(t)$ and $B^*(t) = C^T(t)$ is controllable if and only if the original system equation (8.1) is observable and vice versa. Any of the conditions for controllability of continuous-time systems that we obtained previously when applied to the system (8.36) will give conditions for observability for our actual system (8.1).

Note: If we also take $C^*(t) = B^T(t)$, then we can show that observability of the system (8.36) is necessary and sufficient for controllability of the original system (8.1). The system governed by

$$\begin{aligned} \dot{\boldsymbol{x}}_a(t) &= -A^T(t)\boldsymbol{x}_a(t) + C^T(t)\boldsymbol{u}_a(t) \\ \boldsymbol{y}_a(t) &= B^T(t)\boldsymbol{x}_a(t) + D^T(t)\boldsymbol{u}_a(t) \end{aligned} \tag{8.45}$$

is called the "adjoint" to the original system. The order n of the adjoint system is the same as the original, but where the original system has r inputs and m outputs, the adjoint system has m inputs and r outputs.

For a fixed analog system, the adjoint system (8.45) becomes

$$\begin{aligned} \dot{\boldsymbol{x}}_a(t) &= -A^T\boldsymbol{x}_a(t) + C^T\boldsymbol{u}_a(t) \\ \boldsymbol{y}_a(t) &= B^T\boldsymbol{x}_a(t) + D^T\boldsymbol{u}_a(t). \end{aligned} \tag{8.46}$$

When we form the controllability matrix $\widetilde{\Gamma}_n$ for the fixed adjoint system (8.46) we find that

$$\widetilde{\Gamma}_n = [C^T \quad -A^T C^T \quad (A^T)^2 C^T \ldots (-1)^{(n-1)}(A^T)^{n-1} C^T]. \tag{8.47}$$

Applying Condition 3 we see that (8.46) is controllable, and thus the fixed analog system

$$\begin{aligned} \dot{\boldsymbol{x}}(t) &= A\boldsymbol{x}(t) + B\boldsymbol{u}(t) \\ \boldsymbol{y}(t) &= C\boldsymbol{x}(t) + D\boldsymbol{u}(t) \end{aligned} \tag{8.48}$$

is observable, if and only if the rank of the matrix in (8.47) is n. Since the rank of a matrix does not change when the sign of all the elements of any column is changed, this shows that the rank of $\widetilde{\Gamma}_n$ in (8.47) equals the rank Γ_n^* in (8.35), so Condition 6 holds for observability of fixed analog systems too.

For fixed analog systems it is usually more convenient to consider the system "dual" to (8.48)

$$\begin{aligned} \dot{\boldsymbol{x}}_d(t) &= A^T \boldsymbol{x}_d(t) + C^T \boldsymbol{u}_d(t) \\ \boldsymbol{y}_d(t) &= B^T \boldsymbol{x}_d(t) + D^T \boldsymbol{u}_d(t) \end{aligned} \tag{8.49}$$

instead of the adjoint. The only difference is the sign of A^T. We see that the controllability matrix Γ_n of (8.49) is the observability matrix Γ_n^* of the original system (8.48) and vice versa. Thus the system (8.48) is observable if and only if (8.49) is controllable, and (8.48) is controllable if and only if (8.49) is observable.

8.8 Sufficient conditions for time-varying analog systems

For fixed systems it was not necessary to find the transition matrix of the system (equivalent to obtaining solutions of the differential equations) in order to determine controllability. It would be desirable to obtain such a condition for time-varying systems also, especially for continuous-time systems, since it is for such systems that it is most difficult if not impossible to obtain the transition matrix. We will thus seek a controllability criterion for analog time-varying systems that will enable us to test for controllability without first obtaining the transition matrix. The following development uses the approach in the paper by Silverman and Meadows (1967).

First assume that the system

$$\dot{\boldsymbol{x}}(t) = A(t)\boldsymbol{x}(t) + B(t)\boldsymbol{u}(t) \tag{8.50}$$

is not controllable. Then by Condition 2, there exists a t_0 and some constant $\boldsymbol{\mu} \neq \mathbf{0}$ such that for every $t > t_0$, $\boldsymbol{z}^T(\tau) = \boldsymbol{\mu}^T \phi(t_0, \tau)B(\tau) = \mathbf{0}^T$ for all $\tau \in [t_0, t]$. Equivalently for the vector $\boldsymbol{v}(\tau)$ of (8.6), there exists a t_0 and some constant $\boldsymbol{\lambda} \neq \mathbf{0}$ such that for every $t > t_0$,

$$\boldsymbol{v}^T(\tau) = \boldsymbol{\lambda}^T \phi(t, \tau)B(\tau) = \mathbf{0}^T \tag{8.51}$$

for all $\tau \in [t_0, t]$. We know that we can write $\phi(t, \tau) = Q(t)Q^{-1}(\tau)$ where

$$\dot{Q}(t) = A(t)Q(t).$$

Thus we see that

$$\boldsymbol{v}^T(\tau) = \boldsymbol{\lambda}^T Q(t)Q^{-1}(\tau)B(\tau) = \mathbf{0}^T. \tag{8.52}$$

Now if this expression is differentiated k times with respect to τ, it becomes

$$\boldsymbol{\lambda}^T Q(t)\frac{d^k}{d\tau^k}\left[Q^{-1}(\tau)B(\tau)\right] = \mathbf{0}^T. \tag{8.53}$$

Note that

$$\frac{d}{d\tau}[Q^{-1}(\tau)B(\tau)] = Q^{-1}(\tau)\frac{dB(\tau)}{d\tau} + \frac{dQ^{-1}(\tau)}{d\tau}B(\tau). \tag{8.54}$$

We can use the result (8.39) obtained above in (8.54), which shows that

$$\frac{d}{d\tau}[Q^{-1}(\tau)B(\tau)] = Q^{-1}(\tau)\left[\frac{dB(\tau)}{d\tau} - A(\tau)B(\tau)\right] = Q^{-1}(\tau)M_1(\tau) \tag{8.55}$$

where we have defined

$$M_1(\tau) \triangleq \frac{dB(\tau)}{d\tau} - A(\tau)B(\tau). \tag{8.56}$$

By proceeding in the same way we find that

$$\frac{d^2}{d\tau^2}[Q^{-1}(\tau)B(\tau)] = \frac{d}{d\tau}[Q^{-1}(\tau)M_1(\tau)] \triangleq Q^{-1}(\tau)M_2(\tau) \tag{8.57}$$

where

$$M_2(\tau) \triangleq \frac{dM_1(\tau)}{d\tau} - A(\tau)M_1(\tau). \tag{8.58}$$

Continuing as above gives the general result

$$\frac{d^k}{d\tau^k}[Q^{-1}(\tau)B(\tau)] = Q^{-1}(\tau)M_k(\tau) \tag{8.59}$$

where

$$M_k(\tau) \triangleq \frac{dM_{k-1}(\tau)}{d\tau} - A(\tau)M_{k-1}(\tau) \quad \text{for } k = 1, 2, \ldots \tag{8.60}$$

when we define $M_0(\tau) = B(\tau)$. Thus if the system (8.50) is not controllable, then there exists a $\boldsymbol{\lambda} \neq \mathbf{0}$ such that for all $t > t_0$

$$\boldsymbol{\lambda}^T Q(t)Q^{-1}(\tau)M_k(\tau) = \mathbf{0}^T \tag{8.61}$$

for all $\tau \in [t_0, t]$ and all $k = 0, 1, 2, \ldots$. By setting $\tau = t$ in (8.61), it becomes

$$\boldsymbol{\lambda}^T M_k(t) = \mathbf{0}^T \quad \text{for } k = 0, 1, 2, \ldots. \tag{8.62}$$

Combining into a matrix, this is

$$\boldsymbol{\lambda}^T \begin{bmatrix} M_0(t) & M_1(t) & M_2(t) & \ldots \end{bmatrix} = \mathbf{0}^T. \tag{8.63}$$

In particular, if we take k only up to $n-1$, we see that when a system is not controllable, then there exists a $\boldsymbol{\lambda} \neq \mathbf{0}$ such that for all $t > t_0$,

$$\boldsymbol{\lambda}^T \begin{bmatrix} M_0(t) & M_1(t) & M_2(t) & \ldots & M_{n-1}(t) \end{bmatrix} = \mathbf{0}^T. \tag{8.64}$$

The condition (8.64) says that the rank of the $n \times (rn)$ matrix

$$\Gamma_n(t) = \begin{bmatrix} M_0(t) & M_1(t) & M_2(t) & \dots & M_{n-1}(t) \end{bmatrix} \tag{8.65}$$

is less than n for all $t > t_0$. Therefore if the rank of the matrix $\Gamma_n(t)$ is equal to n for some $t > t_0$ then no such vector $\boldsymbol{\lambda} \neq \mathbf{0}$ can exist, so the system is controllable. Thus we have the sufficient condition

Condition 7 *A time-varying continuous-time system is controllable if rank $\Gamma_n(t)$ is n for some $t > t_0$.*

Actually, (8.64) gives even a weaker sufficient condition since there might not exist a constant $\boldsymbol{\lambda} \neq \mathbf{0}$ which makes $\boldsymbol{\lambda}^T \Gamma_n(t) = \mathbf{0}^T$ for all $t > t_0$ even when the rank of $\Gamma_n(t)$ is less than n for all $t > t_0$. We can express this as

Condition 8 *A time-varying continuous-time system is controllable if there does not exist a constant vector $\boldsymbol{\lambda} \neq \mathbf{0}$ such that $\boldsymbol{\lambda}^T \Gamma_n(t) = \mathbf{0}^T$ for all $t > t_0$.*

It should be noted that Conditions 7 and 8 are only sufficient for controllability (they are not necessary). When either one does not hold the system might still be controllable (see Problem 8.8e where Condition 7 fails).

By applying Conditions 7 and 8 to the adjoint system (8.45) we obtain sufficient conditions for observability.

Condition 9 *A time-varying continuous-time system is observable if rank $\Gamma_n^*(t)$ is n for some $t > t_0$ where*

$$\Gamma_n^*(t) = \begin{bmatrix} M_0^*(t) & M_1^*(t) & M_2^*(t) & \dots & M_{n-1}^*(t) \end{bmatrix} \tag{8.66}$$

and

$$M_k^*(\tau) \triangleq \frac{dM_{k-1}^*(\tau)}{d\tau} + A^T(\tau) M_{k-1}^*(\tau) \quad \text{for } k = 1, 2, \dots \tag{8.67}$$

with $M_0^*(t) = C^T(t)$.

Condition 10 *A time-varying continuous-time system is observable if there does not exist a constant vector $\boldsymbol{\eta} \neq \mathbf{0}$ such that $\boldsymbol{\eta}^T \Gamma_n^*(t) = \mathbf{0}^T$ for all $t > t_0$.*

8.9 The structure of noncontrollable and nonobservable systems

We saw in Section 8.1, that for controllability of a system, the ability to go from any initial state to any final state in finite time is equivalent to the ability to go from the zero state $\mathbf{0}$ to any final state in finite time. For observability, we require that for any initial time t_0, any initial state $\boldsymbol{x}(t_0)$ can be determined from observation of the output $y(t)$ over a finite time interval $[t_0, t]$ when the input is known on this interval. Kalman (1963) showed that systems which are not controllable or not observable can be put in a special structure that reveals these properties clearly.

8.9.1 The controllable subspace

We will now restrict our attention to time-invariant systems and determine, for systems that are not controllable but are observable, what set of states *can* be reached from the zero state $\mathbf{0}$ in finite time.

Definition 8.4 *The set of states that can be reached from the zero state* $\mathbf{0}$ *in finite time is called the controllable subspace and is denoted as* $\mathcal{C}$.

For a discrete system

$$\boldsymbol{x}(k+1) = A\boldsymbol{x}(k) + B\boldsymbol{u}(k) \tag{8.68}$$

we have seen that for $\boldsymbol{x}(0) = \mathbf{0}$,

$$\boldsymbol{x}(k) = \sum_{i=0}^{k-1} A^{k-1-i} B\boldsymbol{u}(i).$$

Since any power A^k for $k > n-1$ can be expressed as a linear combination of the powers $A^{n-1}, A^{n-2}, \ldots, A, I$, we see that the states that can be reached from $\mathbf{0}$ must be expressible as a linear combination of the columns of $\Gamma_n = [B \quad AB \quad A^2B \ldots A^{n-1}B]$. When these columns span the the entire n-dimensional state space $\mathcal{S}$ (i.e., there are n linearly independent columns so rank $\Gamma_n = n$), then any state $\boldsymbol{x}^* \in \mathcal{S}$ can be reached from $\mathbf{0}$, and the system is controllable. But if rank $\Gamma_n = \ell < n$, then only those states which lie in the range space of Γ_n (i.e., the ℓ-dimensional subspace spanned by any ℓ linearly independent columns of Γ_n) can be reached from $\mathbf{0}$. Thus we have shown that the controllable subspace $\mathcal{C}$ is the ℓ-dimensional subspace spanned by any ℓ linearly independent columns of Γ_n.

For analog systems the result is exactly the same. This follows from the fact that for $\boldsymbol{x}(0) = \mathbf{0}$, the solution to

$$\dot{\boldsymbol{x}}(t) = A\boldsymbol{x}(t) + B\boldsymbol{u}(t) \tag{8.69}$$

is given by

$$\boldsymbol{x}(t) = \int_0^t e^{A(t-\tau)} B\boldsymbol{u}(\tau)\, d\tau$$

so by expanding $e^{A(t-\tau)}$ in its power series, we get

$$\boldsymbol{x}(t) = \sum_{i=0}^{\infty} \frac{A^i}{i!} B \int_0^t (t-\tau)^i \boldsymbol{u}(\tau)\, d\tau.$$

Thus $\boldsymbol{x}(t)$ can only lie in the space spanned by the columns of the series of matrices $B, AB, A^2B, \ldots, A^kB, \ldots$. As before, we know that for $k \geq n$, there can be no additional linearly independent columns, so $\boldsymbol{x}(t)$ must lie in the subspace spanned by the columns of Γ_n.

Definition 8.5 *A subspace $\mathcal{X}$ is called invariant under the transformation T, if for any $\boldsymbol{x} \in \mathcal{X}$ the vector $T\boldsymbol{x}$ is also in $\mathcal{X}$.*

Property 1: $\mathcal{C}$ is invariant under the transformation A.

Proof: Since any $\boldsymbol{x} \in \mathcal{C}$ can be expressed as $\boldsymbol{x} = \Gamma_n \boldsymbol{\alpha}$ for some $\boldsymbol{\alpha}$, we see that $A\boldsymbol{x} = A\Gamma_n\boldsymbol{\alpha} = [AB \quad A^2B \quad A^3B \ldots A^nB]\boldsymbol{\alpha}$. But the columns of A^nB can be expressed as a linear combination of the columns of Γ_n, so we see that $A\boldsymbol{x}$ can be expressed as a linear combination of the columns of Γ_n, thus it lies in $\mathcal{C}$. QED.

As a result of Property 1, it follows that any initial state $\boldsymbol{x}_0$ in $\mathcal{C}$ can be moved to any final state $\boldsymbol{x}_f$ in $\mathcal{C}$, in finite time. That is, all states lying in $\mathcal{C}$ can be controlled and only those states can be. We can show this explicitly if we make a transformation of our state variables.

For rank $\Gamma_n = \ell < n$, let us choose any ℓ linearly independent columns $\boldsymbol{g}_1, \boldsymbol{g}_2, \ldots, \boldsymbol{g}_\ell$ of Γ_n. This forms a basis for the controllable subspace $\mathcal{C}$ (i.e., any vector $\boldsymbol{v}$ in $\mathcal{C}$ can be expressed as $\boldsymbol{v} = \sum_{i=1}^{\ell} \alpha_i \boldsymbol{g}_i$ for some ℓ vector $\boldsymbol{\alpha}$). Next, choose any $(n - \ell)$ vectors $\boldsymbol{g}_{\ell+1}, \boldsymbol{g}_{\ell+2}, \ldots, \boldsymbol{g}_n$ which together with $\boldsymbol{g}_1, \boldsymbol{g}_2, \ldots, \boldsymbol{g}_\ell$ span the entire state space $\mathcal{S}$. Let us define the matrix $T_1 = [\boldsymbol{g}_1 \mid \boldsymbol{g}_2 \mid \ldots \mid \boldsymbol{g}_\ell]$ and $T_2 = [\boldsymbol{g}_{\ell+1} \mid \boldsymbol{g}_{\ell+2} \mid \ldots \mid \boldsymbol{g}_n]$. The matrix $T = [T_1 \mid T_2]$ has n linearly independent columns, so its inverse exists. This inverse will be written as

$$T^{-1} = \begin{bmatrix} \hat{T}_1 \\ \hat{T}_2 \end{bmatrix}$$

so

$$T^{-1}T = \left[\begin{array}{c:c} \hat{T}_1 T_1 & \hat{T}_1 T_2 \\ \hdashline \hat{T}_2 T_1 & \hat{T}_2 T_2 \end{array}\right] = \left[\begin{array}{c:c} I_\ell & \Theta \\ \hdashline \Theta & I_{n-l} \end{array}\right] \tag{8.70}$$

where Θ is a zero matrix. Now make the change of state variables as defined by the transformation $\hat{\boldsymbol{x}} = T^{-1}\boldsymbol{x}$. In terms of this new state, the state differential equations become

$$\dot{\hat{\boldsymbol{x}}}(t) = \hat{A}\,\hat{\boldsymbol{x}}(t) + \hat{B}\,\boldsymbol{u}(t) \tag{8.71}$$

where $\hat{A} = T^{-1}AT$ and $\hat{B} = T^{-1}B$ (see Section 2.4). This gives

$$\hat{A} = \begin{bmatrix} \hat{T}_1 \\ \hat{T}_2 \end{bmatrix} A[T_1 \ \vdots \ T_2] = \left[\begin{array}{c:c} \hat{T}_1AT_1 & \hat{T}_1AT_2 \\ \hdashline \hat{T}_2AT_1 & \hat{T}_2AT_2 \end{array}\right]. \tag{8.72}$$

Let $\hat{A}_{11} = \hat{T}_1AT_1$, $\hat{A}_{12} = \hat{T}_1AT_2$ and $\hat{A}_{22} = \hat{T}_2AT_2$. Note that since the columns of T_1 form a basis for $\mathcal{C}$ which is invariant under the transformation A, this means that the columns of AT_1 lie in $\mathcal{C}$. But from (8.70) we see that $\hat{T}_2T_1 = \Theta$, so any vector $\boldsymbol{v}$ in $\mathcal{C}$ makes $\hat{T}_2\boldsymbol{v} = \mathbf{0}$. Thus $\hat{T}_2AT_1 = \Theta$. By the same reasoning,

$$\hat{B} = \begin{bmatrix} \hat{T}_1B \\ \hat{T}_2B \end{bmatrix} = \begin{bmatrix} \hat{B}_1 \\ \Theta \end{bmatrix} \tag{8.73}$$

where $\hat{T}_2B = \Theta$ because the columns of B lie in $\mathcal{C}$. Thus in terms of the new state vector $\hat{\boldsymbol{x}} = \begin{bmatrix} \hat{\boldsymbol{x}}_1 \\ \hat{\boldsymbol{x}}_2 \end{bmatrix}$, the state differential equations can be written as

$$\begin{aligned} \dot{\hat{\boldsymbol{x}}}_1(t) &= \hat{A}_{11}\hat{\boldsymbol{x}}_1(t) + \hat{A}_{12}\hat{\boldsymbol{x}}_2(t) + \hat{B}_1\boldsymbol{u}(t) \\ \dot{\hat{\boldsymbol{x}}}_2(t) &= \qquad\qquad\quad \hat{A}_{22}\hat{\boldsymbol{x}}_2(t) \end{aligned} \tag{8.74}$$

and the output equation is

$$\boldsymbol{y}(t) = \hat{C}_1\hat{\boldsymbol{x}}_1(t) + \hat{C}_2\hat{\boldsymbol{x}}_2(t) + D\boldsymbol{u}(t). \tag{8.75}$$

We see that the input does not affect $\hat{\boldsymbol{x}}_2$ at all, so this part of the state vector is uncontrollable. It lies in the uncontrollable subspace $\overline{\mathcal{C}} = \mathcal{S} - \mathcal{C}$.

The system described by (8.74) and (8.75) and whose structure is shown in Figure 8.3 is said to be the controllability decomposition for the system. The eigenvalues of $\hat{A}_{11}$ are called the controllable eigenvalues, and the eigenvalues of $\hat{A}_{22}$ are called the uncontrollable eigenvalues.

For a discrete system, we would find exactly the same controllability decomposition except that the dynamic elements would be delays instead of integrators.

Example 8.2

Consider a system governed by the state equations

$$\begin{aligned} \dot{x}_1(t) &= -x_1(t) + x_2(t) - x_3(t) + u(t) \\ \dot{x}_2(t) &= -x_3(t) + 2u(t) \\ \dot{x}_3(t) &= x_1(t) - x_3(t) + x_4(t) + 2u(t) \\ \dot{x}_4(t) &= x_1(t) - u(t) \\ y_1(t) &= x_1(t) \\ y_2(t) &= x_3(t) - u(t). \end{aligned}$$

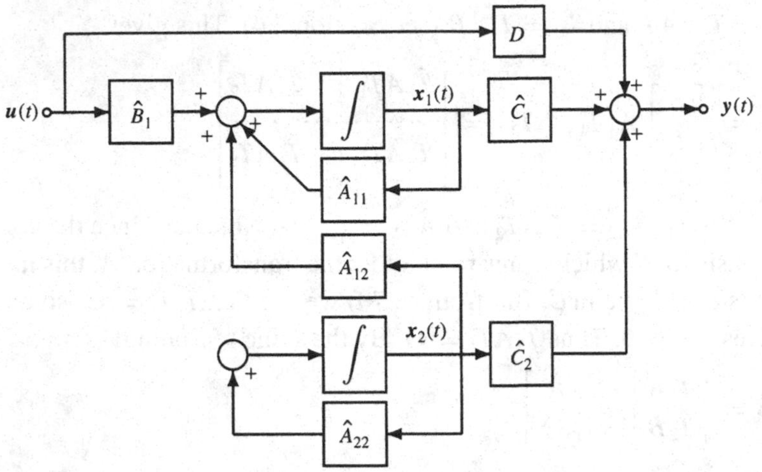

Fig. 8.3. The controllable and uncontrollable subsystems

This is the fourth-order system of Example 2.10 that is shown in Figure 2.14. We see that

$$A = \begin{bmatrix} -1 & 1 & -1 & 0 \\ 0 & 0 & -1 & 0 \\ 1 & 0 & -1 & 1 \\ 1 & 0 & 0 & 0 \end{bmatrix} \qquad B = \begin{bmatrix} 1 \\ 2 \\ 2 \\ -1 \end{bmatrix}$$

$$C = \begin{bmatrix} 1 & 0 & 0 & 0 \\ 0 & 0 & 1 & 0 \end{bmatrix} \qquad D = \begin{bmatrix} 0 \\ -1 \end{bmatrix}.$$

For this system we find that the controllability matrix is

$$\Gamma_4 = \begin{bmatrix} 1 & -1 & 1 & -1 \\ 2 & -2 & 2 & -2 \\ 2 & -2 & 2 & -2 \\ -1 & 1 & -1 & 1 \end{bmatrix}$$

which has rank equal to 1. The observability matrix Γ_4^* is found to be

$$\Gamma_4^* = \begin{bmatrix} 1 & 0 & -1 & 1 & 0 & -1 & 0 & 0 \\ 0 & 0 & 1 & 0 & -1 & 1 & 0 & -1 \\ 0 & 1 & -1 & -1 & 1 & 0 & 0 & 0 \\ 0 & 0 & 0 & 1 & -1 & -1 & 1 & 0 \end{bmatrix}$$

which has rank equal to 4, so the system is observable. To put the system into controllability form, we choose the first column of Γ_4 as g_1. This vector spans the one-dimensional controllable subspace. We can choose any three other vectors that together with $\boldsymbol{g}_1$ form a linearly independent set. Thus we can choose

$$T = \begin{bmatrix} 1 & 0 & 0 & 0 \\ 2 & 1 & 0 & 0 \\ 2 & 0 & 1 & 0 \\ -1 & 0 & 0 & 1 \end{bmatrix}$$

so

$$T^{-1} = \begin{bmatrix} 1 & 0 & 0 & 0 \\ -2 & 1 & 0 & 0 \\ -2 & 0 & 1 & 0 \\ 1 & 0 & 0 & 1 \end{bmatrix}.$$

When we change the state vector to $\hat{\boldsymbol{x}} = T^{-1}\boldsymbol{x}$ we get

$$\dot{\hat{\boldsymbol{x}}}(t) = T^{-1}AT\hat{\boldsymbol{x}}(t) + T^{-1}Bu(t) = \begin{bmatrix} -1 & 1 & -1 & 0 \\ 0 & -2 & 1 & 0 \\ 0 & -2 & 1 & 1 \\ 0 & 1 & -1 & 0 \end{bmatrix} \hat{\boldsymbol{x}}(t) + \begin{bmatrix} 1 \\ 0 \\ 0 \\ 0 \end{bmatrix} u(t).$$

We see that $\dot{\hat{x}}_1(t) = -\hat{x}_1(t) + \hat{x}_2(t) - \hat{x}_3(t) + u(t)$ is the controllable portion of the system, and

$$\begin{bmatrix} \dot{\hat{x}}_2(t) \\ \dot{\hat{x}}_3(t) \\ \dot{\hat{x}}_4(t) \end{bmatrix} = \begin{bmatrix} -2 & 1 & 0 \\ -2 & 1 & 1 \\ 1 & -1 & 0 \end{bmatrix} \begin{bmatrix} \hat{x}_2(t) \\ \hat{x}_3(t) \\ \hat{x}_4(t) \end{bmatrix}$$

is the uncontrollable portion. The output is given by

$$\boldsymbol{y}(t) = CT\hat{\boldsymbol{x}}(t) + Du(t) = \begin{bmatrix} 1 & 0 & 0 & 0 \\ 2 & 0 & 1 & 0 \end{bmatrix} \hat{\boldsymbol{x}}(t) + \begin{bmatrix} 0 \\ -1 \end{bmatrix} u(t).$$

When the system starts in the zero state then $\hat{x}_2(t) = \hat{x}_3(t) = \hat{x}_4(t) = 0$ for all t, so the system can be described by the first-order system $\dot{\hat{x}}_1(t) = -\hat{x}_1(t) + u(t)$, and the output becomes

$$\boldsymbol{y}(t) = \begin{bmatrix} 1 \\ 2 \end{bmatrix} \hat{x}_1(t) + \begin{bmatrix} 0 \\ -1 \end{bmatrix} u(t).$$

This is the minimal realization found in Example 2.10. We see that to get a minimal realization (i.e., a simulation diagram of lowest order) for a fixed system, we merely take the controllable and observable subsystem of any realization that is found.

8.9.2 The observable subspace

For nonobservable time-invariant systems that are controllable, we can find the set of initial states $\boldsymbol{x}(0)$ which can be determined from observation of the zero-input response $\boldsymbol{y}(t)$ over a finite time interval $[0, t]$.

Definition 8.6 *The observable subspace $\mathcal{O}$ of a fixed linear system is the set of states $\mathbf{x}(0)$ that can be determined from observation of the zero-input response $\mathbf{y}(t)$ over a finite time interval* $[0, t]$.

At this point we could proceed directly, as in the preceding section, by first showing that the observable subspace $\mathcal{O}$ is the space spanned by the columns of the observability matrix Γ_n^* and that $\mathcal{O}$ is invariant under the transformation A^T. Since the detailed steps are identical to those carried out in Section 8.9.1, we will instead obtain the observability decomposition by applying the results obtained there to the dual system and taking the dual of the resulting decomposition of the system.

Before doing this, we note that any vector $\boldsymbol{z}$ that lies in the unobservable subspace $\bar{\mathcal{O}} = \mathcal{S} - \mathcal{O}$ is orthogonal to any vector $\boldsymbol{v}$ that lies in the observable subspace $\mathcal{O}$ (i.e., $\boldsymbol{v}^T\boldsymbol{z} = 0$ for such vectors). We can also see that $\bar{\mathcal{O}}$ is invariant under the transformation A since any vector $\boldsymbol{v} \in \mathcal{O}$ can be expressed as a linear combination of the columns of Γ_n^* (i.e., $\boldsymbol{v} = \Gamma_n^*\boldsymbol{\alpha}$ for some vector $\boldsymbol{\alpha}$). Thus for any $\boldsymbol{z} \in \overline{\mathcal{O}}$,

$$\boldsymbol{\alpha}^T\Gamma_n^{*T}\boldsymbol{z} = \boldsymbol{\alpha}^T \begin{bmatrix} C \\ CA \\ CA^2 \\ \vdots \\ CA^{n-1} \end{bmatrix} \boldsymbol{z} = 0 \tag{8.76}$$

which shows that $\boldsymbol{z}$ is orthogonal to the rows of Γ_n^{*T}. If we now take

$$\boldsymbol{\alpha}^T\Gamma_n^{*T}A\boldsymbol{z} = \boldsymbol{\alpha}^T \begin{bmatrix} CA \\ CA^2 \\ CA^3 \\ \vdots \\ CA^{n} \end{bmatrix} \boldsymbol{z}$$

then since A^n can be expressed as a linear combination of the powers $I, A, \ldots, A^{n-1}$, we see that the rows of $\Gamma_n^{*T}A$ can be expressed as a linear combination of the rows of Γ_n^{*T}. Thus $\boldsymbol{\alpha}\Gamma_n^{*T}A\boldsymbol{z} = 0$ for any $\boldsymbol{\alpha}$, so $A\boldsymbol{z}$ also lies in $\overline{\mathcal{O}}$. We have thus proved:

Property 2: $\overline{\mathcal{O}}$ is invariant under the transformation A.

This property will be used below.

To get the observability form for the analog system (8.48), we start with the dual system (8.49). Let $\mathcal{C}^d$ denote the controllable subspace of the dual system (8.49). We know that $\mathcal{C}^d$ is spanned by the columns of its controllability matrix $\hat{\Gamma}_n = \Gamma_n^*$. Since the subspace $\mathcal{O}$ of (8.48) is the subspace $= \mathcal{C}^d$ of (8.49), we see that $\mathcal{O}$ is spanned by the columns of Γ_n^*, and since $\mathcal{C}^d$ is invariant under the transformation A^T, so is $\mathcal{O}$. When rank $\Gamma_n^* = p < n$, (8.49) is not controllable, and (8.48) is not observable. Form the matrix $V = [V_1 \mid V_2]$, where the columns $v_1, v_2, \ldots, v_p$ of the $(n \times p)$ matrix V_1 are chosen from any p linearly independent columns of Γ_n^*. The columns $v_{p+1}, v_{p+2}, \ldots, v_n$ of V_2 are $(n-p)$ linearly independent vectors which together with $v_1, v_2, \ldots, v_p$ span the entire state space $\mathcal{S}$. Then for the new state $\widetilde{x}_d = V^{-1}x_d$, we find as before that in terms of the new state vector $\widetilde{x}_d = \begin{bmatrix} \widetilde{x}_{d1} \\ \widetilde{x}_{d2} \end{bmatrix}$, the state differential equations can be written as

$$\dot{\widetilde{x}}_d(t) = \left[\begin{array}{c:c} \hat{V}_1 A^T V_1 & \hat{V}_1 A^T V_2 \\ \hdashline \hat{V}_2 A^T V_1 & \hat{V}_2 A^T V_2 \end{array}\right] \begin{bmatrix} \widetilde{x}_{d1}(t) \\ \widetilde{x}_{d2}(t) \end{bmatrix} + \begin{bmatrix} \hat{V}_1 C^T \\ \Theta \end{bmatrix} u_d(t), \tag{8.77}$$

where

$$V^{-1} = \begin{bmatrix} \hat{V}_1 \\ \hat{V}_2 \end{bmatrix}.$$

Also,

$$y_d(t) = B^T V \widetilde{x}_d(t) + D^T u_d(t). \tag{8.78}$$

Taking the dual of the system (8.77) and (8.78) gives the observability decomposition of the original system (8.48). This is

$$\begin{aligned} \dot{\widetilde{x}}_1(t) &= \widetilde{A}_{11}\widetilde{x}_1(t) && + \widetilde{B}_1 u(t) \\ \dot{\widetilde{x}}_2(t) &= \widetilde{A}_{21}\widetilde{x}_1(t) + \widetilde{A}_{22}\widetilde{x}_2(t) && + \widetilde{B}_2 u(t) \\ y(t) &= \widetilde{C}_1 \widetilde{x}_1(t) && + Du(t) \end{aligned} \tag{8.79}$$

where $\widetilde{A}_{11} = V_1^T A \hat{V}_1^T$, $\widetilde{A}_{21} = \hat{V}_2^T A \hat{V}_1^T$, $\widetilde{A}_{22} = V_2^T A \hat{V}_2^T$, $\widetilde{B}_1 = V_1^T B$, $\widetilde{B}_2 = V_2^T B$, and $\widetilde{C}_1 = C \hat{V}_1^T$. The structure of this system is shown in Figure 8.4.

We see that the part of the state denoted by $\widetilde{x}_2(t)$ has no affect on the output, so this part is unobservable. It lies in the unobservable subspace $\overline{\mathcal{O}}$. The eigenvalues of $\widetilde{A}_{11}$ are called the observable eigenvalues, and the eigenvalues of $\widetilde{A}_{22}$ are called the unobservable eigenvalues.

For a discrete system that is not observable, here too we would find exactly the same observability form except that the dynamic elements would be delays instead of integrators.

8.9.3 Kalman decomposition

To get the most general form of the decomposition, we consider a time-invariant system that is neither controllable nor observable. In Section 8.9.1, we saw that a

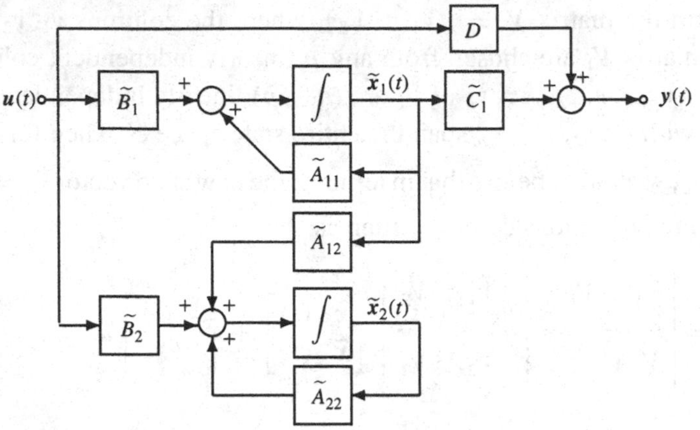

Fig. 8.4. The observable and unobservable subsystems

basis $\boldsymbol{g}_1, \boldsymbol{g}_2, \ldots, \boldsymbol{g}_\ell$ for the subspace $\mathcal{C}$ could be chosen from the linearly independent columns of Γ_n and a basis $\boldsymbol{v}_1, \boldsymbol{v}_2, \ldots, \boldsymbol{v}_p$ for the subspace $\mathcal{O}$ could be chosen from the linearly independent columns of Γ_n^*.

We now take all those vectors $\boldsymbol{g}_i$ or linear combinations of them that also lie in the unobservable subspace $\bar{\mathcal{O}}$ (say there are $\ell_1 \leq \ell$ of them) and call these $z_1, z_2, \ldots, z_{\ell_1}$. Then these vectors span the subspace $\mathcal{C}\bar{\mathcal{O}} = \mathcal{C} \bigcap \bar{\mathcal{O}}$ which is the intersection of $\mathcal{C}$ and $\bar{\mathcal{O}}$. In other words, $\mathcal{C}\bar{\mathcal{O}}$ is the subspace of states that are controllable but not observable. Form the matrix $Z_1 = [z_1 \mid z_2 \mid \ldots \mid z_{\ell_1}]$ and the matrix Z_2 such that its columns together with those of Z_1 form a basis for the subspace of states that are controllable. Thus $Z_2 = [z_{\ell_1+1} \mid z_{\ell_1+2} \mid \ldots \mid z_\ell]$. Note that the matrix T_1 of Section 8.9.1 is $T_1 = [\,Z_1 \mid Z_2\,]$.

We next choose a set of vectors that together with the set $z_1, z_2, \ldots, z_{\ell_1}$ span the unobservable subspace $\bar{\mathcal{O}}$. There are $n - p - \ell_1 = p_1$ such vectors. We use these vectors as the columns of the matrix $Z_3 = [z_{\ell+1} \mid z_{\ell+2} \mid \ldots \mid z_{\ell+p_1}]$. If $\ell + p_1 < n$, we must choose $(n - \ell - p_1)$ more linearly independent vectors $z_{\ell+p_1+1}, z_{\ell+p_1+2}, \ldots, z_n$ which together with the previously chosen vectors z_i span the entire state space. These vectors are the columns of the matrix Z_4. If $\ell + p_1 = n$, then there is no matrix Z_4. Thus the matrix T_2 of Section 8.9.1 is $T_2 = [\,Z_3 \mid Z_4\,]$.

The $(n \times n)$ matrix $Z = [\,Z_1 \mid Z_2 \mid Z_3 \mid Z_4\,]$ has n linearly independent columns, so its inverse exists. This inverse will be written as

$$Z^{-1} = \begin{bmatrix} \hat{T}_1 \\ \hdashline \hat{T}_2 \end{bmatrix} = \begin{bmatrix} \hat{Z}_1 \\ \hat{Z}_2 \\ \hdashline \hat{Z}_3 \\ \hat{Z}_4 \end{bmatrix}$$

so

$$Z^{-1}Z = \left[\begin{array}{c:c:c:c} \hat{Z}_1 Z_1 & \hat{Z}_1 Z_2 & \hat{Z}_1 Z_3 & \hat{Z}_1 Z_4 \\ \hdashline \hat{Z}_2 Z_1 & \hat{Z}_2 Z_2 & \hat{Z}_2 Z_3 & \hat{Z}_2 Z_4 \\ \hdashline \hat{Z}_3 Z_1 & \hat{Z}_3 Z_2 & \hat{Z}_3 Z_3 & \hat{Z}_3 Z_4 \\ \hdashline \hat{Z}_4 Z_1 & \hat{Z}_4 Z_2 & \hat{Z}_4 Z_3 & \hat{Z}_4 Z_4 \end{array}\right] = \left[\begin{array}{c:c:c:c} I_{\ell_1} & \Theta & \Theta & \Theta \\ \hdashline \Theta & I_{\ell-\ell_1} & \Theta & \Theta \\ \hdashline \Theta & \Theta & I_{p_1} & \Theta \\ \hdashline \Theta & \Theta & \Theta & I_{n-\ell-p_1} \end{array}\right] \tag{8.80}$$

Now change the state variables using the transformation $\hat{\boldsymbol{x}} = Z^{-1}\boldsymbol{x}$. In terms of this new state, the state differential equations become

$$\dot{\hat{\boldsymbol{x}}}(t) = \hat{A}\hat{\boldsymbol{x}}(t) + \hat{B}\hat{\underline{\boldsymbol{u}}}(t) \tag{8.81}$$

where $\hat{A} = Z^{-1}AZ$ and $\hat{B} = Z^{-1}B$. This gives

$$\hat{A} = \left[\begin{array}{c:c:c:c} \hat{Z}_1 A Z_1 & \hat{Z}_1 A Z_2 & \hat{Z}_1 A Z_3 & \hat{Z}_1 A Z_4 \\ \hdashline \hat{Z}_2 A Z_1 & \hat{Z}_2 A Z_2 & \hat{Z}_2 A Z_3 & \hat{Z}_2 A Z_4 \\ \hdashline \hat{Z}_3 A Z_1 & \hat{Z}_3 A Z_2 & \hat{Z}_3 A Z_3 & \hat{Z}_3 A Z_4 \\ \hdashline \hat{Z}_4 A Z_1 & \hat{Z}_4 A Z_2 & \hat{Z}_4 A Z_3 & \hat{Z}_4 A Z_4 \end{array}\right] \qquad \hat{B} = \begin{bmatrix} \hat{Z}_1 B \\ \hat{Z}_2 B \\ \hat{Z}_3 B \\ \hat{Z}_4 B \end{bmatrix}. \tag{8.82}$$

Let $\hat{A}_{ij} = \hat{Z}_i A Z_j$ and $\hat{B}_i = \hat{Z}_i B$. Note that since the columns of $T_1 = [Z_1 \vdots Z_2]$ form a basis for $\mathcal{C}$ which is invariant under the transformation A, this means that the columns of AZ_1 and AZ_2 lie in $\mathcal{C}$. From (8.80) we see that

$$\hat{T}_2 T_1 = \left[\begin{array}{c:c} \hat{Z}_3 Z_1 & \hat{Z}_3 Z_2 \\ \hdashline \hat{Z}_4 Z_1 & \hat{Z}_4 Z_2 \end{array}\right] = \left[\begin{array}{c:c} \Theta & \Theta \\ \hdashline \Theta & \Theta \end{array}\right]$$

so any vector $\boldsymbol{v}$ in $\mathcal{C}$ makes $\hat{T}_2\boldsymbol{v} = \boldsymbol{0}$. Thus

$$\hat{T}_2 A T_1 = \left[\begin{array}{c:c} \hat{Z}_3 A Z_1 & \hat{Z}_3 A Z_2 \\ \hdashline \hat{Z}_4 A Z_1 & \hat{Z}_4 A Z_2 \end{array}\right] = \left[\begin{array}{c:c} \Theta & \Theta \\ \hdashline \Theta & \Theta \end{array}\right].$$

By the same reasoning

$$\hat{T}_2 B = \begin{bmatrix} \hat{Z}_3 B \\ \hat{Z}_4 B \end{bmatrix} = \begin{bmatrix} \Theta \\ \Theta \end{bmatrix}.$$

We note further that the columns of Z_1 and Z_3 together span the unobservable subspace $\overline{\mathcal{O}}$ (we chose them that way), and $\bar{\mathcal{O}}$ is invariant under the transformation A (by Property 2 of Section 8.9.2). Therefore we see that since (8.80) shows that $\hat{Z}_2[Z_1 \ \vdots \ Z_3] = [\Theta \ \vdots \ \Theta]$ then $\hat{Z}_2 z = \mathbf{0}$ for any $z \in \bar{\mathcal{O}}$. Since Az is in $\bar{\mathcal{O}}$ this shows that $\hat{Z}_2[AZ_1 \ \vdots \ AZ_3] = [\hat{Z}_2 AZ_1 \ \vdots \ \hat{Z}_2 AZ_3] = [\Theta \ \vdots \ \Theta]$. For the same reason since $\hat{Z}_4[Z_1 \ \vdots \ Z_3] = [\Theta \ \vdots \ \Theta]$ it follows that $[\hat{Z}_4 AZ_1 \ \vdots \ \hat{Z}_4 AZ_3] = [\Theta \ \vdots \ \Theta]$ and also $[CZ_1 \ \vdots \ CZ_3] = [\Theta \ \vdots \ \Theta]$. Thus the matrices $\hat{A}$ and $\hat{B}$ have the form

$$\hat{A} = \left[\begin{array}{c:c:c:c} \hat{A}_{11} & \hat{A}_{12} & \hat{A}_{13} & \hat{A}_{14} \\ \hdashline \Theta & \hat{A}_{22} & \Theta & \hat{A}_{24} \\ \hdashline \Theta & \Theta & \hat{A}_{33} & \hat{A}_{34} \\ \hdashline \Theta & \Theta & \Theta & \hat{A}_{44} \end{array}\right] \qquad \hat{B} = \begin{bmatrix} \hat{B}_1 \\ \hat{B}_2 \\ \Theta \\ \Theta \end{bmatrix}. \tag{8.83}$$

In terms of the new state vector, the output equation becomes

$$\begin{aligned} \boldsymbol{y}(t) &= \hat{C}\hat{\boldsymbol{x}}(t) + D\boldsymbol{u}(t) \\ &= CZ\hat{\boldsymbol{x}}(t) + D\boldsymbol{u}(t) = \left[\Theta \ \vdots \ CZ_2 \ \vdots \ \Theta \ \vdots \ CZ_4\right]\hat{\boldsymbol{x}}(t) + D\boldsymbol{u}(t) \end{aligned} \tag{8.84}$$

since $CZ_1 = \Theta$ and $CZ_3 = \Theta$. Let CZ_2 and CZ_4 be denoted as $\hat{C}_2$ and $\hat{C}_4$ respectively.

Therefore in terms of the new state vector

$$\hat{\boldsymbol{x}} = \begin{bmatrix} \hat{\boldsymbol{x}}_{c\bar{o}} \\ \hat{\boldsymbol{x}}_{co} \\ \hat{\boldsymbol{x}}_{\overline{co}} \\ \hat{\boldsymbol{x}}_{\bar{c}o} \end{bmatrix}$$

partitioned to show the different substates, the state equations can be written as

$$\begin{aligned} \dot{\hat{\boldsymbol{x}}}_{c\bar{o}}(t) &= \hat{A}_{11}\hat{\boldsymbol{x}}_{c\bar{o}}(t) + \hat{A}_{12}\hat{\boldsymbol{x}}_{co}(t) + \hat{A}_{13}\hat{\boldsymbol{x}}_{\overline{co}}(t) + \hat{A}_{14}\hat{\boldsymbol{x}}_{\bar{c}o}(t) + \hat{B}_1\boldsymbol{u}(t) \\ \dot{\hat{\boldsymbol{x}}}_{co}(t) &= \hat{A}_{22}\hat{\boldsymbol{x}}_{co}(t) + \hat{A}_{24}\hat{\boldsymbol{x}}_{\bar{c}o}(t) + \hat{B}_2\boldsymbol{u}(t) \\ \dot{\hat{\boldsymbol{x}}}_{\overline{co}}(t) &= \hat{A}_{33}\hat{\boldsymbol{x}}_{\overline{co}}(t) + \hat{A}_{34}\hat{\boldsymbol{x}}_{\bar{c}o}(t) \\ \dot{\hat{\boldsymbol{x}}}_{\bar{c}o}(t) &= \hat{A}_{44}\hat{\boldsymbol{x}}_{\bar{c}o}(t) \\ \boldsymbol{y}(t) &= \hat{C}_2\hat{\boldsymbol{x}}_{co}(t) + \hat{C}_4\hat{\boldsymbol{x}}_{\bar{c}o}(t) + D\boldsymbol{u}(t). \end{aligned} \tag{8.85}$$

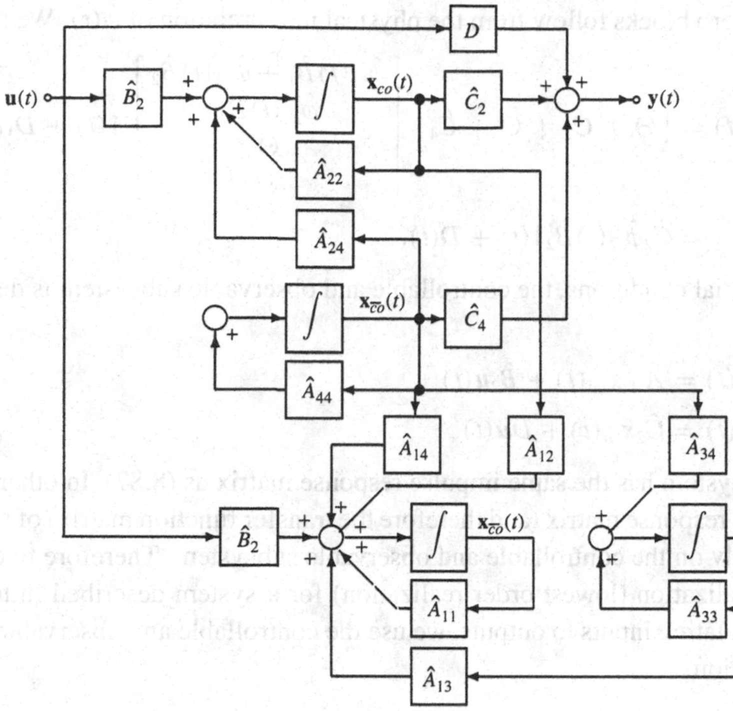

Fig. 8.5. The Kalman decomposition

The structure of this system is shown in Figure 8.5. This system is called the Kalman decomposition of the original system (8.48). It shows explicitly the parts of the state that lie in $\mathcal{C} \cap \mathcal{O}$ (both controllable and observable), $\mathcal{C} \cap \overline{\mathcal{O}}$ (controllable but not observable), $\overline{\mathcal{C}} \cap \mathcal{O}$ (not controllable but observable), and $\overline{\mathcal{C}} \cap \overline{\mathcal{O}}$ (neither controllable nor observable). For discrete systems, the structure is the same with delays replacing integrators.

Before concluding this topic, let us observe that the impulse response matrix for the system (8.85) is

$$H(t) = \hat{C}\hat{\phi}(t)\hat{B}1(t) + D\delta(t). \tag{8.86}$$

The transition matrix $\hat{\phi}(t)$ has the form

$$\hat{\phi}(t) = \begin{bmatrix} \hat{\phi}_{11}(t) & \hat{\phi}_{12}(t) & \hat{\phi}_{13}(t) & \hat{\phi}_{14}(t) \\ \Theta & \hat{\phi}_{22}(t) & \Theta & \hat{\phi}_{24}(t) \\ \Theta & \Theta & \hat{\phi}_{33}(t) & \hat{\phi}_{34}(t) \\ \Theta & \Theta & \Theta & \hat{\phi}_{44}(t) \end{bmatrix}$$

where the zero blocks follow from the physical interpretation of $\hat{\phi}(t)$. We find that

$$H(t) = \begin{bmatrix} \Theta & \vdots & \hat{C}_2 & \vdots & \Theta & \vdots & \hat{C}_4 \end{bmatrix} \begin{bmatrix} \hat{\phi}_{11}(t)\hat{B}_1 + \hat{\phi}_{12}(t)\hat{B}_2 \\ \hat{\phi}_{22}(t)\hat{B}_2 \\ \Theta \\ \Theta \end{bmatrix} 1(t) + \underset{\cdot}{D}(t)$$

$$= \hat{C}_2\hat{\phi}_{22}(t)\hat{B}_2 1(t) + \underset{\cdot}{D}(t). \tag{8.87}$$

For zero initial conditions, the controllable and observable subsystem is described by

$$\dot{\hat{\boldsymbol{x}}}_{co}(t) = \hat{A}_{22}\hat{\boldsymbol{x}}_{co}(t) + \hat{B}_2\boldsymbol{u}(t)$$
$$\boldsymbol{y}(t) = \hat{C}_2\hat{\boldsymbol{x}}_{co}(t) + D\boldsymbol{u}(t)$$

so this subsystem has the same impulse response matrix as (8.87). In other words, the impulse response matrix (and therefore the transfer function matrix) of a system depends only on the controllable and observable subsystem. Therefore to obtain a minimal realization (lowest order realization) for a system described in terms of equations relating inputs to outputs, we use the controllable and observable part of any realization.

Example 8.3

For the system governed by the state equations

$$\dot{x}_1(t) = x_2(t) + u(t)$$
$$\dot{x}_2(t) = x_1(t) - u(t)$$
$$\dot{x}_3(t) = x_1(t) + x_2(t) - x_3(t) + 2u(t)$$
$$y(t) = x_1(t)$$

we see that

$$A = \begin{bmatrix} 0 & 1 & 0 \\ 1 & 0 & 0 \\ 1 & 1 & -1 \end{bmatrix} \quad B = \begin{bmatrix} 1 \\ -1 \\ 2 \end{bmatrix} \quad C = \begin{bmatrix} 1 & 0 & 0 \end{bmatrix}.$$

We find that the controllability matrix is

$$\Gamma_3 = \begin{bmatrix} 1 & -1 & 1 \\ -1 & 1 & -1 \\ 2 & -2 & 2 \end{bmatrix}$$

which has rank equal to 1, so the system is not controllable. The observability matrix is

$$\Gamma_3^* = \begin{bmatrix} 1 & 0 & 1 \\ 0 & 1 & 0 \\ 0 & 0 & 0 \end{bmatrix}$$

which has rank equal to 2, so the system is not observable either. To get the Kalman decomposition for this system, we first note that $\mathcal{C}$ is spanned by the vector

$$g_1 = \begin{bmatrix} 1 \\ -1 \\ 2 \end{bmatrix}$$

(one of the columns of Γ_3) and $\mathcal{O}$ is spanned by the vectors

$$\begin{bmatrix} 1 \\ 0 \\ 0 \end{bmatrix} \quad \text{and} \quad \begin{bmatrix} 0 \\ 1 \\ 0 \end{bmatrix}$$

(two linearly independent columns of Γ_3^*). The unobservable subspace $\bar{\mathcal{O}}$ is thus spanned by one vector orthogonal to both of these, such as

$$\begin{bmatrix} 0 \\ 0 \\ 1 \end{bmatrix}.$$

We can see that the vector g_1 does not lie in $\bar{\mathcal{O}}$, therefore $\mathcal{C}\overline{\mathcal{O}}$ is empty and $\ell_1 = 0$, and Z_1 does not exist. Z_2 is just the single column g_1. Since there are no columns in Z_1, the columns of Z_3 must span $\overline{\mathcal{O}}$. We saw that $\overline{\mathcal{O}}$ is one-dimensional, so $p_1 = 1$ and

$$Z_3 = \begin{bmatrix} 0 \\ 0 \\ 1 \end{bmatrix}.$$

We need to choose one more column to span the state space. We will arbitrarily pick

$$Z_4 = \begin{bmatrix} 0 \\ 1 \\ 0 \end{bmatrix}.$$

This gives

$$Z = \begin{bmatrix} Z_2 \;\vdots\; Z_3 \;\vdots\; Z_4 \end{bmatrix} = \begin{bmatrix} 1 & 0 & 0 \\ -1 & 0 & 1 \\ 2 & 1 & 0 \end{bmatrix}.$$

Taking the inverse gives

$$Z^{-1} = \begin{bmatrix} 1 & 0 & 0 \\ -2 & 0 & 1 \\ 1 & 1 & 0 \end{bmatrix}$$

which results in

$$\hat{A} = Z^{-1}AZ = \begin{bmatrix} -1 & 0 & 1 \\ 0 & -1 & -1 \\ 0 & 0 & 1 \end{bmatrix}; \qquad \hat{B} = Z^{-1}B = \begin{bmatrix} 1 \\ 0 \\ 0 \end{bmatrix};$$

$$\hat{C} = CZ = \begin{bmatrix} 1 & 0 & 0 \end{bmatrix}.$$

The state equations for this decomposed system are thus

$$\begin{aligned} \dot{\hat{x}}_1(t) &= -\hat{x}_1(t) + \hat{x}_3(t) + u(t) \\ \dot{\hat{x}}_2(t) &= \hat{x}_2(t) - \hat{x}_3(t) \\ \dot{\hat{x}}_3(t) &= \hat{x}_3(t) \\ y(t) &= \hat{x}_1(t). \end{aligned}$$

We can now easily see that $\hat{x}_1(t)$ is the only state variable that is affected by the input and that $\hat{x}_2(t)$ has no effect on the output. That is, $\hat{x}_1(t)$ is both controllable, and observable, $\hat{x}_3(t)$ is observable (since it affects $y(t)$ through $\hat{x}_1(t)$) but not controllable, and $\hat{x}_2(t)$ is neither controllable nor observable.

In this example the subspace $\mathcal{C}\bar{\mathcal{O}}$ is empty, so there is no controllable but unobservable subsystem. The following example illustrates the procedure when $\mathcal{C}\bar{\mathcal{O}}$ is not empty.

Example 8.4

We consider a system (either analog or discrete) with system matrices

$$A = \begin{bmatrix} 1 & 2.5 & 0.5 & 1.5 \\ 1 & -0.5 & 1.5 & 0.5 \\ 1 & 1.5 & -0.5 & 0.5 \\ -3 & -3 & -2 & -3 \end{bmatrix}; \qquad B = \begin{bmatrix} 1.5 \\ -0.5 \\ 1.5 \\ -1 \end{bmatrix}; \quad C = \begin{bmatrix} 0 & -1 & 1 & 0 \end{bmatrix}.$$

This system has the controllability and observability matrices

$$\Gamma_4 = \begin{bmatrix} 1.5 & -0.5 & 3.5 & -4.5 \\ -0.5 & 3.5 & -4.5 & 11.5 \\ 1.5 & -0.5 & 3.5 & -4.5 \\ -1 & -3 & 1 & -7 \end{bmatrix}; \qquad \Gamma_4^* = \begin{bmatrix} 0 & 0 & 0 & 0 \\ -1 & 2 & -4 & 8 \\ 1 & -2 & 4 & -8 \\ 0 & 0 & 0 & 0 \end{bmatrix}.$$

We find that rank $\Gamma_4 = 2$ and rank $\Gamma_4^* = 1$, so the controllable subspace $\mathcal{C}$ is two-dimensional, and the observable subspace $\mathcal{O}$ is one-dimensional. We can choose the first two columns of Γ_4 as basis vectors for $\mathcal{C}$, so

$$g_1 = \begin{bmatrix} 1.5 \\ -0.5 \\ 1.5 \\ -1 \end{bmatrix}; \qquad g_2 = \begin{bmatrix} -0.5 \\ 3.5 \\ -0.5 \\ -3 \end{bmatrix}.$$

The first column of Γ_4^* is a basis vector for $\mathcal{O}$, so we must choose three linearly independent vectors orthogonal (perpendicular) to this column as a basis for $\bar{\mathcal{O}}$. Let us choose these to be

$$\bar{v}_1 = \begin{bmatrix} 1 \\ 0 \\ 0 \\ 0 \end{bmatrix}; \qquad \bar{v}_2 = \begin{bmatrix} 0 \\ 0 \\ 0 \\ 1 \end{bmatrix}; \qquad \bar{v}_3 = \begin{bmatrix} 0 \\ 1 \\ 1 \\ 0 \end{bmatrix}.$$

Obviously neither $\boldsymbol{g}_1$ nor $\boldsymbol{g}_2$ lies in the space $\bar{\mathcal{O}}$, but we must determine if a linear combination of them does. We note that $\bar{v}_1$ permits the first component of a vector in $\bar{\mathcal{O}}$ to take any value and $\bar{v}_2$ permits the fourth component to take any value. However, $\bar{v}_3$ requires the second and third components of a vector in $\bar{\mathcal{O}}$ to be equal. We see that the linear combination $2\boldsymbol{g}_1 + \boldsymbol{g}_2$ has equal second and third components, so it lies in $\bar{\mathcal{O}}$. We can scale this to any convenient level. Therefore we choose

$$Z_1 = z_1 = \frac{2}{5}\left[2\boldsymbol{g}_1 + \boldsymbol{g}_2\right] = \begin{bmatrix} 1 \\ 1 \\ 1 \\ -2 \end{bmatrix}.$$

For $Z_2 = z_2$ we can choose any vector that together with z_1 spans $\mathcal{C}$. This could be either of the $\boldsymbol{g}_i$ but just to show that it really does not matter, let us take

$$Z_2 = z_2 = \boldsymbol{g}_1 - \frac{1}{2}z_1 = \begin{bmatrix} 1 \\ -1 \\ 1 \\ 0 \end{bmatrix}.$$

We next must choose Z_3. Since z_1 lies in the three-dimensional subspace $\bar{\mathcal{O}}$ we need two more basis vectors, so let us choose $\bar{v}_1$ and $\bar{v}_2$, which are linearly independent of z_1. At this point we already have four linearly independent vectors that span our four-dimensional state space, so there is no Z_4 matrix. Therefore we have

$$Z = \left[Z_1 \mid Z_2 \mid Z_3\right] = \left[\begin{array}{c:c:cc} 1 & 1 & 1 & 0 \\ 1 & -1 & 0 & 0 \\ 1 & 1 & 0 & 0 \\ -2 & 0 & 0 & 1 \end{array}\right]$$

so

$$Z^{-1} = \begin{bmatrix} \hat{Z}_1 \\ \hat{Z}_2 \\ \hat{Z}_3 \end{bmatrix} = \left[\begin{array}{cccc} 0 & 0.5 & 0.5 & 0 \\ \hdashline 0 & -0.5 & 0.5 & 0 \\ \hdashline 1 & 0 & -1 & 0 \\ 0 & 1 & 1 & 1 \end{array}\right].$$

This results in

$$\hat{A} = Z^{-1}AZ = \begin{bmatrix} 1 & 1 & 1 & 0.5 \\ 0 & -2 & 0 & 0 \\ 0 & 0 & 0 & 1 \\ 0 & 0 & -1 & -2 \end{bmatrix}; \quad \hat{B} = Z^{-1}B = \begin{bmatrix} 0.5 \\ 1 \\ 0 \\ 0 \end{bmatrix};$$

$$\hat{C} = CZ = \begin{bmatrix} 0 & 2 & 0 & 0 \end{bmatrix}.$$

If this is a discrete system, then the controllable and observable subsystem is first-order and is described by

$$\hat{x}_2(k+1) = -2\hat{x}_2(k) + u(k)$$
$$y(k) = 2\hat{x}_2(k).$$

The uncontrollable and unobservable subsystem is second-order and is described by

$$\hat{x}_3(k+1) = \hat{x}_4(k)$$
$$\hat{x}_4(k+1) = -\hat{x}_3(k) - 2\hat{x}_4(k),$$

and the controllable but unobservable subsystem is first-order and is described by

$$\hat{x}_1(k+1) = \hat{x}_1(k) + \hat{x}_2(k) + \hat{x}_3(k) + 0.5\hat{x}_4(k) + 0.5u(k).$$

Problems

8.1 Consider the system below.

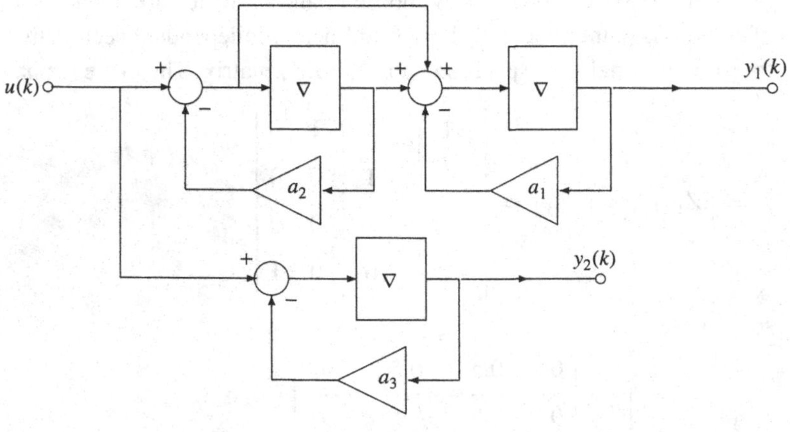

a) For what condition on the constants a_1, a_2, a_3 is this system controllable?

b) Replace the delay elements by integrators and repeat part (a).

8.2 Determine whether each of the systems below is controllable and/or observable.

a)

$$\begin{aligned}\dot{x}_1(t) &= -x_1(t) + u(t)\\ \dot{x}_2(t) &= -x_2(t) + x_3(t)\\ \dot{x}_3(t) &= -2x_3(t) + u(t)\\ y(t) &= x_1(t) + x_2(t) + u(t)\end{aligned}$$

b)

$$\begin{aligned}x_1(k+1) &= -4x_1(k) + \frac{3}{2}x_2(k) + x_3(k) + u_1(k) + u_2(k)\\ x_2(k+1) &= -x_1(k) \qquad + x_3(k) \qquad + u_2(k)\\ x_3(k+1) &= -5x_1(k) + 3x_2(k) + 3x_3(k) + 2u_1(k) + u_2(k)\\ y(k) &= x_1(k) \qquad - x_3(k) \qquad + u_1(k)\end{aligned}$$

c) A system with

$$A = \begin{bmatrix} 0 & 1 & 0 & \dots & 0 \\ 0 & 0 & 1 & \ddots & \vdots \\ \vdots & \vdots & \vdots & \ddots & \vdots \\ 0 & 0 & 0 & \dots & 1 \\ -\alpha_1 & -\alpha_2 & -\alpha_3 & \dots & -\alpha_n \end{bmatrix}; \qquad \boldsymbol{b} = \begin{bmatrix} 0 \\ 0 \\ \vdots \\ 0 \\ 1 \end{bmatrix}$$

$$C = \begin{bmatrix} 1 & 0 & 0 & \dots & 0 \end{bmatrix}$$

Note: For $i > 1$, the ith column of A is $\boldsymbol{a}_i = \boldsymbol{e}_{i-1} - \alpha_i \boldsymbol{e}_n$, where $\boldsymbol{e}_i$ is a unit vector with 1 in the ith position and 0 everywhere else.

8.3 For the system described by the state equations

$$\begin{aligned}\dot{x}_1(t) &= -x_1(t) + x_2(t) + x_3(t)\\ \dot{x}_2(t) &= \qquad x_2(t) - x_3(t) + u(t)\\ \dot{x}_3(t) &= \qquad - x_2(t) + x_3(t) - u(t)\end{aligned}$$

$$\begin{aligned}y_1(t) &= x_1(t) \qquad + x_3(t)\\ y_2(t) &= \qquad x_2(t) - x_3(t)\end{aligned}$$

a) Determine whether the system is controllable.

b) Determine whether the system is observable.

Justify your answers.

8.4 An analog system has the A, B, and C matrices

$$A = \begin{bmatrix} -2 & -3 & -4 \\ 0 & -2 & 2 \\ -2 & -1 & 2 \end{bmatrix}; B = \begin{bmatrix} 1 & 1 \\ 0 & 1 \\ 1 & 0 \end{bmatrix}; C = [1 \quad 0 \quad 0].$$

a) Determine whether the system is controllable and/or observable.
b) If these are the matrices of a discrete system, are the controllability or observability properties changed?

8.5 A system is governed by the state equations

$$\begin{aligned} \dot{q}_1(t) &= -2q_1(t) + q_2(t) + x_1(t) - x_2(t) \\ \dot{q}_2(t) &= -2q_2(t) + 2q_3(t) + x_2(t) \\ \dot{q}_3(t) &= -2q_1(t) - q_2(t) + 2q_3(t) + x_1(t) \end{aligned}$$

$$y(t) = q_1(t) + q_3(t).$$

a) Determine whether the system is controllable.
b) Determine whether the system is observable.

8.6 Determine whether the system below is controllable and/or observable.

$$\begin{aligned} q_1(k+1) &= \frac{3}{4}q_1(k) - \frac{1}{4}q_2(k) - \frac{3}{2}q_3(k) + 3x_2(k) \\ q_2(k+1) &= -\frac{1}{4}q_1(k) + \frac{3}{4}q_2(k) + \frac{5}{2}q_3(k) + 2x_1(k) - 2x_2(k) \\ q_3(k+1) &= q_3(k) + \frac{1}{2}x_2(k) \end{aligned}$$

$$\begin{aligned} y_1(k) &= \frac{1}{2}q_1(k) + \frac{1}{2}q_2(k) - q_3(k) \\ y_2(k) &= 2q_3(k). \end{aligned}$$

8.7 A time-varying system has the transition matrix

$$\phi(t, \tau) = \begin{bmatrix} e^{-(t-\tau)} & e^{-(t-\tau)} - \dfrac{t+1}{\tau+1} \\ 0 & \dfrac{t+1}{\tau+1} \end{bmatrix}$$

and

$$B = \begin{bmatrix} 1 \\ 0 \end{bmatrix}; C = \begin{bmatrix} e^{-t} & 0 \end{bmatrix}; D = 0.$$

a) Is this system controllable?
b) Is this system observable?

8.8 Consider the system below.

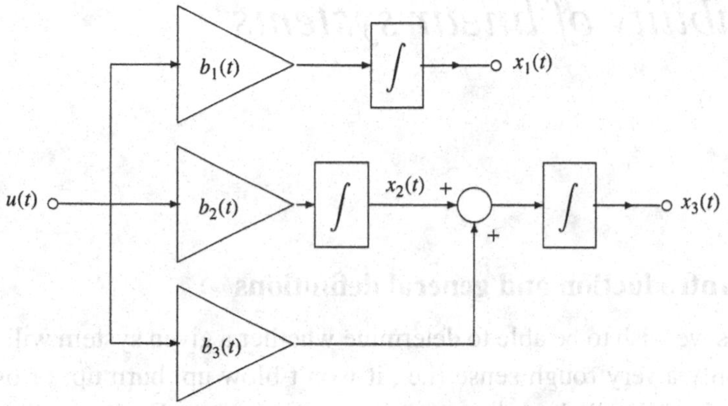

Determine, for each of the following conditions, whether or not the system is completely controllable.
a) b_1, b_2, b_3 are constant for all time.
b) $b_2(t) = t$ and b_1 and b_3 are constant for all time.
c) b_1 and b_2 are constant and $b_3(t) = t$.
d) $b_1(t) = t$, $b_2(t) = 2t$, and b_3 is constant.
e) $b_1(t) = \begin{cases} \sin \pi t & \text{for } 2k \leq t < 2k+1 \\ 0 & \text{for } 2k+1 \leq t < 2k+2 \end{cases}$ where k is an integer.

$b_2(t) = b_1(t+1)$ and $b_3 = \text{constant}$.

8.9 Find the Kalman decomposition for a system having the system matrices

$$A = \begin{bmatrix} 1 & 2.5 & 0.5 & 1.5 \\ 1 & -0.5 & 1.5 & 0.5 \\ 1 & 1.5 & -0.5 & 0.5 \\ -3 & -3 & -2 & -3 \end{bmatrix}; \quad B = \begin{bmatrix} 1.5 \\ -0.5 \\ 1.5 \\ -1 \end{bmatrix};$$

$$C = \begin{bmatrix} 2 & 1 & 1 & 2 \end{bmatrix}.$$

Note: This system has the same A and B matrices as the system of Example 8.4.

9

Stability of linear systems

9.1 Introduction and general definitions

In many cases we wish to be able to determine whether a given system will operate properly in only a very rough sense (i.e., it won't blow up, burn up, or oscillate) without obtaining a detailed solution of the system equations. Even more important, since we could obtain solutions for only a few typical inputs, we might wish to know if there is *any* input which could cause improper behavior. Such questions are considered in the general topic of system stability.

We begin by defining certain types of stability in terms of the system's state response with no inputs applied (zero-input or Lyapunov stability). These definitions given below are applicable to nonlinear as well as linear systems.

Recall that for an unforced system governed by

$$\dot{\boldsymbol{x}} = \boldsymbol{f}(\boldsymbol{x}(t), \boldsymbol{0}, t) \tag{9.1}$$

we call the state $\boldsymbol{x}_e$ an "*equilibrium*" state if $\boldsymbol{f}(\boldsymbol{x}_e, \boldsymbol{0}, t) \equiv \boldsymbol{0}$ for all t. For linear systems $\boldsymbol{0}$ is always an equilibrium state.

Definition 9.1 *For equation (9.1), an equilibrium state $\boldsymbol{x}_e$ is called* **bounded** *at t_0 if and only if there exist some numbers $\delta(t_0) > 0$ and $K(t_0) < \infty$ such that if $\|\boldsymbol{x}(t_0) - \boldsymbol{x}_e\| < \delta(t_0)$ then $\|\boldsymbol{x}(t) - \boldsymbol{x}_e\| < K(t_0)$ for all $t \geq t_0$. It is called uniformly bounded on $[t_0, \infty]$ if δ and K do not depend on t_0.*

Note: The symbol $\|\boldsymbol{z}\|$ is used to denote some measure of the magnitude of a vector. It is called the *norm* of the vector $\boldsymbol{z}$, and it must satisfy the norm conditions:

$$\begin{aligned} &\text{a} \quad \|\boldsymbol{z}\| > 0 \quad \textit{for} \quad \boldsymbol{z} \neq \boldsymbol{0} \quad \textit{where } \|\boldsymbol{0}\| = 0 \\ &\text{b} \quad \|\alpha \boldsymbol{z}\| = |\alpha| \|\boldsymbol{z}\| \\ &\text{c} \quad \|\boldsymbol{z}_1 + \boldsymbol{z}_2\| \leq \|\boldsymbol{z}_1\| + \|\boldsymbol{z}_2\|. \end{aligned} \tag{9.2}$$

Examples of norms

1 $\|z\| = \left[\sum_{i=1}^{n} z_i^2\right]^{\frac{1}{2}}$ (the Euclidean norm)

2 $\|z\| = \sum_{i=1}^{n} |z_i|$

3 $\|z\| = \max_i |z_i|$. (9.3)

Definition 9.2 *An equilibrium state* $\boldsymbol{x}_e$ *of (9.1) is called* **stable** *(S) at* t_0 *if and only if for any* $\epsilon > 0$, *there exists a* $\delta(\epsilon, t_0) > 0$ *such that*

$$\|\boldsymbol{x}(t_0) - \boldsymbol{x}_e\| < \delta(\epsilon, t_0) \Longrightarrow \|\boldsymbol{x}(t) - \boldsymbol{x}_e\| < \epsilon \quad \text{for all } t \geq t_0.$$

Again the stability is uniform if δ *does not depend on* t_0. *If* $\boldsymbol{x}_e$ *is not stable it is called* **unstable** *(US).*

Note: Stability implies boundedness, but boundedness does not necessarily imply stability.

Definition 9.3 *An equilibrium state* $\boldsymbol{x}_e$ *of (9.1) is called* **asymptotically stable** *(AS) if*

a *It is stable and*

b *Every solution* $\boldsymbol{x}(t)$ *which starts sufficiently close to* $\boldsymbol{x}_e$ *converges to* $\boldsymbol{x}_e$ *as* $t \to \infty$.

In other words, there exists a $\gamma > 0$ *such that if* $\|\boldsymbol{x}(t_0) - \boldsymbol{x}_e\| < \gamma$ *then* $\lim_{t\to\infty} \|\boldsymbol{x}(t) - \boldsymbol{x}_e\| = 0$.

Definition 9.4 *An equilibrium state* $\boldsymbol{x}_e$ *of (9.1) is called* **asymptotically stable in the large** *(ASIL) or* **globally asymptotically stable** *if it is asymptotically stable for any finite initial condition (i.e.,* γ *is any finite number).*

The same general stability definitions hold for discrete systems but the time variable t_k is discrete, and the equilibrium state $\boldsymbol{x}_e$ satisfies $\boldsymbol{x}_e \equiv \boldsymbol{f}(\boldsymbol{x}_e, \boldsymbol{0}, t_k)$ for all t_k.

9.2 Zero-input stability of linear systems

Consider the unforced linear analog system

$$\dot{\boldsymbol{x}}(t) = A(t)\boldsymbol{x}(t) \tag{9.4}$$

which has the solution

$$\boldsymbol{x}(t) = \phi(t, t_0)\boldsymbol{x}(t_0). \tag{9.5}$$

We note that the origin is an equilibrium state.

Theorem 9.1 *For a linear analog system, the equilibrium state at the origin is*

a *Stable (thus also bounded) if and only if $\|\phi(t, t_0)\|$ is bounded for all $t \geq t_0$.*

b *Asymptotically stable if and only if in addition to (a), $\|\phi(t, t_0)\| \to 0$ as $t \to \infty$.*

Proof: (a) (Sufficiency) Taking the norm on both sides of (9.5) gives

$$\|\boldsymbol{x}(t)\| = \|\phi(t, t_0)\boldsymbol{x}(t_0)\| \leq \|\phi(t, t_0)\| \, \|\mathbf{x}(t_0)\| \tag{9.6}$$

where the *norm of a matrix* is defined as

$$\|T\| = \max_{x \neq 0} \frac{\|T\boldsymbol{x}\|}{\|\boldsymbol{x}\|}. \tag{9.7}$$

Equation (9.6) follows from the definition (9.7) since

$$\|T\boldsymbol{x}\| = \|T\boldsymbol{x}\| \frac{\|\boldsymbol{x}\|}{\|\boldsymbol{x}\|} \leq \max_{x \neq 0} \frac{\|T\boldsymbol{x}\|}{\|\boldsymbol{x}\|} \|\boldsymbol{x}\| = \|T\| \, \|\boldsymbol{x}\|. \tag{9.8}$$

We see that if $\|\phi(t, t_0)\|$ is bounded for all $t \geq t_0$, then there exists a $K < \infty$ such that $\|\phi(t, t_0)\| < K$, so by taking δ in the definition of stability to be $\delta = \epsilon/K$ we see that when $\|\boldsymbol{x}(t_0)\| < \delta$ then $\|\boldsymbol{x}(t)\| < \epsilon$ for all $t \geq t_0$. Thus the equilibrium state state $\boldsymbol{x}_e = \mathbf{0}$ is stable (and thus is also bounded).

(Necessity) If $\|\phi(t, t_0)\|$ is not bounded, then there is at least one element for which $|\phi_{ij}(t, t_0)|$ goes to infinity as $t \to \infty$ (Problem 9.2b is to prove this). Therefore the initial state $x_j(t_0) = \delta/2 > 0$ and $x_k(t_0) = 0$ for $k \neq j$ will make $|x_i(t)| \to \infty$ as $t \to \infty$, so the equilibrium state at $\mathbf{0}$ is not bounded and thus is not stable. QED.

The proof of (b) is left as an exercise (Problem 9.3).

It is easier to apply the conditions of Theorem 9.1 when they are expressed directly in terms of the elements of $\phi(t, t_0)$.

Corollary to Theorem 9.1 *For a linear analog system, the equilibrium state is S if and only if $|\phi_{ij}(t, t_0)|$ is bounded for all i, j, and all $t \geq t_0$. It is AS if and only if in addition for all i and j, $\phi_{ij}(t, t_0) \to 0$ as $t \to \infty$.*

Note: The conditions for stability and asymptotic stability are imposed on $\phi(t, t_0)$ and therefore are independent of $\boldsymbol{x}(t_0)$. This means that for linear systems all stability properties are valid globally (i.e., they are true for any finite initial condition).

For linear systems with constant A matrix, we know that $\phi(t) = e^{At}$ and that in general (see (4.44))

$$e^{At} = \sum_{i=1}^{s} \sum_{j=0}^{m_i - 1} \left[\frac{d^j \, e^{\lambda t}}{d\lambda^j} \right]_{\lambda = \lambda_i} Z_{ij} \tag{9.9}$$

where m_i is the multiplicity of the eigenvalue λ_i of the minimum polynomial $m(\lambda)$. Since

$$\left[\frac{d^j e^{\lambda t}}{d\lambda^j}\right]_{\lambda=\lambda_i} = t^j e^{\lambda_i t} \tag{9.10}$$

we see that

$$e^{At} = \sum_{i=1}^{s}\sum_{j=0}^{m_i-1} t^j e^{\lambda_i t} Z_{ij}. \tag{9.11}$$

Each term $t^j e^{\lambda_i t}$ is bounded if $\Re e\{\lambda_i\} < 0$ regardless of the finite power j (for $j = 0$ it is bounded even when $\Re e\{\lambda_i\} = 0$), and $t^j e^{\lambda_i t} \to 0$ as $t \to \infty$ if and only if $\Re e\{\lambda_i\} < 0$. This result can be expressed as

Theorem 9.2 *The equilibrium state at the origin for a linear, continuous-time system with constant A matrix is*

a *Asymptotically stable if and only if $\Re e\{\lambda_i\} < 0$ for all i.*
b *Stable if and only if $\Re e\{\lambda_i\} \leq 0$ for all i and those λ_i for which $\Re\{\lambda_i\} = 0$ are simple (i.e., $m_i = 1$ so the multiplicity in the minimum polynomial is one).*
c *Unstable if $\Re e\{\lambda_i\} > 0$ for any i or $\Re e\{\lambda_i\} = 0$ and $m_i > 1$.*

Using the same approach for fixed discrete systems with $\phi(k) = A^k$ we find that taking

$$\left[\frac{d^j \lambda^k}{d\lambda^j}\right]_{\lambda=\lambda_i} = p_j(k)\lambda^{(k-j)}$$

where $p_j(k)$ denotes a polynomial in k of degree j. Here each term in the expansion for A^k is bounded and goes to zero as $k \to \infty$ for any finite power j if and only if $|\lambda_i| < 1$. When $j = 0$ it is bounded even for $|\lambda_i| = 1$. This gives

Theorem 9.3 *The equilibrium state at the origin of a linear, discrete system with constant A matrix is*

a *Asymptotically stable if and only if $|\lambda_i| < 1$ for all i.*
b *Stable if and only if $|\lambda_i| \leq 1$ for all i and all λ_i with $|\lambda_i| = 1$ are simple (i.e., $m_i = 1$, so λ_i is not repeated in $m(\lambda)$).*
c *Unstable if $|\lambda_i| > 1$ for any i or $|\lambda_i| = 1$ and $m_i > 1$.*

Prove this theorem as an exercise (see Problem 9.6).

9.3 Zero-state stability of linear systems

The stability discussed in the preceding sections involved the unforced system. In this section we introduce a type of stability which involves the response of the system to inputs.

Definition 9.5 *A forced linear system is bounded input bounded output (BIBO) stable if the zero-state response is bounded for all $t \geq t_0$ when $\boldsymbol{x}(t)$ is bounded for all $t \geq t_0$.*

For causal, linear analog systems the zero-state response is

$$\begin{aligned} \boldsymbol{y}(t) &= \int_{t_0}^{t} H(t,\tau)\boldsymbol{x}(\tau)d\tau \\ &= \int_{t_0}^{t} [C(t)\phi(t,\tau)B(\tau)1(t-\tau) + D(\tau)\delta(t-\tau)]\boldsymbol{x}(\tau)d\tau \end{aligned} \tag{9.12}$$

and we have

Theorem 9.4 *A linear analog system is BIBO stable if and only if*

$$\int_{t_0}^{t} |h_{ij}(t,\tau)|d\tau < \infty \quad \textit{for all} \quad i, j, \textit{ and } t \geq t_0. \tag{9.13}$$

Proof: (Sufficiency) $\|\boldsymbol{y}(t)\| \leq \int_{t_0}^{t} \|H(t,\tau)\| \, \|\boldsymbol{x}(\tau)\| d\tau$ for all $t \geq t_0$. For some $K_1 < \infty$

$$\|H(t,\tau)\| \leq K_1 \sum_{i=1}^{m} \sum_{j=1}^{r} |h_{ij}(t,\tau)|$$

(Problem 9.2d is to prove this), and for bounded inputs $\|\boldsymbol{x}(\tau)\| \leq K_2$ for some $K_2 < \infty$ and for all $\tau \geq t_0$. Letting $K = K_1K_2$, we get

$$\|\boldsymbol{y}(t)\| \leq K \sum_{i=1}^{m} \sum_{j=1}^{r} \int_{t_0}^{t} |h_{ij}(t,\tau)|d\tau \quad \text{for all } t \geq t_0.$$

Thus when the conditions of the theorem hold we see that $\|\boldsymbol{y}(t)\| < \infty$ for all $t \geq t_0$, so the system is BIBO stable.

(Necessity) If any term becomes unbounded (say, the lk term) then at some $t = t_1$ by choosing $x_j(\tau) = \text{sgn } h_{lj}(t_1,\tau)$ for all $j = 1 \ldots r$, $\boldsymbol{x}(\tau)$ is a bounded input, but

$$y_l(t_1) = \sum_{j=1}^{r} \int_{t_0}^{t_1} |h_{lj}(t_1,\tau)|d\tau \geq \int_{t_0}^{t_1} |h_{lk}(t_1,\tau)|d\tau \longrightarrow \infty$$

so $\|\boldsymbol{y}(t_1)\| \to \infty$ as $t_1 \to \infty$. Thus the output is unbounded, so the system is not BIBO stable. QED.

Corollary to Theorem 9.4 *If $B(t)$, $C(t)$, and $D(t)$ are bounded for $t \geq t_0$ then the system is BIBO stable if*

$$\int_{t_0}^{t} |\phi_{ij}(t,\tau)|\, d\tau < \infty \quad \text{for all } i,\ j, \text{ and } t \geq t_0. \tag{9.14}$$

Note: This is sufficient but not necessary as will be seen later.

Proof: We use the fact that

$$\|T_1 T_2\| = \max_{x \neq 0} \frac{\|T_1 T_2 x\|}{\|x\|} = \max_{x \neq 0} \frac{\|T_1 T_2 x\|}{\|T_2 x\|} \frac{\|T_2 x\|}{\|x\|} \leq \|T_1\|\,\|T_2\|. \tag{9.15}$$

Then

$$\begin{aligned} \|y(t)\| &\leq \left\| \int_{t_0}^{t} C(t)\phi(t,\tau)B(\tau)x(\tau)d\tau \right\| + \|D(t)\|\,\|x(t)\| \\ &\leq \int_{t_0}^{t} \|C(t)\|\,\|\phi(t,\tau)\|\,\|B(\tau)\|\,\|x(\tau)\| d\tau + \|D(t)\|\,\|x(t)\| \end{aligned}$$

so when $B(t)$, $C(t)$, and $D(t)$ are finite and $x(t)$ is bounded there exist constants K_1 and K_2 such that

$$\|y(t)\| \leq K_1 \int_{t_0}^{t} \|\phi(t,\tau)\|\, d\tau + K_2.$$

The remainder of the proof is identical to the sufficiency part of the preceding proof. QED.

Example 9.1

This example will illustrate that the condition (9.14) of the corollary is not necessary for BIBO stability. We see from Figure 9.1 that $H(t-\tau) = e^{-(t-\tau)}$ since the input only affects the state x_1 not x_2 (i.e., the system is not controllable), so

$$\int_{t_0}^{t} |H(t-\tau)| d\tau = \int_{t_0}^{t} |e^{-(t-\tau)}| d\tau = 1 - e^{-(t-t_0)} < \infty \quad \text{for all } t \geq t_0.$$

The system is thus BIBO stable, but $\phi_{22}(t-\tau) = e^{(t-\tau)}$, so $\int_{t_0}^{t} |e^{(t-\tau)}|\, d\tau = e^{(t-t_0)} - 1$ which $\longrightarrow \infty$ as $t \to \infty$. Thus we see that for uncontrollable systems

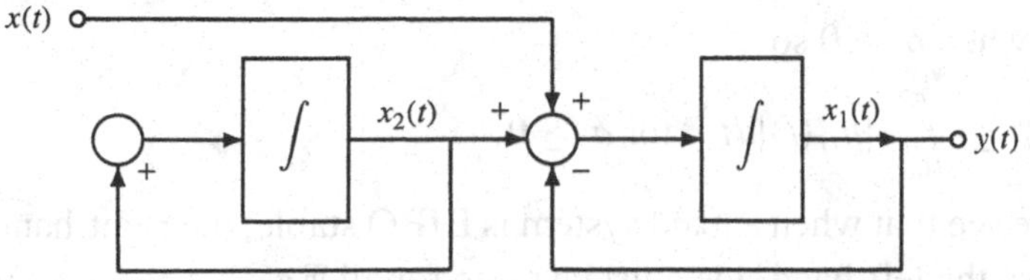

Fig. 9.1. An unstable system that is BIBO stable

the condition (9.14) is not necessary for BIBO stability, so the system may be BIBO stable and the unforced system may still be unstable.

One may ask whether stability of the unforced system is enough to guarantee that the system is BIBO stable. The following example shows that for time-varying systems we may have asymptotic stability but not BIBO stability.

Example 9.2

$$\dot{x}(t) = -\frac{x(t)}{t+1} + x(t)$$

We found that $\phi(t, t_0) = \dfrac{t_0 + 1}{t+1}$ for $t \geq t_0$ so we see that $\phi(t, t_0) \longrightarrow 0$ as $t \to \infty$ so the equilibrium point at the origin is AS if $t_0 > -1$. The response to $x(t) = 1(t)$ is

$$x(t) = \int_0^t \frac{1+\tau}{1+t} d\tau = \frac{1}{1+t}\left(t + \frac{t^2}{2}\right) = t\left(\frac{1+(1/2)t}{1+t}\right)$$

which $\longrightarrow \infty$ as $t \to \infty$, so the system is not BIBO stable. The difficulty here is that $\phi(t, t_0)$ does not go to zero fast enough to make $\int_{t_0}^t \phi(t, \tau)d\tau$ finite.

We shall see that for fixed systems there is a close relation between AS of the equilibrium points and BIBO stability of the system.

9.4 Conditions for BIBO stability of fixed linear systems

For BIBO stability of fixed systems Theorem 9.4 requires that

$$\int_0^t |h_{ij}(\eta)|d\eta < \infty \quad \text{for all } t \geq 0 \text{ and all } i, j.$$

Note that the Laplace transform of $h_{ij}(t)$ is

$$H_{ij}(s) = \int_0^\infty h_{ij}(t)e^{-st}dt$$

so

$$|H_{ij}(s)| \leq \int_0^\infty |h_{ij}(t)|e^{-\sigma t}dt \quad \text{where } \sigma = \Re e\{s\}.$$

Now $e^{-\sigma t} \leq 1$ when $\sigma \geq 0$ so

$$|H_{ij}(s)| \leq \int_0^\infty |h_{ij}(t)|dt \quad \text{for } \sigma \geq 0. \tag{9.16}$$

Therefore we see that when a fixed system is BIBO stable, the right-hand side of (9.16) is finite so the left-hand side must be finite for all $\Re e\{s\} \geq 0$. In other words, a *necessary* condition for BIBO stability is that no element of the transfer function matrix have poles in the right half plane (RHP) or on the $j\omega$-axis of the s-plane. This condition is necessary even for systems not governed by ordinary differential equations (i.e., nonlumped systems).

Is this condition also sufficient for BIBO stability? It can easily be shown that it is, since a typical element of the transfer function matrix $\mathrm{H}(s)$ has the form $H_{ij}(s) = P(s)/Q(s)$ where in this case $Q(s)$ has zeros only in the left half plane (LHP). Thus for $t \geq 0$

$$h_{ij}(t) = \sum_{l=1}^{k} \sum_{m=0}^{n_l - 1} \alpha_{lm} t^m e^{\lambda_l t}. \tag{9.17}$$

This goes to 0 as $t \to \infty$ for $\Re e\{\lambda_l\} < 0$, and since it goes to zero exponentially it is absolutely integrable over any interval for $t > 0$.

This result can be expressed as

> **Theorem 9.5** *A necessary and sufficient condition for BIBO stability of a fixed analog system is that every element $H_{ij}(s)$ of $\mathrm{H}(s)$ be finite for $\Re e\{s\} \geq 0$ (i.e., no poles at which $\mathrm{H}(s)$ is infinite are allowed in the right half of the s-plane (RHP)).*

When the system is governed by the state equations

$$\begin{aligned} \dot{\boldsymbol{x}}(t) &= A\boldsymbol{x}(t) + B\boldsymbol{u}(t) \\ \boldsymbol{y}(t) &= C\boldsymbol{x}(t) + D\boldsymbol{u}(t) \end{aligned} \tag{9.18}$$

then $\mathrm{H}(s) = \mathcal{L}\{H(t)\} = C(sI - A)^{-1}B + D = C\,\dfrac{\operatorname{adj}(sI - A)}{\det(sI - A)}B + D$. We see that

1 The only singularities which can possibly appear in $H_{ij}(s)$ are poles, and
2 these poles can occur only at the zeros of $\det(sI - A)$ (i.e., only at the eigenvalues of A).

Thus if A has eigenvalues λ_l only in the left half of the s-plane (LHP), $\mathcal{L}\{H(t)\} = \mathrm{H}(s)$ is finite for $\Re e\{s\} \geq 0$. This satisfies the necessary and sufficient conditions of Theorem 9.5.

We have shown

1 A fixed, analog linear system governed by state equations (9.18) is BIBO stable if and only if every element $H_{ij}(s)$ of the transfer function matrix $\mathrm{H}(s)$ has poles only in the LHP.
2 For a fixed, continuous-time linear system governed by state equations (9.18)

   AS $\Longrightarrow$ the system is BIBO stable.

3 We note again that in general BIBO stability of the system does not imply AS, however, *if the system is fixed and also controllable and observable (with B, C, and D finite) then AS does follow from BIBO stability.*

It is a bit complicated to prove this rigorously. The argument is essentially as follows:

a Because of controllability, the system can be moved (using a bounded input) from the origin to any state in a finite time. If $\phi(t)$ is not bounded, then the equilibrium point at zero is unstable, so we can get to a state for which $\boldsymbol{x}(t) \to \infty$ as $t \to \infty$ even when the input is removed. Because of observability, we would see $\boldsymbol{y}(t) \to \infty$. This cannot happen for a system which is BIBO stable so the equilibrium point must at least be stable.

b Suppose $\|\phi(t)\|$ does not go to zero as $t \to \infty$, then because the system is controllable and observable there is at least one component of $H(t)$ that does not go to zero as $t \to \infty$. Since this cannot happen in a BIBO stable system, the contradiction shows that $\|\phi(t)\| \to 0$ so the system is AS.

For linear discrete-time systems governed by the state equations

$$\begin{aligned} \boldsymbol{x}(k+1) &= A(k)\boldsymbol{x}(k) + B(k)\boldsymbol{u}(k) \\ \boldsymbol{y}(k) &= C(k)\boldsymbol{x}(k) + D(k)\boldsymbol{u}(k) \end{aligned} \tag{9.19}$$

completely analogous results are obtained. They can be stated as follows:

1 A linear discrete-time system is BIBO stable if and only if

$$\sum_{l=k_0}^{k} |h_{ij}(k,l)| < \infty \quad \text{for all } i,\ j, \text{ and all } k \geq k_0.$$

2 A necessary and sufficient condition for BIBO stability of a fixed discrete system is that $\mathcal{H}_{ij}(z)$ be finite for $|z| \geq 1$. In other words, a fixed discrete system is BIBO stable if and only if every element $\mathcal{H}_{ij}(z)$ of the transfer function matrix $\mathcal{H}(z)$ has all its poles inside the unit circle.

3 For a fixed discrete system governed by state equations (9.19)

$$\text{AS} \begin{cases} \Longrightarrow \text{ system is BIBO stable} \\ + \\ \Longleftarrow \text{ observability and controllability.} \end{cases}$$

9.5 Lyapunov's second method

This is an approach to the study of the stability of equilibrium states for a system governed by state equations (linear or nonlinear) which does not require that the solutions of the equations be known explicitly (Lyapunov 1892; Hahn 1963).

The basic idea

Let $\boldsymbol{x}(t)$ be the state of the system at time t and $E(\boldsymbol{x})$ the energy of the system as a function of the state. If the rate of change of energy $\dot{E}(\boldsymbol{x}(t))$ is negative for every possible $\boldsymbol{x}(t)$ except for the equilibrium state $\boldsymbol{x}^*$, then the system will lose energy (i.e., $E(\boldsymbol{x})$ decreases with time) until it assumes its minimum value at $\boldsymbol{x}^*$. In other words, a dissipative system when disturbed from equilibrium will always return to equilibrium. In order to develop this idea into a useful tool the method should be made independent of the concept of energy. Lyapunov (1892) did this by using a real, scalar, positive definite function which can be thought of as a generalized measure of energy or distance.

Definition 9.6 *A real scalar function $V(\boldsymbol{x})$ is called* **positive definite** *if $V(\boldsymbol{x}) > 0$ for $\boldsymbol{x} \neq \boldsymbol{0}$ and $V(\boldsymbol{0}) = 0$. It is called* **positive semidefinite** *if $V(\boldsymbol{x}) \geq 0$. $V(\boldsymbol{x})$ is* **negative definite** *if $-V(\boldsymbol{x})$ is positive definite. A time-varying function $V(\boldsymbol{x}, t)$ is called* **positive definite** *if there exists a positive definite function $W(\boldsymbol{x})$, independent of t, such that $V(\boldsymbol{x}, t) \geq W(\boldsymbol{x})$ for all t and $\boldsymbol{x}$ with $V(\boldsymbol{0}, t) \equiv 0$.*

Definition 9.7 *A real scalar function $V(\boldsymbol{x}, t)$*

a *which is continuous and has continuous partials with respect to its arguments,*

b *which is positive definite in a region $\mathcal{R}$ containing the origin,*

c *whose time derivative along a solution is negative semidefinite, that is,*

$$\dot{V}(\boldsymbol{x}, t) = \frac{\partial V(\boldsymbol{x}, t)}{\partial t} + \sum_{i=1}^{n} \frac{\partial V(\boldsymbol{x}, t)}{\partial x_i} \frac{dx_i}{dt}$$

is called a **Lyapunov function** *for the system $\dot{\boldsymbol{x}}(t) = \boldsymbol{f}(\boldsymbol{x}(t), t)$ which has an equilibrium state at the origin.*

Theorem 9.6

a *If there exists a Lyapunov function $V(\boldsymbol{x}, t)$ in some neighborhood $\mathcal{R}$ of the origin, then the equilibrium state at the origin is stable.*

b *If in addition $\dot{V}(\boldsymbol{x}, t)$ is negative definite in this neighborhood $\mathcal{R}$ then the equilibrium state at the origin is asymptotically stable.*

Note: We only require that these conditions hold in a sufficiently small region $\mathcal{R}$ containing the origin.

Proof:

a $\mathcal{R}$ is the region in which a Lyapunov function exists. Since

$$V(\boldsymbol{x}, t) = V(\boldsymbol{x}(t_0), t_0) + \int_{t_0}^{t} \dot{V}(\boldsymbol{x}, \tau)\, d\tau$$

and as long as $\boldsymbol{x}(t)$ stays in the region $\mathcal{R}$, we know that $\dot{V}(\boldsymbol{x}, t) \leq 0$, so we see that

$$V_0 \triangleq V(\boldsymbol{x}_0, t_0) \geq V(\boldsymbol{x}, t) > W(\boldsymbol{x}).$$

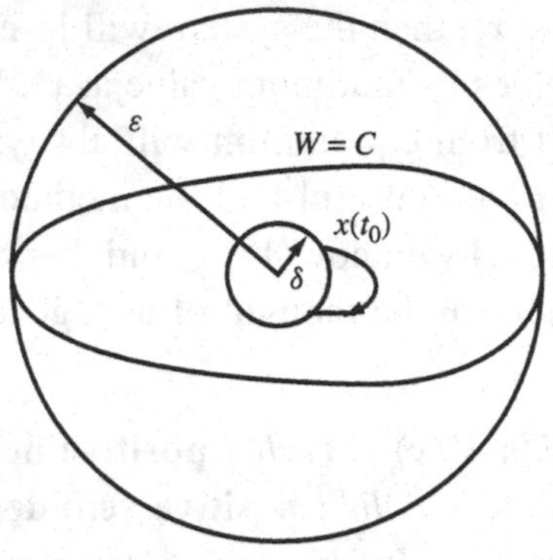

Fig. 9.2.

To prove stability we must show that for any $\varepsilon > 0$ we can guarantee $\|\boldsymbol{x}(t)\| < \varepsilon$ when $\|\boldsymbol{x}(t_0)\| < \delta$. Now for any $\varepsilon > 0$ we can find a number $C > 0$ which is the minimum value of $W(\boldsymbol{x})$ on the set $\|\boldsymbol{x}\| = \varepsilon$ (see Figure 9.2), a continuous function on a compact set assumes the minimum and maximum on the set, but the greatest lower bound could be used in any case; thus

$$W(\boldsymbol{x}) < C \Longrightarrow \|\boldsymbol{x}(t)\| < \varepsilon.$$

For $t = t_0$ we can find a $\delta > 0$ such that $\|\boldsymbol{x}\| < \delta$ makes $V(\boldsymbol{x}, t_0) < C$. (This is always possible since $V(\boldsymbol{0}, t) \equiv 0$ and $V(\boldsymbol{x}, t)$ is continuous in $\boldsymbol{x}$.) Thus we see that as long as we use an ε which makes the set $\|\boldsymbol{x}\| < \varepsilon$ lie entirely in the region $\mathcal{R}$ we have

$$\|\boldsymbol{x}(t_0)\| < \delta \Longrightarrow C > V_0 > V(\boldsymbol{x}, t) > W(\boldsymbol{x}) \Longrightarrow \|\boldsymbol{x}(t)\| < \varepsilon \quad \text{for all } t.$$

This proves stability of the system.

b When part (a) holds we proved that the system is stable. For AS we must also show that $\lim_{t\to\infty} \|\boldsymbol{x}(t)\| = 0$. From condition (b) of the theorem we see that if $\|\boldsymbol{x}(t)\| \neq 0$ then $\dot{V}(t) < 0$ so V is monotonically decreasing in time and it must be ≥ 0. Thus $V(\boldsymbol{x}, t)$ approaches a limit γ as $t \to \infty$. If $\gamma > 0$ then $\|\boldsymbol{x}(t)\|$ does not go to zero, and consequently $\dot{V}(\boldsymbol{x}, t)$ cannot go to zero since it is negative definite. But if $\dot{V}(\boldsymbol{x}, t)$ does not go to zero then $V(\boldsymbol{x}, t)$ cannot approach the nonzero constant γ. Thus we see that γ must be zero, and consequently since $V(\boldsymbol{x}, t)$ can be zero only when $\boldsymbol{x} = \boldsymbol{0}$, we see that $\|\boldsymbol{x}(t)\| \to 0$ as $t \to \infty$. QED.

Alternate proof of (b): Since V is monotonically decreasing, it follows that $V(\boldsymbol{x}(t), t) \geq \gamma$ for all $t \geq t_0$, where $\lim_{t\to\infty} V = \gamma$. If $\gamma > 0$ this means that

$\|\boldsymbol{x}(t)\| > \eta(\gamma) > 0$ for all $t \geq t_0$. Since $\dot{V}$ is negative definite, $\dot{V} \leq -\lambda$ for all $t \geq t_0$. This now implies that

$$V(\boldsymbol{x}, t) = V_0 + \int_{t_0}^{t} \dot{V}\, d\tau \leq V_0 - \lambda(t - t_0)$$

which says that V can go negative for large enough time. Since this is impossible, it means that $\|\boldsymbol{x}(t)\| \to 0$ as $t \to \infty$. QED.

Note: This theorem gives a *sufficient condition* for stability and asymptotic stability of the equilibrium point. This means that when a Lyapunov function can be found, the appropriate stability is guaranteed. However, when one cannot be found then we cannot tell anything about stability.

In many applications it is not possible to find a Lyapunov function which satisfies condition (b) of Theorem 9.6, but it is possible to find one which satisfies weaker conditions.

Theorem 9.7 *If in addition to part (a) of Theorem 9.6 the set of states $\mathcal{X}$ for which $\dot{V}(\boldsymbol{x}, t) = 0$ contains no trajectory other than $\boldsymbol{x}(t) \equiv \boldsymbol{0}$ (i.e., $\dot{V}(\boldsymbol{x}, t)$ is not identically zero along a solution of $\dot{\boldsymbol{x}} = \boldsymbol{f}(\boldsymbol{x}, t)$ except for $\boldsymbol{x} = \boldsymbol{0}$) then the origin is AS.*

Proof: The proof of Theorem 9.6 showed that $\dot{V}$ cannot remain negative as $t \to \infty$ but must approach 0. Thus $\boldsymbol{x}(t)$ must approach a point $\hat{\boldsymbol{x}} \in \mathcal{X}$ or if it starts at $\hat{\boldsymbol{x}} \in \mathcal{X}$ it must remain in $\mathcal{X}$. On the other hand, the condition here rules this out unless $\hat{\boldsymbol{x}} = \boldsymbol{0}$. QED.

To establish AS, Theorem 9.7 permits use of a Lyapunov function with a negative semidefinite derivative as long as no solution other than $\boldsymbol{x}(t) \equiv \boldsymbol{0}$ can remain on a surface of constant V.

Example 9.3

Consider

$$\begin{aligned} \dot{x}_1 &= x_2 - a x_1 (x_1^2 + x_2^2) \\ \dot{x}_2 &= x_1 - a x_2 (x_1^2 + x_2^2) \quad \text{where } a > 0. \end{aligned} \tag{9.20}$$

Let us try $V(\boldsymbol{x}) = x_1^2 + x_2^2$ as a Lyapunov function; then

$$\begin{aligned} \dot{V}(\boldsymbol{x}) &= \frac{\partial V}{\partial x_1}\dot{x}_1 + \frac{\partial V}{\partial x_2}\dot{x}_2 \\ &= 2x_1\left[x_2 - a x_1 (x_1^2 + x_2^2)\right] + 2x_2\left[- x_1 - a x_2 (x_1^2 + x_2^2)\right] \\ &= -2a\left(x_1^2 + x_2^2\right)^2. \end{aligned} \tag{9.21}$$

This is negative definite, so the system is asymptotically stable.

9.6 Lyapunov functions for fixed linear systems

Although for nonlinear systems the techniques for constructing Lyapunov functions can be quite complicated, for fixed linear systems we can always construct a Lyapunov function.

Consider the unforced fixed linear system

$$\dot{\boldsymbol{x}}(t) = A\boldsymbol{x}(t) \tag{9.22}$$

and choose as a Lyapunov function

$$V(\boldsymbol{x}(t)) = \boldsymbol{x}^T(t)M\boldsymbol{x}(t) \tag{9.23}$$

where M is a symmetric positive definite matrix. Now take the derivative with respect to t,

$$\dot{V}(\boldsymbol{x}(t)) = \dot{\boldsymbol{x}}^T(t)M\boldsymbol{x}(t) + \boldsymbol{x}^T(t)M\dot{\boldsymbol{x}}(t) = \boldsymbol{x}^T(t)[A^T M + MA]\boldsymbol{x}(t) \tag{9.24}$$

and let

$$-Q = A^T M + MA. \tag{9.25}$$

Then

$$\dot{V}(\boldsymbol{x}(t)) = -\boldsymbol{x}^T(t)Q\boldsymbol{x}(t) \tag{9.26}$$

where we observe that Q is symmetric. Equation (9.25) is called the Lyapunov equation. We see that if for any symmetric positive definite matrix Q the symmetric matrix M which satisfies (9.25) is positive definite, then, by Theorem 9.6, the system is asymptotically stable.

On the other hand, if the system is asymptotically stable, then for any symmetric positive definite Q we can choose

$$V(\boldsymbol{x}(t)) = \int_t^{\infty} \boldsymbol{x}^T(\tau)Q\boldsymbol{x}(\tau)d\tau > 0 \tag{9.27}$$

and since a solution to (9.22) can be written as

$$\boldsymbol{x}(\tau) = e^{A(\tau-t)}\boldsymbol{x}(t) \tag{9.28}$$

we find that

$$\begin{aligned} V(\boldsymbol{x}(t)) &= \int_t^{\infty} \boldsymbol{x}^T(\tau)Q\boldsymbol{x}(\tau)d\tau \\ &= \boldsymbol{x}^T(t)\int_t^{\infty} (e^{A(\tau-t)})^T Q e^{A(\tau-t)} d\tau \boldsymbol{x}(t) \\ &= \boldsymbol{x}^T(t)\int_0^{\infty} (e^{A\tau})^T Q e^{A\tau} d\tau \boldsymbol{x}(t) \triangleq \boldsymbol{x}^T(t)M\boldsymbol{x}(t) \end{aligned} \tag{9.29}$$

where M is symmetric and finite since we are integrating exponentially decreasing time functions. Also, (9.27) shows that $\boldsymbol{x}(t)M\boldsymbol{x}(t) > 0$ for $\boldsymbol{x}(t) \neq \boldsymbol{0}$. We find that

$$\dot{V}(\boldsymbol{x}(t)) = \boldsymbol{x}^T(t)[A^T M + MA]\boldsymbol{x}(t) \tag{9.30}$$

as before. But by taking the derivative of

$$V(\boldsymbol{x}(t)) = \int_t^\infty \boldsymbol{x}^T(\tau)Q\boldsymbol{x}(\tau)d\tau \tag{9.31}$$

we see that

$$\dot{V}(\boldsymbol{x}(t)) = -\boldsymbol{x}^T(t)Q\boldsymbol{x}(t). \tag{9.32}$$

Thus we have shown that when the system is AS the matrix M which solves $-Q = A^T M + MA$ for any positive definite Q is symmetric and positive definite. This can be summarized as

Theorem 9.8 *The equilibrium point at the origin of a fixed continuous-time system $\dot{\boldsymbol{x}}(t) = A\boldsymbol{x}(t)$ is AS if and only if for any symmetric positive definite Q, there exists a positive definite matrix M which satisfies the Lyapunov equation (9.25).*

Note: When Q is only positive semidefinite but $\boldsymbol{x}^T(\tau)Q\boldsymbol{x}(\tau)$ is not identically zero along a solution of $\dot{\boldsymbol{x}} = A\boldsymbol{x}$, then by Theorem 9.7 the equilibrium point at the origin is AS. On the other hand, when this is true and the equilibrium point is AS then the function

$$V(\boldsymbol{x}(t)) = \int_t^\infty \boldsymbol{x}(\tau)^T Q\boldsymbol{x}(\tau)d\tau = \boldsymbol{x}^T(t)M\boldsymbol{x}(t)$$

is positive definite, because the integrand cannot be zero along a solution, so it must become positive for some time interval; thus $V(\boldsymbol{x}(t) > 0$ for any $\boldsymbol{x}(t) \neq \boldsymbol{0}$. We have the corollary:

Corollary to Theorem 9.8 *The equilibrium point at zero is AS if and only if for any symmetric positive semidefinite matrix Q there exists a positive definite M satisfying (9.25) and $\dot{V} = -\boldsymbol{x}^T(t)Q\boldsymbol{x}(t)$ is not identically zero along a trajectory.*

Discrete-time systems

Analogous results for discrete-time systems may be obtained in a similar fashion and stated as follows:

Definition 9.8 *A real scalar function $V(\boldsymbol{x}, k)$*

a Which is continuous in $\boldsymbol{x}$,

b Which is positive definite in a region $\mathcal{R}$ containing the origin, *c Whose first forward difference satisfies*

$$\Delta V(\boldsymbol{x}, k) \triangleq V(\boldsymbol{x}(k+1), k+1) - V(\boldsymbol{x}(k), k) \leq 0 \tag{9.33}$$

(i.e., ΔV is negative semidefinite) is called a Lyapunov function for the discrete-time system

$$\boldsymbol{x}(k+1) = \boldsymbol{f}(\boldsymbol{x}(k), k) \tag{9.34}$$

which has an equilibrium point at $\mathbf{0}$.

Theorem 9.9 *If a Lyapunov function $V(\boldsymbol{x}, k)$ for the discrete system (9.34) exists in some neighborhood $\mathcal{R}$ of the origin then the equilibrium state at the origin is stable. If in addition $\Delta V(\boldsymbol{x}, k)$ is negative definite or if the set $\mathcal{X}$, of states for which $\Delta V(\boldsymbol{x}, k) = 0$, contains no solution of the system other than the trivial solution, then the equilibrium state is AS.*

Finally,

Theorem 9.10 *The equilibrium point at the origin of a fixed discrete-time linear system $\boldsymbol{x}(k+1) = A\boldsymbol{x}(k)$ is AS if and only if for any symmetric positive definite matrix Q, there exists a positive definite matrix M satisfying*

$$A^T M A - M = -Q. \tag{9.35}$$

9.7 Stability tests for fixed analog systems

We have seen that if $\det(sI - A) = 0$ has roots only in the LHP then the equilibrium point at the origin is AS, and in addition the system is BIBO stable. Thus we must be able to determine when a polynomial equation

$$D_n(s) = a_n s^n + a_{n-1} s^{n-1} + \cdots + a_1 s + a_0 = 0 \tag{9.36}$$

has all its roots in the LHP. Such a polynomial is called a Hurwitz polynomial. Since it is not practical to actually solve for the root locations of high-order polynomials, it is desirable to find a test which will tell us whether or not a polynomial is Hurwitz. This problem (called the Routh-Hurwitz problem because of the work of Hurwitz 1895 and Routh 1877) has been studied extensively since the late nineteenth century.

Theorem 9.7 could be used to give a test of the type we seek, but it requires that a set of $n(n+1)/2$ linear algebraic equations be solved. We will use this theorem to derive some simpler stability tests.

9.7.1 The continued fraction test

Step 1 Form a rational function $G_0(s)$ (i.e., a ratio of polynomials) from $D_n(s)$ as shown.

$$G_0(s) = \frac{P_1(s)}{P_2(s)} = \begin{cases} \dfrac{D_n(s) + D_n(-s)}{D_n(s) - D_n(-s)} = \dfrac{\text{even part of } D_n(s)}{\text{odd part of } D_n(s)} & \text{if } n \text{ is even} \\ \dfrac{D_n(s) - D_n(-s)}{D_n(s) + D_n(-s)} = \dfrac{\text{odd part of } D_n(s)}{\text{even part of } D_n(s)} & \text{if } n \text{ is odd.} \end{cases} \tag{9.37}$$

Note: The numerator polynomial $P_1(s)$ is one degree higher than the denominator polynomial $P_2(s)$ so that

$$G_0(s) = \frac{P_1(s)}{P_2(s)} = \frac{a_n s^n + a_{n-2}s^{n-2} + \cdots}{a_{n-1}s^{n-1} + a_{n-3}s^{n-3} + \cdots}. \tag{9.38}$$

Step 2 Divide the denominator $P_2(s)$ of $G_0(s)$ into the numerator $P_1(s)$ once and call the remainder polynomial $P_3(s)$. This gives

$$G_0(s) = \alpha_1 s + \frac{P_3(s)}{P_2(s)} \triangleq \alpha_1 s + 1/G_1(s) \tag{9.39}$$

where $\alpha_1 = a_n/a_{n-1}$ and $G_1(s) = P_2(s)/P_3(s)$ is a rational function with numerator $P_2(s)$ of degree $(n-1)$ and denominator $P_3(s)$ of degree $(n-2)$.

Step 3 Divide the denominator $P_3(s)$ of $G_1(s)$ into the numerator $P_2(s)$ once and call the remainder polynomial $P_4(s)$. This gives

$$G_1(s) = \alpha_2 s + \frac{P_4(s)}{P_3(s)} \triangleq \alpha_2 s + 1/G_2(s) \tag{9.40}$$

where $G_2(s) = P_3(s)/P_4(s)$ is a rational function with numerator $P_3(s)$ of degree $(n-2)$ and denominator $P_4(s)$ of degree $(n-3)$. Note that α_2 is not equal to a_{n-1}/a_{n-2}.

Step 4 Continue in this way to get the rational functions $G_i(s)$ and the α_{i+1} for i up to $(n-1)$. The process stops with $G_{n-1}(s) = \alpha_n s$. This procedure gives what is called the *continued fraction* expansion of $G_0(s)$.

$$G_0(s) = \alpha_1 s + \cfrac{1}{\alpha_2 s + \cfrac{1}{\alpha_3 s + \cfrac{1}{\ddots + \cfrac{1}{\alpha_{n-1}s + 1/\alpha_n s}}}}. \tag{9.41}$$

Note: If at any point we find that a numerator polynomial P_i has degree greater than one plus the degree of the denominator polynomial P_{i+1} this test terminates and the polynomial $D_n(s)$ is not Hurwitz.

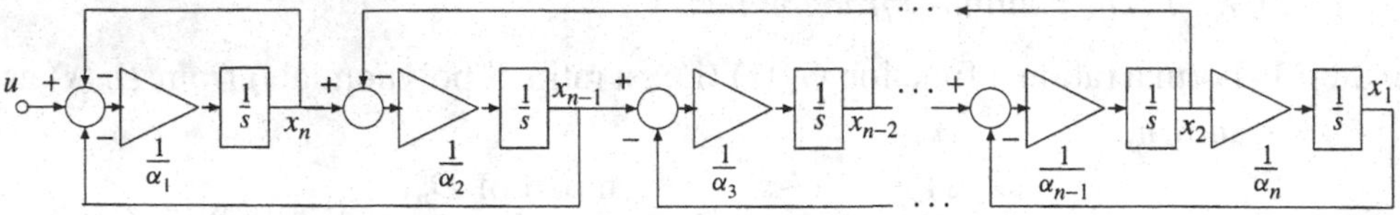

Fig. 9.3. System for stability proof

We can prove the following theorem.

Theorem 9.11 Continued fraction theorem *$D_n(s)$ is a Hurwitz polynomial if and only if $\alpha_i > 0$ for all $i = 1 \ldots n$.*

Proof: Consider the system of Figure 9.3 when none of the α_i are zero. This limitation is due to this method of proof (Parks 1962). Agashe's theorem, proved later in this chapter, would give the same result.

The transfer function of the system in Figure 9.3 from $U(s)$ to $X_n(s)$ is

$$H(s) \triangleq \frac{X_n(s)}{U(s)} = \frac{1/\alpha_1 s}{1 + (1/\alpha_1 s) + (H_1(s)/\alpha_1 s)} = \frac{1}{1 + \alpha_1 s + H_1(s)} \tag{9.42}$$

where

$$H_1(s) \triangleq \frac{X_{n-1}(s)}{X_n(s)} = \frac{1/\alpha_2 s}{1 + (H_2(s)/\alpha_2 s)} = \frac{1}{\alpha_2 s + H_2(s)}. \tag{9.43}$$

This operation can be continued so that in general

$$H_k(s) \triangleq \frac{X_{n-k}(s)}{X_{n+1-k}(s)} = \frac{1/\alpha_{k+1} s}{1 + (H_{k+1}(s)/\alpha_{k+1} s)} = \frac{1}{\alpha_{k+1} s + H_{k+1}(s)} \quad \text{for } 1 < k \le n-1 \tag{9.44}$$

but $H_n(s) = 1/(\alpha_n s)$. Thus we see that the transfer function $H(s)$ has the form

$$H(s) = \cfrac{1}{1 + \alpha_1 s + \cfrac{1}{\alpha_2 s + \cfrac{1}{\alpha_3 s + \cfrac{1}{\ddots + \cfrac{1}{\alpha_{n-1} s + 1/\alpha_n s}}}}} = \frac{1}{1 + G_0(s)}. \tag{9.45}$$

Using (9.37), we see that for n even

$$H(s) = \frac{D_n(s) - D_n(-s)}{2D_n(s)} \tag{9.46}$$

while for n odd

$$H(s) = \frac{D_n(s) + D_n(-s)}{2D_n(s)}. \tag{9.47}$$

In either case we see that if this system is AS then $D_n(s)$ is Hurwitz, and since it can be shown that the system is observable and controllable it can be AS only if $D_n(s)$ is Hurwitz. The unforced state equations of the system are

$$\dot{x} = \begin{bmatrix} 0 & 1/\alpha_n & 0 & \cdots & 0 \\ -1/\alpha_{n-1} & 0 & 1/\alpha_{n-1} & \cdots & 0 \\ 0 & -1/\alpha_{n-2} & 0 & \ddots & 0 \\ \cdots & \cdots & \cdots & \cdots & \cdots \\ 0 & \cdots & -1/\alpha_2 & 0 & -1/\alpha_2 \\ 0 & \cdots & 0 & -1/\alpha_1 & -1/\alpha_1 \end{bmatrix} x \triangleq Ax. \tag{9.48}$$

Let us define the square matrix M to be

$$M \triangleq \begin{bmatrix} \alpha_n & 0 & \cdots & 0 \\ 0 & \alpha_{n-1} & \cdots & 0 \\ \vdots & \vdots & \ddots & \vdots \\ 0 & 0 & \cdots & \alpha_1 \end{bmatrix} \tag{9.49}$$

so we find that

$$MA = \begin{bmatrix} 0 & 1 & 0 & 0 & \ldots & 0 \\ -1 & 0 & 1 & 0 & \ldots & 0 \\ 0 & -1 & 0 & 1 & \ddots & 0 \\ 0 & 0 & -1 & 0 & \ddots & 0 \\ \vdots & \ddots & \ddots & \ddots & \ddots & \vdots \\ 0 & 0 & \ldots & \ldots & -1 & -1 \end{bmatrix} \tag{9.50}$$

which gives

$$A^T M + MA = (MA)^T + MA = -Q \triangleq \begin{bmatrix} 0 & 0 & \cdots & 0 \\ 0 & 0 & \cdots & 0 \\ \vdots & \vdots & \ddots & \vdots \\ 0 & 0 & \cdots & -2 \end{bmatrix}. \tag{9.51}$$

Thus if all the $\alpha_1, \alpha_2, \ldots, \alpha_n$ are positive, we can choose $V(\boldsymbol{x}) = \boldsymbol{x}^T M \boldsymbol{x}$ as a Lyapunov function and get

$$\dot{V} = -\boldsymbol{x}^T (A^T M + M A)\boldsymbol{x} = -\boldsymbol{x}^T Q \boldsymbol{x} = -2x_n^2 \tag{9.52}$$

which is negative semidefinite.

We next show that it is not possible for $x_n(t) \equiv 0$ along a solution of $\dot{\boldsymbol{x}}(t) = A\boldsymbol{x}(t)$ unless all $x_i(t) \equiv 0$, so the system is AS. To do this, note that $\dot{V}(\boldsymbol{x}) \equiv 0$ if and only if $x_n(t) \equiv 0$, but from (9.48) we see that $\dot{x}_n(t) \equiv 0$ if and only if $x_{n-1}(t) \equiv 0$ also. Continuing in this way we see that $\dot{x}_{n-1}(t) \equiv 0$ if and only if $x_{n-2}(t) \equiv 0$ also. Repeat this argument until we show that $x_1(t) \equiv 0$ must also hold. Thus only the trivial solution $\boldsymbol{x}(t) = 0$ makes $\dot{V}(\boldsymbol{x}) \equiv 0$. We have thus shown that $\alpha_i > 0$ for all i is sufficient for AS.

To show that this condition is also necessary for AS, we use the corollary to Theorem 9.8 and a result from matrix theory which says that if a matrix A has no zero or equal and opposite roots then the Lyapunov equation (9.25) has a unique solution M for any Q. We choose Q as that of (9.51), and thus for the matrix A of (9.48) with all its roots in the LHP we know that (9.51) has a unique solution for M. Since the M of (9.49) is a solution it must be the only one, and the corollary to Theorem 9.8 tells us that M must be positive definite. Thus we see that all $\alpha_i > 0$. QED.

9.7.2 The Routh test

Much of the writing of the continued fraction algorithm can be eliminated by proper organization of the computations (Routh 1877). In forming the continued fraction expansion of $G_0(s)$ in (9.38) we divide the denominator $P_2(s)$ into the numerator $P_1(s)$ giving

$$\begin{array}{r|l}
 & \dfrac{a_n}{a_{n-1}} s = d_1 s \\
\hline
a_{n-1}s^{n-1} + a_{n-3}s^{n-3} + a_{n-5}s^{n-5} + \cdots & a_n s^n + \quad a_{n-2}s^{n-2} + \quad a_{n-4}s^{n-4} + \cdots \\
 & s^n + a_{n-3}\alpha_1 s^{n-2} + a_{n-5}\alpha_1 s^{n-4} + \cdots \\
\cline{2-2}
 & P_3(s) \longrightarrow \; b_1 s^{n-2} + \quad b_2 s^{n-4} + \cdots
\end{array}$$

where

$$b_1 = a_{n-2} - a_{n-3}\alpha_1 = \frac{a_{n-1}a_{n-2} - a_{n-3}}{a_{n-1}}$$

$$b_2 = a_{n-4} - a_{n-5}\alpha_1 = \frac{a_{n-1}a_{n-4} - a_{n-5}}{a_{n-1}}.$$

$$\ldots\ldots\ldots\ldots .$$

Similarly, dividing the denominator $P_3(s)$ of $G_1(s)$ into the numerator $P_2(s)$ gives

$$\begin{array}{r|l} & \frac{a_{n-1}}{b_1}s = \alpha_2 s \\ \hline b_1 s^{n-2} + b_2 s^{n-4} + b_3 s^{n-6} + \cdots & a_{n-1}s^{n-1} + a_{n-3}s^{n-3} + a_{n-5}s^{n-5} + \cdots \\ & a_{n-1}s^{n-1} + b_2\alpha_2 s^{n-3} + b_3\alpha_2 s^{n-5} + \cdots \\ \hline & P_4(s) \longrightarrow \quad c_1 s^{n-3} + \quad c_2 s^{n-5} + \cdots \end{array}$$

where

$$c_1 = a_{n-3} - b_2\alpha_2 = \frac{b_1 a_{n-3} - b_2 a_{n-1}}{b_1}$$

$$c_2 = a_{n-5} - b_3\alpha_2 = \frac{b_1 a_{n-5} - b_3 a_{n-1}}{b_1}.$$

$$\cdots\cdots\cdots\cdots .$$

These calculations are carried out until the n values for α_i are obtained. Actually it is not necessary to write out the long divisions since these results can be obtained more easily by forming the Routh array shown below.

$$\begin{array}{lccccl} \textit{row } 1 & a_n & a_{n-2} & a_{n-4} & a_{n-6} & \cdots \longleftarrow \text{coefficients of } P_1(s) \\ \textit{row } 2 & a_{n-1} & a_{n-3} & a_{n-5} & a_{n-7} & \cdots \longleftarrow \text{coefficients of } P_2(s) \\ \textit{row } 3 & b_1 & b_2 & b_3 & b_4 & \cdots \longleftarrow \text{coefficients of } P_3(s) \\ \textit{row } 4 & c_1 & c_2 & c_3 & c_4 & \cdots \longleftarrow \text{coefficients of } P_4(s) \\ \textit{row } 5 & d_1 & d_2 & d_3 & d_4 & \cdots \longleftarrow \text{coefficients of } P_5(s) \\ \vdots & \vdots & \vdots & \vdots & \vdots & \vdots \\ \textit{row } n & v_1 & 0 & & & \\ \textit{row } (n+1) & w_1 & 0. & & & \end{array} \tag{9.53}$$

The first two rows of the Routh array are formed from the coefficients of the polynomial $D_n(s)$ to be tested. The first row contains the coefficients of the dividend (numerator) polynomial $P_1(s)$ and the second row contains the coefficients of the divisor (denominator) polynomial $P_2(s)$. If we denote the entry in the ith row and jth column of the array by r_{ij}, then the array is formed by using

$$r_{ij} = \frac{r_{(i-1)1}r_{(i-2)(j+1)} - r_{(i-2)1}r_{(i-1)(j+1)}}{r_{(i-1)1}}. \tag{9.54}$$

A blank entry is considered to be zero.

Note that the third row of the array contains the coefficients of the polynomial $P_3(s)$, the fourth row contains the coefficients of $P_4(s)$, and so forth. Thus the calculations (9.54) used to produce the ith row of the array actually generate the coefficients of the remainder polynomial $P_i(s)$ resulting from the division of $P_{(i-1)}(s)$ into $P_{(i-2)}(s)$.

The α_i elements are found to be

$$\begin{aligned}
\alpha_1 &= \frac{r_{11}}{r_{12}} = \frac{a_n}{a_{n-1}} \\
\alpha_2 &= \frac{r_{12}}{r_{13}} = \frac{a_{n-1}}{b_1} \\
\alpha_3 &= \frac{r_{13}}{r_{14}} = \frac{b_1}{c_1} \\
&\ldots\ldots\ldots\ldots
\end{aligned} \tag{9.55}$$

Note that the α_i are all formed from the elements of the first column of the Routh array and that they are all greater than 0 if and only if all $r_{i1} > 0$ for $i = 1, \ldots, (n+1)$. This result can be expressed as

> **Theorem 9.12 The Routh theorem** (Routh 1877) *$D_n(s)$ is a Hurwitz polynomial if and only if all the elements in the first column of the Routh array are positive.*

It is important to note that when $a_n > 0$, neither the continued fraction theorem nor the Routh theorem conditions can be satisfied when any of the other coefficients a_i of $D_n(s)$ are less than or equal to zero. Thus a necessary (but not sufficient) condition for $D_n(s)$ to be a Hurwitz polynomial is that all a_i have the same sign.

Example 9.4

Consider the polynomial $D_4(s) = s^4 + 4s^3 + 7s^2 + 6s + 3$. If some coefficients were not positive we could say immediately that the polynomial was not Hurwitz, but since all the coefficients are positive we must form the Routh array, which becomes

$$\begin{array}{|ccc}
1 & 7 & 3 \\
4 & 6 & 0 \\
11/2 & 3 & \\
42/11 & 0 & \\
3. & &
\end{array}$$

We see that all five terms in the first row are positive, so $D_4(s)$ is a Hurwitz polynomial.

Example 9.5

Consider the polynomial $D_4(s) = s^4 + 3s^3 + 2s + 1$. Here we see that the coefficient of the s^2 term is zero, so $D_4(s)$ cannot be Hurwitz. We do not need to form the

Routh array, but if we did, we would find that it is

$$\begin{array}{|lll} 1 & 0 & 1 \\ 3 & 2 & 0 \\ -2/3 & 1 & \\ 13/2 & 0 & \\ 1\,. & & \end{array}$$

We see that there is a negative term in the first column, which confirms the conclusion that $D_4(s)$ is not Hurwitz; in fact $D_4(s)$ has two zeros in the RHP.

Example 9.6

Suppose we wish to determine the range over which the parameter K can vary and still have $D_4(s) = s^4 + 3s^3 + Ks^2 + 2s + 1$ a Hurwitz polynomial.

$$\begin{array}{|lll} 1 & K & 1 \\ 3 & 2 & \\ \dfrac{3K-2}{3} & 1 & \\ \dfrac{6K-13}{3K-2} & & \\ 1\,. & & \end{array}$$

We see that when $K > 13/6$ then all the entries in the first column are positive, so $D_4(s)$ has all its zeros in the LHP.

Comments on the Routh test

One should be aware that the Routh array actually shows how many of the zeros of $D_n(s)$ in (9.36) lie in the RHP. Specifically, the number of RHP zeros of $D(s)$ equals the number of changes of sign which occur in the first column of the Routh array. When all the elements are positive, then there are no sign changes, so there are no zeros of $D_n(s)$ in the RHP.

We have seen that any real nth-degree polynomial $D_n(s)$ can be expressed in terms of its even and odd parts $D_n(s) = P_1(s) + P_2(s)$, and from these the Routh procedure essentially generates a sequence of polynomials $P_3(s),\ P_4(s), \ldots,\ P_{n+1}(s)$ where $P_{n+1}(s)$ is a constant. The $(n+1)$ rows of the Routh array are formed from the coefficients of the alternately even and odd polynomials. Let us denote the Routh array for $D_n(s)$ by the symbol RA_n. It is obvious that the last n rows of RA_n form a Routh array denoted by RA_{n-1} for the $n-1$ degree polynomial $D_{n-1}(s) = P_2(s) + P_3(s)$. Therefore, if m_n is the number of sign changes in the first column of RA_n and m_{n-1} is the number of sign changes in the first column of RA_{n-1} we see that

$$\begin{aligned} m_n &= m_{n-1} \quad \text{if there is no sign change between row 1 and 2 of } \mathrm{RA}_n \\ m_n &= m_{n-1} + 1 \quad \text{if there is a sign change.} \end{aligned}$$

Thus the Routh procedure can be viewed as repeatedly determining the difference between the number of RHP zeros of a polynomial $D_i(s)$ and those of the polynomial $D_{i-1}(s)$ derived from it for $i = n, n-1, \ldots, 1$ where

$$D_i(s) = P_{n-i+1}(s) + P_{n-i+2}(s) \quad \text{for} \quad i = n, n-1, \ldots, 1$$
$$D_0(s) = P_{n+1}.$$

Note that P_{n+1} is a constant (a zero-degree polynomial) so it has no zeros at all. When the entire first column of RA_n has no zero elements, then the procedure is straightforward and the number of RHP zeros of $D_n(s)$ is the total number of sign changes. What do we do when a zero term appears in the first column? We cannot proceed in the normal way because when a zero appears in the first column of any row, a division by zero would be required in order to produce the next row. A common approach to this dilemma is to replace the zero by ϵ and to proceed normally. At the end let ϵ approach zero and count the number of sign changes. Unfortunately, this does not always give the same result for positive and negative ϵ as is illustrated by the following example from Gantmacher (1959).

Example 9.7

Consider the polynomial $D_6(s) = s^6 + s^5 + 3s^4 + 3s^3 + 3s^2 + 2s + 1$. The first three rows of the Routh array are

$$\begin{vmatrix} 1 & 3 & 3 & 1 \\ 1 & 3 & 2 & \\ 0 & 1 & 1. & \end{vmatrix}$$

Let us replace the 0 in the first column of the third row with a small ϵ. The complete Routh array is then

$$\begin{vmatrix} 1 & 3 & 3 & 1 \\ 1 & 3 & 2 & \\ \epsilon & 1 & 1 & \\ 3 - \frac{1}{\epsilon} & 2 - \frac{1}{\epsilon} & & \\ x & 1 & & \\ y & 1 & & \\ 1 & & & \end{vmatrix}$$

where

$$x = 1 - \frac{2\epsilon - 1}{3 - (1/\epsilon)} = 1 - \frac{1 - 2\epsilon}{1 - 3\epsilon}\epsilon$$

and

$$y = \frac{(2 - (1/\epsilon))(1 - 4\epsilon + 2\epsilon^2) - (3 - (1/\epsilon))(1 - 3\epsilon)}{1 - 4\epsilon + 2\epsilon^2} = -\frac{1 - 4\epsilon}{1 - 4\epsilon + 2\epsilon^2}\epsilon.$$

We see that $x \to 1$ and $y \to -\epsilon$ as $\epsilon \to 0$ for ϵ both positive and negative. Thus, for ϵ positive the first position of rows 1, 2, 3, 5, and 7 is positive, while in rows 4 and 6 it is negative, so there are 4 sign changes. However, when ϵ is negative then the first position of row 3 is negative, while in all the other rows it is positive, so there are only 2 sign changes. In this case the Routh test does not show how many RHP zeros $D_6(s)$ has.

9.7.3 Agashe's modification of the Routh test

Instead of replacing a zero in the first column of the Routh array, Agashe (1985) suggests an alternate procedure. This involves modifying the way the array and the $P_i(s)$ polynomials are formed. Let us illustrate the procedure by an example.

Example 9.8

For the polynomial $D_5(s) = 2s^5 + 4s^3 + s^2 + 5s + 3$ we can immediately note that the s^4 term is missing, so $D_5(s)$ is not Hurwitz. If we try to determine how many RHP zeros there are, we can start to form the Routh array for $D_5(s)$. This results in

$$\begin{vmatrix} 2 & 4 & 5 \\ 0 & 1 & 3 \end{vmatrix}.$$

Because the second row begins with zero, we rewrite it as shown below. We still want to generate the polynomial remainder of the division of $P_1(s)$ by $P_2(s)$, but we want the remainder which is one degree less than the divisor $P_2(s)$.

$$\begin{array}{|ccc l}
2 & 4 & 5 & \longleftarrow P_1(s) = 2s^5 + 4s^3 + 5s \text{ of degree } 5 \\
1 & 3 & & \longleftarrow P_2(s) = s^2 + 3 \text{ of degree } 2 \\
-2 & 5 & & \longleftarrow \hat{P}_2(s) = -2s^3 + 5s \text{ of degree } 3 \\
1 & 3 & & \longleftarrow \text{divide again by } P_2(s) = s^2 + 3 \text{ of degree } 2 \\
11 & & & \longleftarrow P_3(s) = 11s \text{ of degree } 1 \\
3 & & & \longleftarrow P_4 = 3
\end{array}.$$

Note that the third row corresponds to a third-degree polynomial, so we must divide again by $P_2(s)$. This division yields the first-degree remainder $P_3(s)$. The last row is formed in the usual way. We could now cross out the third and fourth rows, since we are only interested in the four rows corresponding to $P_1(s)$ to $P_4(s)$.

Note, that in going from $D_5(s) = P_1(s) + P_2(s)$ to $D^*(s) = P_2(s) + P_3(s)$ the difference in degree is 3 not 1. Therefore, a new, more complicated rule must be applied to determine the difference in the number of their RHP zeros. This is given by the following generalization of the Routh theorem by Agashe. In his paper Agashe (1985) gives a theorem and proof for general polynomials with complex coefficients.

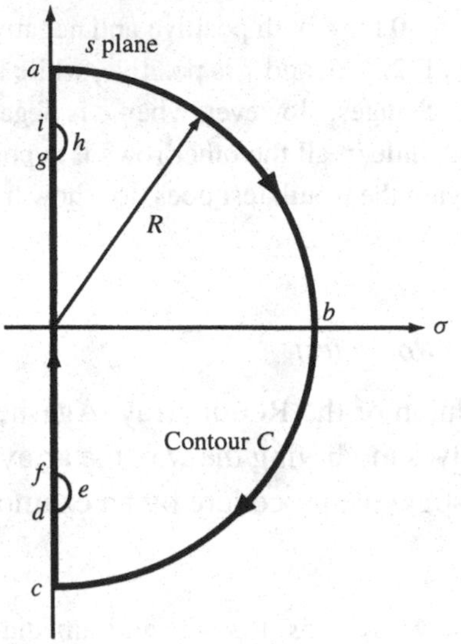

Fig. 9.4. Nyquist contour C

Here they are presented in a simplified form, since we are only interested in real polynomials.

Theorem 9.13 Agashe theorem (for real polynomials) *Let β denote the lead coefficient, d the degree, and m the number of RHP zeros of $D(s) = P_i(s) + P_{i+1}(s)$. Let β^* denote the lead coefficient, d^* the degree, and m^* the number of RHP zeros of $D^*(s) = P_{i+1}(s) + P_{i+2}(s)$. Let $\alpha = \beta/\beta^*$ and express $(d - d^*) = 4k + j$ where $k = 0, 1, 2, 3, \ldots$ and $j = 1$ or 3. (Note that since $D_i(s)$ is a real polynomial, $(d - d^*)$ is always odd.) The difference in the number of RHP zeros is given by*

$$m - m^* = \begin{cases} 2k & \text{if } j = 1 \text{ and } \alpha > 0 \\ 2k + 2 & \text{if } j = 3 \text{ and } \alpha > 0 \\ 2k + 1 & \text{if } \alpha < 0. \end{cases}$$

Proof: Consider the rational function $F(s) = D(s)/D^*(s)$ and a contour C in the s-plane (as shown in Figure 9.4) which encircles the right half-plane once in the clockwise direction while avoiding all zeros of $D(s)$ and $D^*(s)$ on the $j\omega$-axis via small semicircles of radius ρ.

This is the usual Nyquist contour, which for sufficiently large radius R and sufficiently small radius ρ encloses all the zeros of $D(s)$ and $D^*(s)$ that lie in the right half of the s-plane. By Cauchy's Principle of the Argument we know that $(m - m^*)$, the difference between the number of zeros of $D(s)$ and $D^*(s)$ inside C, equals the total number of clockwise encirclements of the origin of the $F(s)$-plane by the image $F(C)$ of the contour C. We also observe that in order for $F(C)$ to encircle the origin it must cross the negative real axis of the $F(s)$-plane and the number of clockwise encirclements equals the net number of clockwise crossings N (count $+1$ for a crossing from below to above the axis, and -1 for a crossing from above to below). Note that there are three kinds of segments that make up the contour C. Segments of C on the $j\omega$-axis such as cd, fg, ia in Figure 9.4 (we call the total number of negative real axis crossings of $F(C)$ due to all of this type of segment, N_ω), segments of C on small semicircles such as def, ghi in Figure 9.4 (we call the total number of negative real axis crossings due to all of this type of segment, N_ρ), and one segment of C consisting of the large semicircle abc (we call the total number of negative real axis crossings due to this N_R), so $N = N_\omega + N_\rho + N_R$. We next investigate each type of segment separately.

1 Imaginary axis segments such as fg

First note that in forming $D^*(s)$ from $D(s)$ we had $D(s) = P_i(s) + P_{i+1}(s)$ where $P_i(s)$ is even and $P_{i+1}(s)$ is odd or vice versa. Also note that $p_i =$ the degree of $P_i(s) > p_{i+1}$, so when $P_i(s)$ is divided by $P_{i+1}(s)$ we get

$$\frac{P_i(s)}{P_{i+1}(s)} = Q(s) + \frac{P_{i+2}(s)}{P_{i+1}(s)} \tag{9.56}$$

where $Q(s)$ is always an odd polynomial and $P_{i+2}(s)$ is even when $P_{i+1}(s)$ is odd or vice versa. We then formed $D^*(s) = P_{i+1}(s) + P_{i+2}(s)$.

Note: When p_i and p_{i+1} differ by 1, then $Q(s) = ks$ and p_{i+2} is one less than p_{i+1}. This occurred in (9.39) and (9.40) of the continued fraction expansion. We see from (9.56) that

$$P_i(s) = Q(s)P_{i+1}(s) + P_{i+2}(s) \tag{9.57}$$

so

$$\begin{aligned} F(s) = \frac{D(s)}{D^*(s)} &= \frac{Q(s)P_{i+1}(s) + P_{i+2}(s) + P_{i+1}(s)}{P_{i+1}(s) + P_{i+2}(s)} \\ &= 1 + \frac{Q(s)P_{i+1}(s)}{P_{i+1}(s) + P_{i+2}(s)}. \end{aligned} \tag{9.58}$$

On the imaginary axis $s = j\omega$, so we get

$$F(j\omega) = 1 + \frac{Q(j\omega)P_{i+1}(j\omega)}{P_{i+1}(j\omega) + P_{i+2}(j\omega)}. \tag{9.59}$$

Now $Q(s)$ is an odd function of s, so $Q(j\omega) = jq(\omega)$ where $q(\omega)$ is a real function of ω. When $P_{i+1}(s)$ is odd then $P_{i+2}(s)$ is even, so $P_{i+1}(j\omega) =$

$j\gamma(\omega)$ and $P_{i+2}(j\omega) = \eta(\omega)$, where $\gamma(\omega)$ and $\eta(\omega)$ are real functions of ω. Therefore we find that

$$\begin{aligned} F(j\omega) &= 1 - \frac{q(j\omega)\gamma(j\omega)}{\eta(s) + j\gamma(s)} \times \frac{\eta(\omega) - j\gamma(\omega)}{\eta(\omega) - j\gamma(\omega)} \\ &= 1 - \frac{q(\omega)\gamma(\omega)\eta(\omega)}{\eta^2(\omega) + \gamma^2(\omega)} + j\frac{q(\omega)\gamma^2(\omega)}{\eta^2(\omega) + \gamma^2(\omega)}. \end{aligned} \tag{9.60}$$

When $P_{i+1}(s)$ is even then $P_{i+2}(s)$ is odd so $P_{i+1}(j\omega) = \gamma(\omega)$ and $P_{i+2}(j\omega) = j\eta(\omega)$. We find that we get exactly the same final equation as in (9.60) for $F(j\omega)$. We see from (9.60) that $F(j\omega)$ can be real only when $q(\omega)\gamma^2(\omega) = 0$. This makes $F(j\omega) = 1$ so $F(C)$ cannot cross the negative real axis for segments of C on the $j\omega$-axis. This shows that $N_\omega = 0$.

2 *Small semicircular segments such as ghi*
For these segments $s = j\omega_0 + \rho e^{j\phi}$, where $j\omega_0$ denotes the location of a zero of $D(s)$ or $D^*(s)$ on the imaginary axis, ϕ varies from $-\pi/2 \to 0 \to \pi/2$ and ρ is extremely small (i.e., $\rho \to 0$). Now when $D(j\omega_0) = P_i(j\omega_0) + P_{i+1}(j\omega_0) = 0$, this requires that both $P_i(j\omega_0) = 0$ and $P_{i+1}(j\omega_0) = 0$ since one is real and the other is imaginary and (9.57) shows that $P_{i+2}(j\omega_0) = 0$. When $D^*(j\omega_0) = P_{i+1}(j\omega_0) + P_{i+2}(j\omega_0) = 0$, both $P_{i+1}(j\omega_0) = 0$ and $P_{i+2}(j\omega_0) = 0$ for the same reason, and (9.57) shows that in this case $P_i(j\omega_0) = 0$. Therefore we see that $P_i(j\omega_0)$, $P_{i+1}(j\omega_0)$, and $P_{i+2}(j\omega_0)$ will all have zeros at $s = j\omega_0$. In fact $D(j\omega_0) = 0$ if and only if $D^*(j\omega_0) = 0$. When the smallest multiplicity factor $(s - j\omega_0)^k$ is divided out between $P_{i+1}(j\omega_0)$ and $P_{i+2}(j\omega_0)$, (9.58) becomes

$$F(s) = 1 + \frac{Q(s)\bar{P}_{i+1}(s)}{\bar{P}_{i+1}(s) + \bar{P}_{i+2}(s)}$$

where $\bar{P}_{i+1}(s)$ and $\bar{P}_{i+2}(s)$ are the polynomials remaining after canceling out the smallest multiplicity of the zero at $j\omega_0$. One is even, the other is odd, and both cannot be zero at $j\omega_0$, so $F(s)$ is continuous at $s = j\omega_0$. We showed above that when $F(j\omega_0)$ is real it must equal 1. Therefore as $\rho \to 0$, the continuity of $F(s)$ at $s = j\omega_0$ shows that $F(C) \approx F(j\omega_0)$ for segments of C on the small semicircle, so $F(C)$ also must equal 1 when it is real, and thus it cannot cross the negative real axis for segments of C on the small semicircle. This shows that $N_\rho = 0$.

3 The large semicircle abc

For this segment of C, $s = Re^{j\theta}$ and θ varies from $\pi/2 \to 0 \to -\pi/2$. Thus since R is very large, $D(s)$ and $D^*(s)$ can be approximated by their highest power terms which gives

$$F(C) \approx \alpha R^{(d-d^*)} e^{j(d-d^*)\theta}.$$

For $\alpha > 0$ we see that the argument of $F(C)$ is just $(d - d^*)\theta$, and since $d - d^*$ is odd, when $d - d^* = 1$ there is no crossing of the negative real axis, so $N_R = 0$. When $d - d^* = 3$, the argument of $F(C)$ is 3θ, so as θ varies from $\pi/2 \to 0 \to -\pi/2$, 3θ varies from $3\pi/2 \to \pi \to 0 \to -\pi \to -3\pi/2$. We see that the negative real axis is crossed twice in the clockwise direction (at π and $-\pi$), so $N_R = 2$. If we continue in this way, we find that $N_R = 2$ again for $d - d^* = 5$, $N_R = 4$ for $d - d^* = 7$ and 9, and in general for $d - d^* = 4k + j$, $N_R = 2k$ when $j = 1$ and $N_R = 2k + 2$ when $j = 3$.

For $\alpha < 0$ we replace the minus sign with $e^{j\pi}$ so

$$F(C) \approx |\alpha| R^{(d-d^*)} e^{j[(d-d^*)\theta+\pi]}$$

and the argument of $F(C)$ is just $(d-d^*)\theta+\pi$. When $d-d^* = 1$, as θ varies from $\pi/2 \to 0 \to -\pi/2$, $\theta + \pi$ varies from $3\pi/2 \to \pi \to \pi/2$, so there is one crossing of the negative real axis and $N_R = 1$. When $d - d^* = 3$, the argument of $F(C)$ is $3\theta + \pi$, so as θ varies from $\pi/2 \to 0 \to -\pi/2$, $3\theta + \pi$ varies from $5\pi/2 \to 2\pi \to \pi \to 0 \to -\pi/2$. We see that the negative real axis is crossed once in the clockwise direction (at π), so again $N_R = 1$. In the same way we find that $N_R = 3$ for $d - d^*$ =5 and 7 and $N_R = 5$ for $d - d^*$ =9 and 11. In general $N_R = 2k + 1$ for $d - d^* = 4k + j$ (it does not depend on j). We see therefore that $N = N_R$, since $N_\omega = N_\rho = 0$. This completes the proof. QED.

Applied to Example 9.8 we found that $d - d^* = 3$, so $k = 0$, $j = 3$, and $\beta/\beta^* > 0$. Thus by Agashe's theorem $m - m^* = 2$. This means that $D_5(s) = P_1(s) + P_2(s)$ has 2 RHP zeros since $D_5^*(s) = P_2(s) + P_3(s) = s^2 + 11s + 3$ has none.

It can also happen that in forming the Routh array, an entire row will have only zero elements. This means that there is no remainder, so the $P_i(s)$ associated with the row above the zero row is a factor of the original polynomial $D_n(s)$. Since this $P_i(s)$ is either an even or an odd polynomial it is not Hurwitz, so the $D_n(s)$ is not Hurwitz either. To determine the number of RHP zeros we could divide $P_i(s)$ into $D_n(s)$ to get the other polynomial factor and test both factors separately. An alternative approach is to replace the zero row by the coefficients of a polynomial which is the derivative of $P_i(s)$. This is a well-known procedure, and its validity follows from the following property.

Theorem 9.14 *If a real polynomial $P(s)$ is either even or odd, then $P(s)$ and $\hat{P}(s) = P(s) + (dP(s)/ds)$ have the same number of RHP zeros.*

Agashe proves this result in the same way as the previous theorem. That is, he shows that $F(s) = (\hat{P}(s)/P(s)) = 1 + (P'(s)/P(s))$ (where $P'(s) = dP(s)/ds$) cannot cross the negative real axis for s on the curve C.

Example 9.9

$D_7(s) = s^7 + 3s^6 + 4s^5 + 2s^4 + 2s^3 + 6s^2 + 8s + 4$. We begin to form the Routh array:

$$\left|\begin{array}{llll} 1 & 4 & 2 & 8 \\ 3 & 2 & 6 & 4 \\ \frac{10}{3} & 0 & \frac{20}{3} & \\ 2 & 0 & 4 & \\ 0 & 0 & & \\ 8 & & & \\ 4 & & & \end{array}\right. \begin{array}{l} \longleftarrow P_1(s) = s^7 + 4s^5 + 2s^3 + 8s \\ \longleftarrow P_2(s) = 3s^6 + 2s^4 + 6s^2 + 4 \\ \longleftarrow P_3(s) = \frac{10}{3}s^5 + \frac{20}{3}s \\ \longleftarrow P_4(s) = 2s^4 + 4 \text{ a factor of } D(s) \\ \longleftarrow \text{identically zero remainder so replace with } P_4'(s) \\ \longleftarrow P_5(s) = P_4'(s) = 8s^3 \\ \longleftarrow P_6 = 4; \text{ note that the degree is reduced by 3 not 1} \end{array}$$

We see that the difference in degree between $D_3(s) = 8s^3 + 4$ and $D_3^*(s) = 4$ is $d - d^* = 3$, and since $\beta/\beta^* > 0$ there are 2 more RHP zeros in $D_3(s)$ than in $D_3^*(s)$ which has none. All the coefficients above the bottom two rows are positive, and the degree change is 1, so $D_7(s)$ has the same number of RHP zeros as $D_3(s)$. Thus $D_7(s)$ has two RHP zeros.

Example 9.10

Now let us apply Agashe's theorem to Gantmacher's Example 9.7, $D_6(s) = s^6 + s^5 + 3s^4 + 3s^3 + 3s^2 + 2s + 1$. The first two rows of the array are the same as before, but from the third row on, the array changes. It becomes

$$\left|\begin{array}{llll} 1 & 3 & 3 & 1 \\ 1 & 3 & 2 & \\ 1 & 1 & & \\ 2 & 2 & & \\ 1 & 1 & & \\ 0 & & & \\ 2 & & & \\ 1 & & & \end{array}\right. \begin{array}{l} \longleftarrow P_1(s) = s^6 + 3s^4 + 3s^2 + 1 \\ \longleftarrow P_2(s) = s^5 + 3s^3 + 2s \\ \longleftarrow P_3(s) = s^2 + 1; \text{ note that the degree is reduced by 3 not 1} \\ \longleftarrow \hat{P}_3(s) = 2s^3 + 2s; \text{ remainder has degree } 3 > 2, \text{ so divide again} \\ \longleftarrow \text{divide again by } P_3(s) = s^2 + 1 \\ \longleftarrow \text{identically zero remainder so replace with } P_3'(s) \\ \longleftarrow P_4(s) = P_3'(s) = 2s \\ \longleftarrow P_5 = 1 \end{array}$$

We see that $D_0 = P_5$ has no zeros in the RHP. In going up the array from P_5 to $P_4(s)$ to $P_3(s)$ the degree increases by one at each step and the sign of the first column remains positive. This says that $D_2(s) = P_3(s) + P_4(s) = s^2 + 2s + 1$ has no RHP zeros. However, since there was a row of zeros (a zero remainder) we know that the polynomial $P_3(s) = s^2 + 1$ is a factor of $D_6(s)$. Observe that $P_3(s)$ has zeros on the $j\omega$-axis at $s = \pm j$, so $D_6(s)$ does also. In going the next step up to $D_5 = P_2(s) + P_3(s) = s^5 + 3s^3 + s^2 + 2s + 1$ we see that the degree increases by 3. Applying Theorem 9.13 to $D_5(s)$ and $D_5^*(s) = D_2(s)$ we see that $d - d^* = 3$ so $k = 0$, $j = 3$, and $\beta/\beta^* > 0$, so by Agashe's theorem $m - m^* = 2$. This means that $D_5(s)$ has 2 RHP zeros since $D_5^*(s) = D_2(s) = s^2 + 2s + 1$ has none. Since $D_6(s)$ and $D_5(s)$ have the same number of RHP zeros (they differ in degree by 1 and their first columns have the same sign), we conclude that $D_6(s)$ has 2 zeros in the RHP. We saw that it has 2 zeros on the $j\omega$-axis, so it must have its remaining 2 zeros in the LHP. Agashe's theorem gives the correct number of RHP zeros for the example.

9.7.4 The Hurwitz test

Another stability test, due to Hurwitz (1895), which is equivalent to the preceding ones, is based on a square $n \times n$ matrix known as the Hurwitz matrix. It is formed from the coefficients of $D_n(s)$ in (9.36) as shown.

$$H = \begin{bmatrix} a_{n-1} & a_{n-3} & a_{n-5} & a_{n-7} & \cdots & 0 \\ a_n & a_{n-2} & a_{n-4} & a_{n-6} & \cdots & 0 \\ 0 & a_{n-1} & a_{n-3} & a_{n-5} & \cdots & 0 \\ 0 & a_n & a_{n-2} & a_{n-4} & \cdots & 0 \\ \cdots & \cdots & \cdots & \cdots & \cdots & \cdots \\ 0 & \cdots & a_5 & a_3 & a_1 & 0 \\ 0 & \cdots & a_6 & a_4 & a_2 & a_0 \end{bmatrix}. \tag{9.61}$$

Note that the second and first rows of H are formed like the first and second rows respectively of the Routh array, and are filled in on the right with zeros to give n columns. A new row i is formed by shifting row $i - 2$ one position to the right, dropping the rightmost element and filling in the column 1 position with a zero. This process is carried out for $i = 3, \ldots, n$.

Theorem 9.15 The Hurwitz theorem *$D_n(s)$ is a Hurwitz polynomial if and only if all principal minors of H are positive. That is,*

$$\Delta_1 = a_{n-1} > 0$$

$$\Delta_2 = \det \begin{bmatrix} a_{n-1} & a_{n-3} \\ a_n & a_{n-2} \end{bmatrix} > 0$$

$$\Delta_3 = \det \begin{bmatrix} a_{n-1} & a_{n-3} & a_{n-5} \\ a_n & a_{n-2} & a_{n-4} \\ 0 & a_{n-1} & a_{n-3} \end{bmatrix} > 0 \tag{9.62}$$

$$\cdots\cdots\cdots\cdots$$

$$\Delta_n = \det H > 0.$$

It turns out that many of the conditions (9.62) are redundant so that not all the principal minors need to be checked (Liénard and Chipart 1914). It is only necessary to check the even or odd minors. The result is stated as

Theorem 9.16 The Liénard-Chipart theorem *$D_n(s)$ is a Hurwitz polynomial if and only if the $\Delta_i > 0$ in (9.62) holds for all i odd or all i even, when $a_i > 0$ for all i.*

Example 9.11

For the $D_4(s)$ of Example 9.4 the Hurwitz matrix is

$$\begin{bmatrix} 4 & 6 & 0 & 0 \\ 1 & 7 & 3 & 0 \\ 0 & 4 & 6 & 0 \\ 0 & 1 & 7 & 3 \end{bmatrix}$$

thus $\Delta_1 = 4 > 0$ and

$$\Delta_3 = \det \begin{bmatrix} 4 & 6 & 0 \\ 1 & 7 & 3 \\ 0 & 4 & 6 \end{bmatrix} = 4(42 - 12) - 36 = 84 > 0$$

so as we found earlier $D_4(s)$ is Hurwitz.

9.8 Stability tests for fixed discrete systems

We have seen that for fixed discrete linear systems if $\det(zI - A) = 0$ has all its roots inside the unit circle, then the equilibrium point at zero is AS, and in this case AS guarantees BIBO stability. Thus the stability problem is to determine whether the polynomial equation

$$\mathcal{D}_n(z) = a_n z^n + a_{n-1} z^{n-1} + \cdots + a_1 z + a_0 = 0 \tag{9.63}$$

has all its roots inside the unit circle. The first work concerned with Routh-Hurwitz–like conditions to determine when all roots lie inside the unit circle was the Schur-Cohn stability test (Cohn 1922). It required forming a $2n \times 2n$ matrix somewhat like the Hurwitz matrix, but this approach is unnecessarily complicated to use, so it will not be covered here.

9.8.1 Mapped Routh-Hurwitz test for discrete systems

Another approach is to transform the inside of the unit circle in the z-plane into the left half of the s-plane by a suitable mapping and just use the Routh or Hurwitz tests.

For example, the bilinear transformation

$$s = \frac{z+1}{z-1} \quad \text{so} \quad z = \frac{s+1}{s-1} \tag{9.64}$$

transforms the inside of the unit circle in the z-plane into the left half of the s-plane and the outside of the unit circle into the right half of the s-plane, so in $\mathcal{D}_n(z)$ of (9.53) we replace z by $(s+1)/(s-1)$ to give

$$a_n\left(\frac{s+1}{s-1}\right)^n + a_{n-1}\left(\frac{s+1}{s-1}\right)^{n-1} + \cdots + a_1\left(\frac{s+1}{s-1}\right) + a_0 = 0 \tag{9.65}$$

which after multiplying by $(s-1)^n$ becomes

$$\hat{D}_n(s) = a_n(s+1)^n + a_{n-1}(s+1)^{n-1}(s-1) + \cdots$$
$$+ a_1(s+1)(s-1)^{n-1} + a_0(s-1)^n = 0.$$

We see that $\hat{D}_n(s)$ is an nth-degree polynomial in s,

$$\hat{D}_n(s) = b_n s^n + b_{n-1}s^{n-1} + \cdots + b_1 s + b_0. \tag{9.66}$$

Now apply the R-H tests to $\hat{D}_n(s)$. If $\hat{D}_n(s)$ has any zeros in the RHP or on the $j\omega$-axis then $\mathcal{D}_n(z)$ has a zero outside or on the unit circle. This is not a particularly good method to use since it requires far more work than the next methods discussed.

9.8.2 *Modified continued fraction test for discrete systems*

Jury (1961) presented a more direct algorithm which is analogous to the continued fraction test. For $\mathcal{D}_n(z)$ given by (9.63) define the reverse polynomial $\mathcal{F}_n(z)$ as

$$\mathcal{F}_n(z) = z^n\mathcal{D}_n(z^{-1}) = a_0 z^n + a_1 z^{n-1} + \cdots + a_{n-1}z + a_n.$$

The zeros of $\mathcal{F}_n(z)$ are the reciprocals of the zeros of $\mathcal{D}_n(z)$.

Procedure

1 Divide $\mathcal{F}_n(z)$ by $\mathcal{D}_n(z)$ starting with the highest power so

$$\frac{\mathcal{F}_n(z)}{\mathcal{D}_n(z)} = \alpha_1 + \frac{\mathcal{F}_{n-1}(z)}{\mathcal{D}_n(z)}$$

where $\mathcal{F}_{n-1}(z)$ is of order $(n-1)$ and $\alpha_1 = a_0/a_n$.

2 Form $\mathcal{D}_{n-1}(z) = z^{n-1}\mathcal{F}_{n-1}(z^{-1})$ and divide $\mathcal{F}_{n-1}(z)$ by $\mathcal{D}_{n-1}(z)$ to get

$$\frac{\mathcal{F}_{n-1}(z)}{\mathcal{D}_{n-1}(z)} = \alpha_2 + \frac{\mathcal{F}_{n-2}(z)}{\mathcal{D}_1(z)}.$$

3 Continue for $i = n-2, n-3, \ldots, 2$ by forming $\mathcal{D}_i(z) = z^i \mathcal{F}_i(z^{-1})$ and dividing so

$$\frac{\mathcal{F}_i(z)}{\mathcal{D}_i(z)} = \alpha_{n-i+1} + \frac{\mathcal{F}_{i-1}(z)}{\mathcal{D}_i(z)}$$

until $\alpha_1, \alpha_2, \ldots, \alpha_{n-1}$ are obtained.

Theorem 9.17 The Jury theorem *The necessary and sufficient conditions for all zeros of $\mathcal{D}_n(z)$ to be inside the unit circle in the z-plane are that the following conditions hold:*

a $\mathcal{D}_n(1) > 0$

b $\left.\begin{array}{ll}\mathcal{D}_n(-1) < 0 & \textit{for n odd}\\ \mathcal{D}_n(-1) > 0 & \textit{for n even}\end{array}\right\} = (-1)^n \mathcal{D}_n(-1) > 0$

c $|\alpha_i| < 1$ *for* $i = 1, 2, \ldots, n-1$.

Note: Checking conditions (a) and (b) first will eliminate the work of checking (c) if either (a) or (b) fails.

Example 9.12

Consider the polynomial $\mathcal{D}_3(z) = z^3 - 3z^2 + 2z + 1/2$. To determine whether any zeros of $\mathcal{D}_3(z)$ lie outside the unit circle, we first apply conditions (a) and (b) of the Jury theorem.

$\mathcal{D}_3(1) = 1 - 3 + 2 + 1/2 = 0.5 > 0$ which satisfies (a).

$\mathcal{D}_3(-1) = -1 - 3 - 2 + 1/2 = -5.5 < 0$ which satisfies (b).

Since (a) and (b) are satisfied we must next find α_1 and α_2. We find

$\mathcal{F}_3(z) = (1/2)z^3 + 2z^2 - 3z + 1.$

Dividing $\mathcal{F}_3(z)$ by $\mathcal{D}_3(z)$ gives

$$\begin{array}{r|l} & \frac{1}{2} = d_1 \\ \hline z^3 - 3z^2 + 2z + 1/2 & 1/2z^3 + \quad 2z^2 - 3z + 1 \\ & \underline{1/2z^3 - 3/2z^2 + \ z + 1/4} \\ & \qquad 7/2z^2 - 4z + 3/4 = \mathcal{F}_2(z)\,. \end{array}$$

From the remainder $\mathcal{F}_2(z)$ we get $\mathcal{D}_2(z) = 3/4z^2 - 4z + 7/2$. We now divide $\mathcal{F}_2(z)$ by $\mathcal{D}_2(z)$ to get

$$\begin{array}{r|l} & \frac{14}{3} = \alpha_2 \\ \hline 3/4z^2 - 4z + 7/2 & 7/2z^2 - \quad 4z + 3/4 \\ & \underline{7/2z^2 - 56/3z + 49/3} \end{array}$$

Since $|\alpha_2| > 1$ condition (c) is violated, so $\mathcal{D}_3(z)$ has some zeros outside the unit circle.

9.8.3 A tabular test for discrete systems

Actually, the calculations for this procedure can be simplified by using a table like the Routh array (Jury and Blanchard 1961). We saw that to divide one polynomial (in this case $\mathcal{F}_i(z)$) by another (in this case $\mathcal{D}_i(z)$), it is convenient to enter the coefficients in an array and then perform the calculations (9.54) to generate a new row. This new row gives the coefficients of the remainder polynomial (in this case $\mathcal{F}_{i-1}(z)$).

Since we always divide the $\mathcal{F}_i$ polynomial by the $\mathcal{D}_i$ polynomial, we form the first row of the array from the coefficients of $\mathcal{F}_n(z)$ and the second row from the coefficients of $\mathcal{D}_n(z)$. The third row is calculated from rows 1 and 2 just like for the Routh array, and the fourth row is just the third row written in reverse order. This is continued giving the array below.

$$\begin{array}{rcccccccc}
row\ 1 & a_0 & a_1 & a_2 & a_3 & \cdots & a_{n-1} & a_n \\
row\ 2 & a_n & a_{n-1} & a_{n-2} & a_{n-3} & \cdots & a_1 & a_0 \\
row\ 3 & b_0 & b_1 & b_2 & b_3 & \cdots & b_{n-1} & \\
row\ 4 & b_{n-1} & b_{n-2} & b_{n-3} & b_{n-4} & \cdots & b_0 & \\
row\ 5 & c_0 & c_1 & c_2 & c_3 & \cdots & & \\
row\ 6 & c_{n-2} & c_{n-3} & c_{n-4} & c_{n-5} & \cdots & & \\
\vdots & \vdots & \vdots & \vdots & \vdots & \vdots & & \\
row\ (2n-3) & v_0 & v_1 & v_2 & & & & \\
row\ (2n-2) & v_2 & v_1 & v_0 & & & &
\end{array} \quad . \qquad (9.67)$$

The entries of each odd-numbered row are obtained using the Routh calculations (9.54), whereas the following even-numbered row is obtained by writing the preceding odd-numbered row in reverse order.

The α_i values are just

$$\alpha_1 = \frac{a_0}{a_n}, \quad \alpha_2 = \frac{b_0}{b_{n-1}}, \quad \alpha_3 = \frac{c_0}{c_{n-2}}, \quad \ldots, \quad \alpha_{n-1} = \frac{v_0}{v_2}.$$

Therefore we only need to check that

$$\begin{aligned}
&|a_0| < |a_n| \\
&|b_0| < |b_{n-1}| \\
&|c_0| < |c_{n-2}| \\
&\vdots \\
&|v_0| < |v_2|.
\end{aligned}$$

Example 9.13

When we form this array for $\mathcal{D}_3(z)$ of Example 9.12 we get

$$\begin{vmatrix} 1/2 & 2 & -3 & 1 \\ 1 & -3 & 2 & 1/2 \\ 7/2 & -4 & 3/4 & \\ 3/4 & -4 & 7/2 & \end{vmatrix}.$$

We see that $|a_0| = 1/2 < 1 = |a_3|$ and $|b_0| = 7/2 > 3/4 = |b_2|$ so, as before, we conclude that $\mathcal{D}(z)$ has zeros outside the unit circle.

Let us note that actually the results do not change if a row and its reversed form are multiplied by any nonzero constant. Thus we can often simplify the table by eliminating fractional values. In the preceding example, we could have multiplied the first and second rows by 2 and the third and fourth rows by 4 to obtain exactly the same results.

Example 9.14

Consider $\mathcal{D}_5(z) = z^5 + 2z^4 + 0.5z^3$. This is of degree 5, but we can save considerable work by noticing that $z = 0$ is a zero of $\mathcal{D}_5(z)$. Thus by rewriting $\mathcal{D}_5(z) = (z^2 + 2z + 0.5)z^3$ we can test the second-degree polynomial $\hat{\mathcal{D}}_2(z) = z^2 + 2z + 0.5$ instead. We find that $\hat{\mathcal{D}}_2(1) = 3.5 > 0$ and $\hat{\mathcal{D}}_2(-1) = -0.5 < 0$ which violates the condition, so there is a zero outside the unit circle.

Problems

9.1 Prove that the norm examples of (9.3) satisfy the norm conditions of (9.2).

9.2 Show that
- a) For a vector $\boldsymbol{x}$, $\|\boldsymbol{x}\| < \infty$ if and only if each component $|x_i| < \infty$.
- b) For a matrix A, $\|A\| < \infty$ if and only if every element $|a_{ij}| < \infty$.
- c) For a matrix A, $\|A\|$ defined by (9.7) satisfies the norm conditions (9.2).
- d) For an $(m \times r)$ matrix B, $\|B\| \leq K \sum_{i=1}^{m} \sum_{j=1}^{r} |b_{ij}|$ for some $K < \infty$.

9.3 Prove part (b) of Theorem 9.1.

9.4 Prove that Theorem 9.1 also holds for discrete systems. (The continuous-time variable t must of course be replaced by the discrete-time variable k.)

9.5 For an $n \times m$ matrix A, use equation (9.7) to find an expression for $\|A\|$ using each norm of (9.3).

9.6 Prove Theorem 9.3.

9.7 Use the conditions of Theorem 9.1 to determine whether the equilibrium point at $\mathbf{0}$ is S, AS, or US for

a) $\dot{x}_1(t) = x_2(t)$

$$\dot{x}_2(t) = -\frac{1}{t+1}x_2(t) + u(t)$$

(*Note:* This is the system of Example 4.1.)

b) The system shown in Problem 4.7.

c) The system shown in Problem 4.15.

9.8 Use the conditions of Theorem 9.2 to determine whether the equilibrium point is S, AS, or US for

a) The system of Problem 4.5.

b) The systems of Problem 4.6.

c) The system of Problem 4.12.

9.9 Use Theorem 9.3 to determine the nature of the equilibrium point for

a) The system of Example 4.14.

b) The system of Problem 4.20.

c) The systems of Problem 7.1.

d) The system of Problem 7.6.

9.10 For each of the systems below, determine whether the system is BIBO stable.

a) $\dot{y}(t) = -ty(t) + u(t)$

b) $\dot{y}(t) = -\dfrac{1}{1+t^2}y(t) + u(t)$

c) The system

$$\dot{x}_1(t) = tx_2(t)$$

$$\dot{x}_2(t) = -x_2(t) + u(t)$$

$$y(t) = x_1(t)$$

d) The system of Problem 4.15

9.11 Determine whether the functions below are positive (or negative) definite or semidefinite or none of these.

a) $V(\boldsymbol{x}) = x_1^2 + 4x_1x_2 + x_2^2$

b) $V(\boldsymbol{x}) = x_1^2 - 2x_1x_2 + x_2^2$

c) $V(\boldsymbol{x}) = x_1^2 - x_1x_2 + x_2^2$

d) $V(\boldsymbol{x}) = -x_1^2 - 2x_1x_2 - x_2^2$
e) $V(\boldsymbol{x}) = -2x_1^2 + 2x_1x_2 - 2x_2^2$

9.12 For the system

$$\dot{x}_1 = x_2$$
$$\dot{x}_2 = -x_1 - x_2 + 4x_1^3$$

a) Find all the equilibrium states of this system.
b) Use Lyapunov's second method to show that the equilibrium state at **0** is AS.

9.13 Use Lyapunov's second method to find conditions on a_1, a_2, and a_3 for which the system below is AS:

$$\dot{x}_1 = x_2$$
$$\dot{x}_2 = x_3$$
$$\dot{x}_3 = -a_1x_1 - a_2x_2 - a_3x_3.$$

9.14 Determine whether each of the following polynomials is a Hurwitz polynomial and indicate how many zeros lie in the RHP.
a) $D(s) = s^2 + 1$
b) $D(s) = s^4 + 2s^2 + 1$
c) $D(s) = s^4 - 1$
d) $D(s) = s^2 + 2s + 1$
e) $D(s) = s^3 + s^2 + s + 1$
f) $D(s) = s^4 + 5s^3 + 2s + 10$
g) $D(s) = s^4 + 2s^3 + 6s^2 + 8s + 8$
h) $D(s) = s^6 + 2s^5 + 8s^4 + 12s^3 + 20s^2 + 16s + 16$

9.15 Find the range of the constant K for which each of the following polynomials is Hurwitz.
a) $D(s) = s^4 + 22s^3 + 10s^2 + 2s + K$
b) $D(s) = s^4 + 20Ks^3 + 5s^2 + (10 + K)s + 15$
c) $D(s) = s^3 + (K + 0.5)s^2 + 4Ks + 50$

9.16 For each of the polynomials below, determine whether or not all its zeros lie inside the unit circle in the z-plane.
a) $\mathcal{D}(z) = 12z^6 + 6z^5 + 20z^4 + 3z^3 + 6z^2$
b) $\mathcal{D}(z) = 12z^5 - 12z^4 + 17z^3 - 9z^2 - z + 1$
c) $\mathcal{D}(z) = 8z^4 - 20z^3 + 25z^2 - 10z + 2$
d) $\mathcal{D}(z) = 12z^5 + 8z^4 + 5z^3 + 2z^2 + 3z$
e) $\mathcal{D}(z) = 108z^6 - 210z^5 + 184z^4 - 81z^3 + 16z^2 - z$

APPENDIX

Review of matrix theory

A.1 Terminology

A system of m linear algebraic equations in n unknowns $x_1, x_2, \ldots, x_n$, namely,

$$\begin{aligned} a_{11}x_1 + a_{12}x_2 + \cdots + a_{1n}x_n &= y_1 \\ a_{21}x_1 + a_{22}x_2 + \cdots + a_{2n}x_n &= y_2 \\ \vdots \qquad\qquad & \quad \vdots \\ a_{m1}x_1 + a_{m2}x_2 + \cdots + a_{mn}x_n &= y_m \end{aligned}$$

may be represented by the matrix equation

$$\begin{bmatrix} a_{11} & a_{12} & \cdots & a_{1n} \\ a_{21} & a_{22} & \cdots & a_{2n} \\ \vdots & \vdots & \ddots & \vdots \\ a_{m1} & a_{m2} & \cdots & a_{mn} \end{bmatrix} \begin{bmatrix} x_1 \\ x_2 \\ \vdots \\ x_n \end{bmatrix} = \begin{bmatrix} y_1 \\ y_2 \\ \vdots \\ y_m \end{bmatrix}$$

which in turn is represented by the simple matrix notation

$$Ax = y.$$

Definition A.1 *A rectangular array of real or complex numbers is called a matrix.*

In the above example $A = [a_{ij}]$ is the matrix of coefficients of the equations. A matrix having just one column is called a column matrix (or vector), and if it has one row it is called a row matrix. The entries a_{ij} in the array are called the elements of the matrix. The subscripts identify the location of each element in the array where the first subscript refers to the row number and the second subscript refers to the column number.

A matrix with m rows and n columns is called an m by n matrix, written as $m \times n$. If $m = n$ the matrix is square and is said to be of order n. For square matrices the elements which have identical subscripts (i.e., $a_{11}, a_{22}, a_{33}, \ldots, a_{nn}$) constitute the main or principal diagonal.

A.2 Equality

Definition A.2 *Two $m \times n$ matrices A and B are equal if and only if corresponding elements are equal; that is,*

$$A = B \quad \text{if and only if} \quad a_{ij} = b_{ij} \quad \text{for all } i, j.$$

Note: This requires that both A and B have the same number of rows and columns.

A.3 Addition

Definition A.3 *Given two $m \times n$ matrices $A = [a_{ij}]$ and $= [b_{ij}]$ we define $A + B = C$ where $C = [c_{ij}]$ is the $m \times n$ matrix whose elements are*

$$c_{ij} = a_{ij} + b_{ij} \quad \text{for all } i, j.$$

Example A.1

$$\begin{bmatrix} 2 & 1 & 2 \\ 2 & 3 & 4 \end{bmatrix} + \begin{bmatrix} 1 & 4 & 0 \\ 1 & 6 & 5 \end{bmatrix} = \begin{bmatrix} 3 & 5 & 2 \\ 3 & 9 & 9 \end{bmatrix}.$$

Note: Two matrices can be added only when they have the same number of rows and the same number of columns. Such matrices are called conformable for addition.

A.4 Multiplication by a scalar

Definition A.4 *The scalar multiple of an $m \times n$ matrix $A = [a_{ij}]$ by a scalar k is the $m \times n$ matrix $B = [b_{ij}]$ where $b_{ij} = ka_{ij}$ for all i, j.*

Example A.2

$$2\begin{bmatrix} 1 & 2 & 3 \\ 4 & 5 & 6 \end{bmatrix} = \begin{bmatrix} 2 & 4 & 6 \\ 8 & 10 & 12 \end{bmatrix}.$$

This definition is suggested by the rule of addition; that is,

$$2\begin{bmatrix} 1 & 2 & 3 \\ 4 & 5 & 6 \end{bmatrix} = \begin{bmatrix} 1 & 2 & 3 \\ 4 & 5 & 6 \end{bmatrix} + \begin{bmatrix} 1 & 2 & 3 \\ 4 & 5 & 6 \end{bmatrix} = \begin{bmatrix} 2 & 4 & 6 \\ 8 & 10 & 12 \end{bmatrix}.$$

A.4.1 Subtraction

This is defined by the rule of addition and scalar multiplication by -1. Thus

$$C = A - B \triangleq A + (-1)B \quad \text{or} \quad c_{ij} = a_{ij} - b_{ij} \quad \text{for all } i, j.$$

A.4.2 Commutative and associative laws for addition and scalar multiplication

The following relations are easily proved from the definitions.

1 $A + B = B + A$ (commutative law for addition).

2 $A + (B + C) = (A + B) + C$ (associative law for addition).

3 $(k_1 + k_2)A = k_1 A + k_2 A$ (multiplication by a scalar is distributive with respect to scalar addition).

4 $k_1(A + B) = k_1 A + k_1 B$ (scalar multiplication is distributive over matrix addition).

5 $k_1(k_2 A) = (k_1 k_2)A$ (associative law for scalar multiplication).

A.5 Matrix multiplication

Definition A.5 *If $A = [a_{ij}]$ is an $m \times n$ matrix and if $B = [b k]$ is an $n \times r$ matrix, the product $C = AB$ is the $m \times r$ matrix $C = [c_{ik}]$ where*

$$c_{ik} = \sum_{j=1}^{n} a_{ij} b_{jk} \quad \textit{for all } i, k.$$

Example A.3

$$\begin{bmatrix} 1 & 2 & 3 \end{bmatrix} \begin{bmatrix} -1 & 4 \\ 3 & 0 \\ 2 & 3 \end{bmatrix} = \begin{bmatrix} 11 & 13 \end{bmatrix}.$$

Note: Two matrices can be multiplied only if the number of columns of the first matrix (A) equals the number of rows of the second matrix (B). In this case A and B are called conformable for multiplication. If $C = AB$ we say that A premultiplies B to give C, or B postmultiplies A to give C. The product has the same number of rows as the first matrix (A) and the same number of columns as the second (B).

The general form of matrix multiplication is illustrated below.

$$\begin{bmatrix} \vdots & \vdots & \ddots & \vdots \\ a_{i1} & a_{i2} & \cdots & a_{in} \\ \vdots & \vdots & \ddots & \vdots \end{bmatrix} \begin{bmatrix} \cdots & b_{1j} & \cdots \\ \cdots & b_{2j} & \cdots \\ \ddots & \vdots & \ddots \\ \cdots & b_{nj} & \cdots \end{bmatrix} = \begin{bmatrix} \ddots & \vdots & \ddots \\ \cdots & c_{ij} & \cdots \\ \ddots & \vdots & \ddots \end{bmatrix}.$$

This definition of matrix multiplication is suggested by the procedure of substituting variables in a set of linear equations.

Consider

$$\begin{aligned} a_{11}x_1 + a_{12}x_2 + a_{13}x_3 &= k_1 \\ a_{21}x_1 + a_{22}x_2 + a_{23}x_3 &= k_2. \end{aligned} \tag{A.1}$$

If we make a change from x variables to y variables, where

$$\begin{aligned} x_1 &= b_{11}y_1 + b_{12}y_2 \\ x_2 &= b_{21}y_1 + b_{22}y_2 \\ x_3 &= b_{31}y_1 + b_{32}y_2 \end{aligned} \tag{A.2}$$

then substituting (A.2) into (A.1) we get

$$\begin{aligned} a_{11}(b_{11}y_1 + b_{12}y_2) + a_{12}(b_{21}y_1 + b_{22}y_2) + a_{13}(b_{31}y_1 + b_{32}y_2) &= k_1 \\ a_{21}(b_{11}y_1 + b_{12}y_2) + a_{22}(b_{21}y_1 + b_{22}y_2) + a_{23}(b_{31}y_1 + b_{32}y_2) &= k_2. \end{aligned}$$

Combining terms in each equation gives

$$\begin{aligned} (a_{11}b_{11} + a_{12}b_{21} + a_{13}b_{31})y_1 + (a_{11}b_{12} + a_{12}b_{22} + a_{13}b_{32})y_2 &= k_1 \\ (a_{21}b_{11} + a_{22}b_{21} + a_{23}b_{31})y_1 + (a_{21}b_{12} + a_{22}b_{22} + a_{23}b_{32})y_2 &= k_2. \end{aligned} \tag{A.3}$$

Call the matrix of coefficients of this new set of equations

$$C = [c_{ij}] \quad \text{where } c_{ij} = a_{i1}b_{ij} + a_{i2}b_{2j} + a_{i3}b_{3j}$$

is the same as defined by the product rule of AB; therefore C has been formed from A and B by a process called multiplication.

Observe that the set of equations (A.1) could be written as $AX = K$ by using the product rule. The equations (A.2) could be written as $X = BY$ which gives $ABY = K$, so $C = AB$ is the coefficient matrix of (A.3).

A.5.1 Properties of multiplication

a Three basic properties of scalar multiplication do not hold for matrix multiplication

1. $AB \neq BA$ (matrix multiplication is not in general commutative).
2. $AB = \Theta$ does not imply that either A or B equals Θ, where we define the zero matrix Θ as the matrix whose elements are all zero. (This could also be written as 0, but this notation is easily confused with the scalar 0.)
3. $AB = AC$ and $A \neq \Theta$ does not imply that $B = C$.

Illustration: Consider $A = \begin{bmatrix} 0 & 1 \\ 0 & 2 \end{bmatrix}$, $B = \begin{bmatrix} 1 & 3 \\ 0 & 0 \end{bmatrix}$, $C = \begin{bmatrix} 2 & 1 \\ 0 & 0 \end{bmatrix}$. Here we see that $AB = \begin{bmatrix} 0 & 0 \\ 0 & 0 \end{bmatrix}$, $BA = \begin{bmatrix} 0 & 7 \\ 0 & 0 \end{bmatrix} \neq \Theta$, and neither A nor $B = \Theta$. Also, $AC = \begin{bmatrix} 0 & 0 \\ 0 & 0 \end{bmatrix}$, so $AB = AC$ but $B \neq C$.

b Matrix multiplication is associative (i.e., $(AB)C = A(BC)$)
To see this assume that the matrices are conformable for multiplication, specifically A is $m \times n$, B is $n \times p$, C is $p \times r$. Then by the definition of matrix multiplication, the ij element of the matrix $(AB)C$ is

$$\{(AB)C\}_{ij} = \sum_{k=1}^{p} \{AB\}_{ik} C_{kj} \quad \text{where } \{AB\}_{ik} = \sum_{\ell=1}^{n} a_{i\ell} b_{\ell k}.$$

Therefore

$$\{(AB)C\}_{ij} = \sum_{k=1}^{p} \left(\sum_{\ell=1}^{n} a_{i\ell} b_{\ell k} \right) c_{kj} = \sum_{k=1}^{p} \sum_{\ell=1}^{n} a_{i\ell} b_{\ell k} c_{kj}$$

and similarly

$$\{A(BC)\}_{ij} = \sum_{\ell=1}^{n} a_{il} \left(\sum_{k=1}^{p} b_{\ell k} c_{kj} \right) = \sum_{\ell=1}^{n} \sum_{k=1}^{p} a_{i\ell} b_{\ell k} c_{kj}.$$

The order of summation can be interchanged (since scalar addition is commutative), so the elements are equal for all i, j, and therefore the matrices are equal.

c Matrix multiplication is distributive over addition
That is, $A(B + C) = AB + AC$ and also $(A + B)C = AC + BC$. This is easily verified as in the proof of (b) by using the definitions of matrix multiplication and addition.

A.6 Transpose

Definition A.6 *If $A = [a_{ij}]$ is an $m \times n$ matrix, the transpose of A is the $n \times m$ matrix $B = [b_{ij}] = A^T$ where $b_{ji} = a_{ij}$ for all i, j. (That is, the rows of A become the columns of B.)*

Example A.4

$$\begin{bmatrix} 1 & 3 & 5 \\ 2 & 4 & 6 \end{bmatrix}^T = \begin{bmatrix} 1 & 2 \\ 3 & 4 \\ 5 & 6 \end{bmatrix}.$$

Properties of transposition

1 $(A^T)^T = A$.
2 $(A + B)^T = A^T + B^T$.
3 $(kA)^T = kA^T$ for any scalar k.
4 $(AB)^T = B^T A^T$.

A.7 Special matrices

1 A is symmetric if and only if (iff) $A = A^T$ (i.e., $a_{ij} = a_{ji}$) for all i and j.
2 A is skew-symmetric iff $A = -A^T$ (i.e., $a_{ij} = -a_{ji}$) for all i and j. Note that this requires that the diagonal elements be always zero.

Example A.5

$$A = \begin{bmatrix} 1 & 2 & 4 \\ 2 & 1 & 3 \\ 4 & 3 & 0 \end{bmatrix} \quad \text{is a symmetric matrix.}$$

$$B = \begin{bmatrix} 0 & 2 & -4 \\ -2 & 0 & 3 \\ 4 & -3 & 0 \end{bmatrix} \quad \text{is a skew symmetric matrix.}$$

3 A is Hermitian if $A = (\bar{A})^T \triangleq A^*$, where the overline indicates the complex conjugate (i.e., $a_{ij} = \bar{a}_{ji}$). *Note:* The diagonal elements are always real, and a real symmetric matrix is Hermitian.

Example A.6

$$\begin{bmatrix} 3 & 2 + j3 \\ 2 - j3 & 4 \end{bmatrix} \quad \text{is a Hermitian matrix.}$$

4 A square matrix $A = [a_{ij}]$ is diagonal if $a_{ij} = 0$ for $i \neq j$.

Example A.7

$$\begin{bmatrix} 3 & 0 & 0 \\ 0 & 2 & 0 \\ 0 & 0 & 1 \end{bmatrix} \quad \text{is a diagonal matrix.}$$

5 The identity matrix I is a diagonal matrix with ones along the main diagonal (i.e., $I = [\delta_{ij}]$ where δ_{ij} is the Kronecker delta). Sometimes this will be written as I_n (identity matrix of order n) when it is important to indicate the order.

Property *If A and B are conformable for multiplication and if the two matrices have been partitioned so that their submatrices are conformable for multiplication, then*

$$AB = \begin{bmatrix} A_1 & A_2 \end{bmatrix} \begin{bmatrix} B_1 \\ B_2 \end{bmatrix} = A_1B_1 + A_2B_2.$$

Proof: Suppose A is $m \times n$ and B is $n \times n$. If we partition A so that A_1 is $m \times p$ then A_2 is $m \times (n - p)$. For conformability B_1 must be $p \times n$ and B_2 is $(n - p) \times n$.

A typical element of AB is $\{AB\}_{ij} = \sum_{k=1}^{n} a_{ik}b_{kj}$. Now, a typical element of A_1B_1 is $\{A_1B_1\}_{ij} = \sum_{k=1}^{p} a_{ik}b_{kj}$ (since the first p columns of A are A_1 and the first p rows of B are B_1).

If we call $A_2 = [\hat{a}_{ij}]$ and $B_2 = [\hat{b}_{ij}]$ then $\{A_2B_2\}_{ij} = \sum_{\ell=1}^{n-p} \hat{a}_{i\ell}\hat{b}_{\ell j}$. Note, however, that $\hat{a}_{i\ell} = a_{i,p+\ell}$ and $\hat{b}_{\ell j} = b_{p+\ell,j}$; therefore, $\{A_2B_2\}_{ij} = \sum_{\ell-1}^{n-p} a_{i,p+\ell}b_{p+\ell j}$. Let $k = p + \ell$, then $\{A_2B_2\}_{ij} = \sum_{k=p+1}^{n} a_{ik}b_{kj}$, and so

$$\{A_2B_2\}_{ij} + \{A_2B_2\}_{ij} = \sum_{k=1}^{p} a_{ik}b_{kj} + \sum_{k=p+1}^{n} a_{ik}b_{kj} = \sum_{k=1}^{n} a_{ik}b_{kj}.$$

QED.

Example A.10

$$\left[\begin{array}{cc|c} 0 & 2 & 3 \\ 1 & 4 & 2 \\ \hline 0 & 1 & 0 \end{array}\right] \left[\begin{array}{cc|c} 4 & 2 & 1 \\ 1 & 0 & 1 \\ \hline 2 & 0 & 2 \end{array}\right] = \left[\begin{array}{cc|c} 8 & 0 & 8 \\ 12 & 2 & 9 \\ \hline 1 & 0 & 1 \end{array}\right].$$

Partition as indicated and call the product $C = \begin{bmatrix} C_1 & C_2 \\ C_3 & C_4 \end{bmatrix}$. Then

$$C_1 = \begin{bmatrix} 0 & 2 \\ 1 & 4 \end{bmatrix} \begin{bmatrix} 4 \\ 1 \end{bmatrix} + \begin{bmatrix} 3 \\ 2 \end{bmatrix} [2] = \begin{bmatrix} 2 \\ 8 \end{bmatrix} + \begin{bmatrix} 6 \\ 4 \end{bmatrix} = \begin{bmatrix} 8 \\ 12 \end{bmatrix}$$

$$C_2 = \begin{bmatrix} 0 & 2 \\ 1 & 4 \end{bmatrix} \begin{bmatrix} 2 & 1 \\ 0 & 1 \end{bmatrix} + \begin{bmatrix} 3 \\ 2 \end{bmatrix} \begin{bmatrix} 0 & 2 \end{bmatrix} = \begin{bmatrix} 0 & 2 \\ 2 & 5 \end{bmatrix} + \begin{bmatrix} 0 & 6 \\ 0 & 4 \end{bmatrix} = \begin{bmatrix} 0 & 8 \\ 2 & 9 \end{bmatrix}$$

$$C_3 = \begin{bmatrix} 0 & 1 \end{bmatrix} \begin{bmatrix} 4 \\ 1 \end{bmatrix} + [0]\,[2] = 1$$

$$C_4 = \begin{bmatrix} 0 & 1 \end{bmatrix} \begin{bmatrix} 2 & 1 \\ 0 & 1 \end{bmatrix} + [0] \begin{bmatrix} 0 & 2 \end{bmatrix} = \begin{bmatrix} 0 & 1 \end{bmatrix} + \begin{bmatrix} 0 & 0 \end{bmatrix} = \begin{bmatrix} 0 & 1 \end{bmatrix}.$$

In general, if A and B are conformable for multiplication and are partitioned as

$$A = \begin{bmatrix} A_{11} & A_{12} & \cdots & A_{1p} \\ A_{21} & A_{22} & \cdots & A_{2p} \\ \vdots & \vdots & \ddots & \vdots \\ A_{m1} & A_{m2} & \cdots & A_{mp} \end{bmatrix} \quad B = \begin{bmatrix} B_{11} & B_{12} & \cdots & B_{1n} \\ B_{21} & B_{22} & \cdots & B_{2n} \\ \vdots & \vdots & \ddots & \vdots \\ B_{p1} & B_{p2} & \cdots & B_{pn} \end{bmatrix}$$

where each partition A_{ij} has q_j columns and B_{jk} has q_j rows, then the product matrix

$$C = AB = \begin{bmatrix} C_{11} & C_{12} & \cdots & C_{1n} \\ C_{21} & C_{22} & \cdots & C_{2n} \\ \vdots & \vdots & \ddots & \vdots \\ C_{m1} & C_{m2} & \cdots & C_{mn} \end{bmatrix}$$

can be computed from $C_{ij} = \sum_{k=1}^{p} A_{ik}B_{kj}$. Note that the submatrices of the product matrix are obtained using the rules of matrix multiplication and addition exactly as the elements of the product matrix are obtained using scalar multiplication and addition.

A.9 Matrix inversion

This operation arises when we are interested in the linear equation $AX = Y$ and want to know how X can be written in terms of Y. We cannot write $X = Y/A$ since matrix division is not defined. However, if we can find a matrix B such that $BA = I$ then premultiplying both sides by B would give $X = BY$.

Definition A.7 *Given a square matrix of order n (i.e., A is $n \times n$), X is called the inverse of A if and only if $AX = XA = I_n$. Then denote X by A^{-1}.*

Note: Only a square matrix can have an inverse. If A is not square when B is conformable for premultiplication it will not necessarily be conformable for post-multiplication. For a nonsquare matrix A which is $m \times n$, if $AB = I_m$ then B is called a right inverse, and if $BA = I_n$ then B is called a left inverse of A.

Theorem A.1 *Uniqueness Theorem: If an inverse exists it is unique.*

Proof: Assume $AB = BA = I$ and also that $AC = CA = I$ (i.e., assume that two different inverses exist). Then $AB = I$ and premultiplying by C gives $CAB = C$ but $CA = I$, so $IB = C|$ and $B = C$. QED.

A.9.1 The adjoint matrix

Definition A.8 *The adjoint of an $n \times n$ matrix $A = [a_{ij}]$ is the $n \times n$ matrix $B = [b_{ij}] = Adj\,A$ where $b_{ij} = |A_{ji}|$ = the cofactor of the element a_{ji} which is $(-1)^{i+j} \times$ the determinant of the matrix resulting from crossing out the jth row and ith column.*

Example A.11

$$A = \begin{bmatrix} 1 & 2 & 0 \\ -1 & 3 & 1 \\ 1 & 0 & 2 \end{bmatrix} \quad \text{then} \quad \text{Adj}\,A = \begin{bmatrix} |A_{11}| & |A_{21}| & |A_{31}| \\ |A_{12}| & |A_{22}| & |A_{32}| \\ |A_{13}| & |A_{23}| & |A_{33}| \end{bmatrix}$$

where

$$|A_{11}| = \det \begin{bmatrix} 3 & 1 \\ 0 & 2 \end{bmatrix} = 6 \; |A_{21}| = (-1) \det \begin{bmatrix} 2 & 0 \\ 0 & 2 \end{bmatrix} = -4$$

$$|A_{31}| = \det \begin{bmatrix} 2 & 0 \\ 3 & 1 \end{bmatrix} = 2 \; |A_{12}| = (-1) \det \begin{bmatrix} -1 & 1 \\ 1 & 2 \end{bmatrix} = 3$$

$$|A_{22}| = \det \begin{bmatrix} 1 & 0 \\ 1 & 2 \end{bmatrix} = 2 \; |A_{32}| = (-1) \det \begin{bmatrix} 1 & 0 \\ -1 & 1 \end{bmatrix} = -1$$

$$|A_{13}| = \det \begin{bmatrix} -1 & 3 \\ 1 & 0 \end{bmatrix} = -3 \; |A_{23}| = (-1) \det \begin{bmatrix} 1 & 2 \\ 1 & 0 \end{bmatrix} = 2$$

$$|A_{33}| = \det \begin{bmatrix} 1 & 2 \\ -1 & 3 \end{bmatrix} = 5.$$

If $A = [a_{ij}]$ is an $n \times n$ matrix and if $\det A \neq 0$ then for $B = \text{Adj}\,A/\det A$ we will show that $AB = I$. Multiplying A and B gives

$$A \frac{\text{Adj}\,A}{\det A} = \frac{1}{\det A} \begin{bmatrix} a_{11} & a_{12} & \cdots & a_{1n} \\ a_{21} & a_{22} & \cdots & a_{2n} \\ \vdots & \vdots & \ddots & \vdots \\ a_{n1} & a_{n2} & \cdots & a_{nn} \end{bmatrix} \begin{bmatrix} |A_{11}| & |A_{21}| & \cdots & |A_{n1}| \\ |A_{12}| & |A_{22}| & \cdots & |A_{n2}| \\ \vdots & \vdots & \cdots & \vdots \\ |A_{1n}| & |A_{2n}| & \cdots & |A_{nn}| \end{bmatrix}.$$

Now consider the typical element along the diagonal

$$\left\{ A \frac{\text{Adj}\,A}{\det A} \right\}_{ii} = \frac{a_{i1}|A_{i1}| + a_{i2}|A_{i2}| + \cdots + a_{in}|A_{in}|}{\det A}.$$

By the Laplace expansion of $\det A$ along the ith row of A we see that the numerator is $\det A$, so $\left\{ A \frac{\text{Adj}\,A}{\det A} \right\}_{ii} = 1$. Next consider the typical element off the diagonal for which $i \neq j$:

$$\left\{ A \frac{\text{Adj}\,A}{\det A} \right\}_{ij} = \frac{a_{i1}|A_{j1}| + a_{i2}|A_{j2}| + \cdots + a_{in}|A_{jn}|}{\det A}.$$

Consider a matrix Z which is equal to A except that its ith row is equal to its jth row. Since this matrix has two identical rows, we know $\det Z = 0$. Expanding $\det Z$ along row j gives

$$\det Z = \det \begin{bmatrix} a_{11} & a_{12} & \cdots & a_{1n} \\ a_{21} & a_{22} & \cdots & a_{2n} \\ \vdots & \vdots & \ddots & \vdots \\ a_{i1} & a_{i2} & \cdots & a_{in} \\ \vdots & \vdots & \ddots & \vdots \\ a_{j1} & a_{j2} & \cdots & a_{jn} \\ \vdots & \vdots & \ddots & \vdots \\ a_{n1} & a_{n2} & \cdots & a_{nn} \end{bmatrix} = a_{j1}|A_{j1}| + a_{j2}|A_{j2}| + \cdots + a_{jn}|A_{jn}| = 0.$$

But since row j is identical to row i then $a_{jk} = a_{ik}$ for all k, and the expression above would give $\det Z = a_{i1}|A_{j1}| + a_{i2}|A_{j2}| + \cdots + a_{in}|A_{jn}| = 0$. Therefore since $a_{ii} = 1$ and $a_{ij} = 0$, we have shown that $AB = I$.

In the same way we can show that $BA = I$; we therefore have the result that

Theorem A.2 *If* $\det A \neq 0$ *for the* $n \times n$ *matrix* A*, then* A^{-1} *(the inverse of* A*) is given by*

$$A^{-1} = \frac{\text{Adj } A}{\det A}.$$

This is an existence theorem since if $\det A \neq 0$ we can always form A^{-1}, so $\det A \neq 0 \leftrightarrow A^{-1}$ exists.

Theorem A.3 *If* A *is a square matrix of order* n*, and if* $\det A = 0$*, then no inverse exists.*

Proof: Assume a square matrix B exists which makes $AB = I$. Then $\det(AB) = (\det A)(\det B) = \det I = 1$. Therefore $\det A \neq 0$ which contradicts the hypothesis of the theorem. That is to say, when $\det A = 0$ then no B can exist which makes $AB = I$, so no inverse of A exists. QED.

Definition A.9 *If* A *is a square matrix with* $\det A = 0$ *the matrix is called singular, and if* A *is a square matrix with* $\det A \neq 0$ *the matrix is called nonsingular.*

Theorem A.4 *If* A *is nonsingular and if* B *is any square matrix satisfying* $AB = I$ *then* $B = A^{-1}$.

Proof: Since A^{-1} exists we see $A^{-1}(AB) = A^{-1}(I)$ shows that $B = A^{-1}$.
QED.

A.10 Linear spaces

Definition A.10 *A linear space $\mathcal{S}$ is a set of elements $\underset{\sim}{x}_i$ and two operations which satisfy the following conditions.*

1. *For any two elements $\underset{\sim}{x}_a$ and $\underset{\sim}{x}_b$ in $\mathcal{S}$ there exists a unique element $\underset{\sim}{x}_a + \underset{\sim}{x}_b$ in $\mathcal{S}$ (called the sum of $\underset{\sim}{x}_a$ and $\underset{\sim}{x}_b$). This addition operation has the properties:*
   a. $\underset{\sim}{x}_a + \underset{\sim}{x}_b = \underset{\sim}{x}_b + \underset{\sim}{x}_a$
   b. $\underset{\sim}{x}_a + (\underset{\sim}{x}_b + \underset{\sim}{x}_c) = (\underset{\sim}{x}_a + \underset{\sim}{x}_b) + \underset{\sim}{x}_c$.
2. *There exists a unique zero element $\underset{\sim}{0}$ such that $\underset{\sim}{0} + \underset{\sim}{x} = \underset{\sim}{x}$ and $0\underset{\sim}{x} = \underset{\sim}{0}$.*
3. *For every $\underset{\sim}{x} \in \mathcal{S}$ there exists an inverse element $-\underset{\sim}{x} \in \mathcal{S}$ such that $\underset{\sim}{x} + (-\underset{\sim}{x}) = \underset{\sim}{0}$.*
4. *For any numbers α and β and any $\underset{\sim}{x} \in \mathcal{S}$ the element $\alpha\underset{\sim}{x}$ exists and belongs to $\mathcal{S}$. Also, $\alpha(\underset{\sim}{x}_a + \underset{\sim}{x}_b) = \alpha\underset{\sim}{x}_a + \alpha\underset{\sim}{x}_b$; $\alpha(\beta\underset{\sim}{x}) = (\alpha\beta)\underset{\sim}{x}$; $(\alpha+\beta)\underset{\sim}{x} = \alpha\underset{\sim}{x} + \beta\underset{\sim}{x}$. (Note: The scalars are members of a field; i.e., it contains the sum, difference, product, and quotient of any two numbers except that division by zero is excluded.)*

The set of all n-vectors with the usual definition for vector addition and multiplication by a scalar form a linear space denoted by $\mathcal{U}^n$.

Definition A.11 *Let $\mathcal{S}$ be a linear space and $\mathcal{W}$ a nonempty set of $\mathcal{S}$. $\mathcal{W}$ is called a subspace iff $\mathcal{W}$ is a linear space.*

Example A.12

Let $\mathcal{S}$ be the linear space of 3×1 matrices (i.e., 3-vectors). Let $\mathcal{W}$ be all such vectors with $x_3 = 0$. It is easy to verify that the elements of $\mathcal{W}$ satisfy conditions 1–4 above.

A.11 Linear independence

Definition A.12 *Let $\underset{\sim}{x}_1, \underset{\sim}{x}_2, \ldots, \underset{\sim}{x}_r$ be a set of elements of a linear space $\mathcal{S}$. An element $\underset{\sim}{x}$ is called a linear combination of the elements $\underset{\sim}{x}_1, \underset{\sim}{x}_2, \ldots, \underset{\sim}{x}_r$, iff $\underset{\sim}{x}$ can be expressed as $\underset{\sim}{x} = \sum_{i=1}^{r} c_i \underset{\sim}{x}_i$ for some choice of scalars c_i.*

Example A.13

Consider the set $\boldsymbol{x}_1 = \begin{bmatrix} 1 \\ 0 \\ 0 \end{bmatrix}$, $\boldsymbol{x}_2 = \begin{bmatrix} 0 \\ 1 \\ 0 \end{bmatrix}$, then any vector $\boldsymbol{x} = \begin{bmatrix} a_1 \\ a_2 \\ 0 \end{bmatrix}$ is a linear combination of $\boldsymbol{x}_1$ and $\boldsymbol{x}_2$. A vector $\boldsymbol{x} = \begin{bmatrix} a_1 \\ a_2 \\ a_3 \end{bmatrix}$ with $a_3 \neq 0$ is not.

Definition A.13 *The set of elements $\underset{\sim}{x}_1, \underset{\sim}{x}_2, \ldots, \underset{\sim}{x}_r$ are called linearly dependent if there exists a set of constants $c_1, c_2, \ldots, c_r$ such that*

$$\sum_{i=1}^{r} c_i \underset{\sim}{x}_i = \underset{\sim}{0},$$

and at least one constant is nonzero. If no such set of c_i constants exist, the elements $\underset{\sim}{x}_1, \underset{\sim}{x}_2, \ldots, \underset{\sim}{x}_r$ are called linearly independent.

Example A.14
Consider

$$\boldsymbol{x}_1 = \begin{bmatrix} 1 \\ 2 \\ 3 \end{bmatrix}; \qquad \boldsymbol{x}_2 = \begin{bmatrix} 2 \\ 3 \\ 4 \end{bmatrix}; \qquad \boldsymbol{x}_3 = \begin{bmatrix} 0 \\ 1 \\ 2 \end{bmatrix}.$$

To see whether this set is linearly independent, we can form $c_1\boldsymbol{x}_1 + c_2\boldsymbol{x}_2 + c_3\boldsymbol{x}_3 = \mathbf{0}$. This gives the equations

$$c_1 + 2c_2 + 0c_3 = 0$$
$$2c_1 + 3c_2 + c_3 = 0$$
$$3c_1 + 4c_2 + 2c_3 = 0$$

and has the solution $c_1 = 2,\ c_2 = -1,\ c_3 = -1$, so we see that the set is linearly dependent.
Note: If a set of elements $\underset{\sim}{x}_1, \underset{\sim}{x}_2, \ldots, \underset{\sim}{x}_k$ contains a linearly dependent subset, then the entire set is linearly dependent. Also, a set of elements $\underset{\sim}{x}_1, \underset{\sim}{x}_2, \ldots, \underset{\sim}{x}_k$ is linearly dependent iff one of the elements can be expressed as a linear combination of the other elements.

Definition A.14 *A set of elements $\underset{\sim}{x}_1, \underset{\sim}{x}_2, \ldots, \underset{\sim}{x}_n$ of a linear space $\mathcal{S}$ is said to span $\mathcal{S}$ if any element in $\mathcal{S}$ can be written as a linear combination of the set $\underset{\sim}{x}_1, \underset{\sim}{x}_2, \ldots, \underset{\sim}{x}_n$.*

Definition A.15 *A set B of elements of a linear space $\mathcal{S}$ is called a basis for B if*

1 The elements of B are linearly independent.
2 The elements of B span the space $\mathcal{S}$.

The number of elements in the basis of a linear space is called the dimension of the space.

Theorem A.5 *Let $\underset{\sim}{x}_1, \underset{\sim}{x}_2, \ldots, \underset{\sim}{x}_n$ be a set of elements of a linear space $\mathcal{S}$. This set is a basis for $\mathcal{S}$ if and only if for every $\underset{\sim}{x} \in \mathcal{S}$ there is a set of unique* *scalars $c_1, c_2, \ldots, c_n$ such that $\underset{\sim}{x} = \sum_{i=1}^{n} c_i \underset{\sim}{x}_i$ (i.e., iff any $\underset{\sim}{x}$ can be written as a linear combination of $\underset{\sim}{x}_1, \underset{\sim}{x}_2, \ldots, \underset{\sim}{x}_n$ in exactly one way).*

The unique scalars $c_1, c_2, \ldots, c_n$ are called the components of $\underset{\sim}{x}$ with respect to the basis.

Proof:

1 (necessity) Let $\underset{\sim}{x}_1, \underset{\sim}{x}_2, \ldots, \underset{\sim}{x}_n$ be a basis and let $\underset{\sim}{x} \in \mathcal{S}$. Then, by the definition of a basis, there exists a set of scalars $c_1, c_2, \ldots, c_n$ such that $\underset{\sim}{x} = \sum_{i=1}^{n} c_i \underset{\sim}{x}_i$. Suppose there is another set of scalars $k_1, k_2, \ldots, k_n$ such that $\underset{\sim}{x} = \sum_{i=1}^{n} k_i \underset{\sim}{x}_i$ (i.e., suppose the c_is are not unique). Then

$$\underset{\sim}{x} = \sum_{i=1}^{n} c_i \underset{\sim}{x}_i = \sum_{i=1}^{n} k_i \underset{\sim}{x}_i \quad \text{or} \quad \sum_{i=1}^{n} (c_i - k_i) \underset{\sim}{x}_i = 0.$$

Since the $\underset{\sim}{x}_i$ are a basis, they are linearly independent, so the linear combination above can equal zero only if $c_i - k_i = 0$ for all i. Thus $c_i = k_i$, so the c_is are unique.

2 (sufficiency) Suppose for every $\underset{\sim}{x} \in \mathcal{S}$ there is a unique set of scalars $c_1, c_2, \ldots, c_n$ such that $\underset{\sim}{x} = \sum_{i=1}^{n} c_i \underset{\sim}{x}_i$. Then the set spans $\mathcal{S}$. To show that the set is linearly independent, consider expressing the zero element as a linear combination of the set. That is, $\underset{\sim}{0} = \sum_{i=1}^{n} c_i \underset{\sim}{x}_i$. But we know that the sum $\sum_{i=1}^{n} 0 \underset{\sim}{x}_i = \underset{\sim}{0}$, so by uniqueness $c_i = 0$ for all i. Thus the set $\underset{\sim}{x}_1, \underset{\sim}{x}_2, \ldots, \underset{\sim}{x}_n$ is linearly independent.

We note that any element of a linear space can be represented by its components with respect to the basis elements. Since these components are unique, the representation is unique. QED.

Example A.15

A vector $\boldsymbol{x} = \begin{bmatrix} 1 \\ 2 \\ 3 \end{bmatrix}$ can be written in terms of the basis vectors

$$\boldsymbol{e}_1 = \begin{bmatrix} 1 \\ 0 \\ 0 \end{bmatrix}; \quad \boldsymbol{e}_2 = \begin{bmatrix} 0 \\ 1 \\ 0 \end{bmatrix}; \boldsymbol{e}_3 = \begin{bmatrix} 0 \\ 0 \\ 1 \end{bmatrix}.$$

As $\boldsymbol{x} = 1 \cdot \boldsymbol{e}_1 + 2 \cdot \boldsymbol{e}_2 + 3 \cdot \boldsymbol{e}_3$, thus it is represented by the components $\{1,\ 2,\ 3\}$ with respect to the basis vectors $\boldsymbol{e}_1, \boldsymbol{e}_2, \boldsymbol{e}_3$. On the other hand, the same vector can be written in terms of the basis vectors

$$u_1 = \begin{bmatrix} 1 \\ 0 \\ 0 \end{bmatrix}; \quad u_2 = \begin{bmatrix} 1 \\ 1 \\ 0 \end{bmatrix}; \quad u_3 = \begin{bmatrix} 1 \\ 1 \\ 1 \end{bmatrix}.$$

As $\boldsymbol{x} = -1 \cdot \boldsymbol{u}_1 - 1 \cdot \boldsymbol{u}_2 + 3 \cdot \boldsymbol{u}_3$, so $\boldsymbol{x}$ is represented by the components $\{-1, \ -1, \ 3\}$ with respect to the basis vectors $\boldsymbol{u}_1, \boldsymbol{u}_2, \boldsymbol{u}_3$.

Note: In an n-dimensional vector space, the set $\boldsymbol{e}_1, \boldsymbol{e}_2, \ldots, \boldsymbol{e}_n$, where $\boldsymbol{e}_i$ is the ith column of the identity matrix I_n (i.e., a unit vector with a 1 in the ith position and zeros everywhere else) is called the standard basis. When a basis is not specified, the standard basis is assumed to be the basis.

Theorem A.6 *A square matrix A of order n has* det $A \neq 0$ *if and only if the columns of A are linearly independent (i.e.,* det $A \neq 0 \Longleftrightarrow$ *columns of A are linearly independent).*

Proof:

a Necessity (only if)

For $A = [\boldsymbol{a}_1 \mid \boldsymbol{a}_2 \mid \cdots \mid \boldsymbol{a}_n]$ assume the columns of A are linearly dependent (i.e., $c_1\boldsymbol{a}_1 + c_2\boldsymbol{a}_2 + \cdots + c_n\boldsymbol{a}_n = \boldsymbol{0}$). Let us say $c_1 \neq 0$. We know that $\det [c_1\boldsymbol{a}_1 \mid \boldsymbol{a}_2 \mid \cdots \mid \boldsymbol{a}_n] = c_1 \det A$ and also that the determinant does not change if a multiple of any column is added to any other. Therefore, add $c_2\boldsymbol{a}_2 + c_3\boldsymbol{a}_3 + \cdots + c_n\boldsymbol{a}_n$ to $c_1 a_1$. From our assumption of linear dependence of $\boldsymbol{a}_1, \boldsymbol{a}_2, \cdots, \boldsymbol{a}_n$ we thus get $\det [\boldsymbol{0} \mid \boldsymbol{a}_2 \mid \cdots \mid \boldsymbol{a}_n] = c_1 \det A = 0$. Therefore, since $c_1 \neq 0$ we have shown that the columns being linearly dependent implies det $A = 0$. Thus if det $A \neq 0$, the columns cannot be dependent (i.e., det $A \neq 0 \longrightarrow$ columns of A are linearly independent).

b Sufficiency (if) Outline of proof.

1. Assume det $A = 0$.
2. Argue that if det $A = 0$ it is possible to obtain a zero column by adding multiples of columns to some column.
3. The manipulations which lead to a zero column give the dependence relation.
4. Thus det $A = 0 \longrightarrow$ columns of A are linearly dependent, so if the columns are linearly independent $\longrightarrow$ det $A \neq 0$.

QED.

Note: A similar theorem can be shown for rows.

Theorem A.7 *Let* $\boldsymbol{x}_1, \boldsymbol{x}_2, \ldots, \boldsymbol{x}_n$ *be a set of linearly independent n-vectors. If* $\boldsymbol{y}$ *is any n-vector then it can be written as a linear combination of the* $\boldsymbol{x}_{is}$ *in exactly one way (i.e., the* $\boldsymbol{x}_{is}$*, span the space* $\boldsymbol{y}^n$ *and are a basis for* $\boldsymbol{y}^n$*).*

Proof: From the $n \times n$ matrix $X = [\boldsymbol{x}_1, \boldsymbol{x}_2, \ldots, \boldsymbol{x}_n]$. Then using the rule for multiplication of partitioned matrices write the equation

$$\boldsymbol{y} = \sum_{i=1}^{n} c_i \boldsymbol{x}_i = [\boldsymbol{x}_1 \quad \boldsymbol{x}_2 \quad \ldots \quad \boldsymbol{x}_n] \begin{bmatrix} c_1 \\ c_2 \\ \vdots \\ c_n \end{bmatrix} = X_c.$$

Now the $\boldsymbol{x}_{is}$ are linearly independent, so det $X \neq 0$, and X^{-1} exists. Therefore $\boldsymbol{c} = X^{-1}\boldsymbol{y}$ is a solution to the equation above, and it is unique since X^{-1} is unique. QED.

Theorem A.8 *The set of n-vectors $\boldsymbol{x}_1, \boldsymbol{x}_2, \ldots, \boldsymbol{x}_k$ is linearly dependent for all $k > n$.*

Proof:

1 If $\boldsymbol{x}_1, \boldsymbol{x}_2, \ldots, \boldsymbol{x}_n$ are linearly dependent, then so are $\boldsymbol{x}_1, \boldsymbol{x}_2, \ldots, \boldsymbol{x}_k$.

2 If $\boldsymbol{x}_1, \boldsymbol{x}_2, \ldots, \boldsymbol{x}_n$ are linearly independent, then the vector $\boldsymbol{x}_k$ can be written uniquely as a linear combination of $\boldsymbol{x}_1, \boldsymbol{x}_2, \ldots, \boldsymbol{x}_n$, so the set $\boldsymbol{x}_1, \boldsymbol{x}_2, \ldots, \boldsymbol{x}_k$, is linearly dependent.

QED.

A.12 Rank of a matrix

Definition A.16 *The rank of a matrix A is the maximum number of linearly independent columns of A.*

Example A.16

$$\begin{bmatrix} 1 & 2 & 0 & 0 \\ 4 & 8 & -1 & -1 \\ 3 & 6 & 2 & 2 \end{bmatrix}.$$

Here the rank of A is less than 4 since the columns are 3-vectors and there cannot be 4 independent columns. We can see that $\boldsymbol{x}_2 = 2\boldsymbol{x}_1$ and $\boldsymbol{x}_3 = \boldsymbol{x}_4$. Therefore $\boldsymbol{x}_1$ and $\boldsymbol{x}_2$ are dependent; $\boldsymbol{x}_3$ and $\boldsymbol{x}_4$ are also dependent. There are only two independent columns of A, so the rank of A is two.

Theorem A.9 *If r' = the maximum number of linearly independent rows and r = the maximum number of linearly independent columns of a matrix A, then $r' = r$.*

Proof: Consider the $m \times n$ matrix A. Let $\boldsymbol{a}_1, \boldsymbol{a}_2, \ldots, \boldsymbol{a}_r$ be the independent columns of A and the first r' rows the independent rows of A:

$$\det \left[\begin{array}{cccc:ccc} a_{11} & a_{12} & \cdots & a_{1r} & a_{1(r+1)} & \cdots & a_{1n} \\ a_{21} & a_{22} & \cdots & a_{1r} & a_{1(r+1)} & \cdots & a_{2n} \\ \vdots & \vdots & \ddots & \vdots & \vdots & \ddots & \vdots \\ a_{r'1} & a_{r'2} & \cdots & a_{r'r} & a_{r'(r+1)} & \cdots & a_{r'n} \\ a_{(r'+1)1} & a_{(r'+1)2} & \cdots & a_{(r'+1)r} & a_{(r'+1)(r+1)} & \cdots & a_{(r'+1)n} \\ \vdots & \vdots & \ddots & \vdots & \vdots & \ddots & \vdots \\ a_{n1} & a_{n2} & \cdots & a_{nr} & a_{n(r+1)} & \cdots & a_{nn} \end{array}\right].$$

1 Let us assume that $r' < r$. We can consider the matrix A partitioned as

$$A = \begin{bmatrix} B_1 & B_2 \\ B_3 & B_4 \end{bmatrix}$$

where the columns of B_1 are $b_i = \begin{bmatrix} a_{1i} \\ \vdots \\ a_{r'i} \end{bmatrix}$ for $i = 1, \ldots, r$. Since the $\boldsymbol{b}_i$ are r'-vectors and $r > r'$ the vectors $\boldsymbol{b}_1, \boldsymbol{b}_2, \ldots, \boldsymbol{b}_r$ are linearly dependent, so

$$c_1\boldsymbol{b}_1 + c_2\boldsymbol{b}_2 + \cdots + c_r\boldsymbol{b}_r = \boldsymbol{0} \quad \text{and not all } c_i \text{ are 0.}$$

In matrix notation

$$[\boldsymbol{b}_1\boldsymbol{b}_2 \ldots \boldsymbol{b}_r] \begin{bmatrix} c_1 \\ \vdots \\ c_r \end{bmatrix} = B_1\boldsymbol{c} = \boldsymbol{0} \quad \text{where } \boldsymbol{c} \neq \boldsymbol{0}.$$

2 Now let $\boldsymbol{u}_i^T = [a_{i1}, a_{i2}, \ldots, a_{in}] =$ the ith row of A, so

$$A = \begin{bmatrix} \boldsymbol{u}_1^T \\ \vdots \\ \boldsymbol{u}_{r'}^T \\ \boldsymbol{u}_{r'+1}^T \\ \vdots \\ \boldsymbol{u}_m^T \end{bmatrix}.$$

Since the first r' rows are linearly independent, any other row can be written as a linear combination of the first r' rows. Thus

$$\boldsymbol{u}_i^T = d_{1i}\boldsymbol{u}_1^T + d_{2i}\boldsymbol{u}_2^T + \cdots + d_{r'i}\boldsymbol{u}_{r'}^T = [d_{1i}d_{2i}\cdots d_{r'i}]\begin{bmatrix}\boldsymbol{u}_1^T\\ \boldsymbol{u}_2^T\\ \vdots\\ \boldsymbol{u}_{r'}^T\end{bmatrix}$$

$$= \boldsymbol{d}_i^T[B_1 \vdots B_2] \quad \text{for } i = r'+1, \ldots, m.$$

This is the same as

$$\begin{bmatrix}\boldsymbol{u}_{r'+1}^T\\ \boldsymbol{u}_{r'+2}^T\\ \vdots\\ \boldsymbol{u}_m^T\end{bmatrix} = [\,B_3 \quad B_4\,] = \begin{bmatrix}\boldsymbol{d}_{r'+1}^T\\ \boldsymbol{d}_{r'+2}^T\\ \vdots\\ \boldsymbol{d}_m^T\end{bmatrix}[\,B_1 \quad B_2\,] \triangleq D[\,B_1 \quad B_2\,].$$

We see therefore that $B_3 = DB_1$, and since $B_1\boldsymbol{c} = \boldsymbol{0}$ we see that $B_3\boldsymbol{c} = \boldsymbol{0}$. We now have

$$\begin{bmatrix}B_1\\ B_3\end{bmatrix}\boldsymbol{c} = \boldsymbol{0} \quad \text{where } \boldsymbol{c} \neq \boldsymbol{0}$$

so the columns of $\begin{bmatrix}B_1\\ B_3\end{bmatrix}$ are linearly dependent. These columns are the first r columns of A which by definition of r are linearly independent. So the assumption that $r > r'$ leads to a contradiction and says that r' cannot be less than r. In the same way, if we assume that $r' > r$ and use the same reasoning as above we also get a contradiction, so r' cannot be greater than r. We thus conclude that $r' = r$.

QED.

Theorem A.10 *If a matrix A has rank r then*

1 *Among all the $r \times r$ square submatrices of A, at least one has a nonzero determinant.*

2 *Every square submatrix of order $k > r$ has a zero determinant.*

Proof:

1 This follows directly from the previous proof. As above, arrange the first r rows and columns of A to be independent ones. Since the matrix B_1 was shown to be $r \times r$, if $\det B_1 = 0$ then $B_{1c} = \boldsymbol{0}$ for some $\boldsymbol{c} \neq \boldsymbol{0}$, and as before we show that this says the first r columns of A are linearly dependent. Since by definition these columns are linearly independent we conclude that $\det B_1$ cannot be zero.

2 We know that every set of $k > r$ columns (or rows) is linearly dependent, since by definition r is the maximum number of independent columns (or rows). Thus if we form a matrix A_k from any k columns of a matrix A we have that $A_{kc} = \mathbf{0}$ where $\boldsymbol{c} \neq \mathbf{0}$ is a k-vector. Let $\hat{A}_k$ be a $k \times k$ matrix formed from any k rows of A_k. We see that $\hat{A}_{kc} = \mathbf{0}$ with $\boldsymbol{c} \neq \mathbf{0}$. If det $\hat{A}_k \neq 0$ then $\hat{A}_k^{-1}$ exists, and this would imply that $\boldsymbol{c} = 0$. Since $\boldsymbol{c} \neq \mathbf{0}$ this say that det $\hat{A}_k = 0$.

QED.

A.13 Simultaneous linear equations

A.13.1 Motivation and definitions

For the three sets of equations shown, can we obtain a solution?

$$a \quad \begin{matrix} x_1 + x_2 = 1 \\ x_1 - x_2 = 0 \end{matrix} \qquad b \quad \begin{matrix} x_1 + x_2 = 1 \\ 2x_1 + 2x_2 = 2 \end{matrix} \qquad c \quad \begin{matrix} x_1 + x_2 = 1 \\ x_1 + x_2 = 2 \end{matrix}$$

Set (a) has exactly one solution; set (b) has an infinite number of solutions; set (c) has no solution. In general we would like to investigate the existence of solutions of a system of m linear equations in n unknowns whether $m > n$, $m = n$, or $m < n$. Let us write this system as

$$\begin{aligned} a_{11}x_1 + a_{12}x_2 + \cdots + a_{1n}x_n &= b_1 \\ a_{21}x_1 + a_{22}x_2 + \cdots + a_{2n}x_n &= b_2 \\ \vdots \qquad & \quad \vdots \\ a_{m1}x_1 + a_{m2}x_2 + \cdots + a_{mn}x_n &= b_m. \end{aligned}$$

Any set of values $x_1, x_2, \ldots, x_n$ (or vector $\boldsymbol{x}$) which satisfy these equations is called a *solution*.

If a system has one or more solutions it is called *consistent*, otherwise it is *inconsistent*. A is called the *coefficient matrix* and $[A \mid \boldsymbol{b}]$ is called the *augmented matrix* of the system.

If a consistent system has only one solution (i.e., a unique solution) it is called *determinate*; otherwise it is *indeterminate*.

The basic problems which arise in studying a system of linear equations $A\boldsymbol{x} = \boldsymbol{b}$ are:

1 Determine whether the system is consistent or inconsistent.
2 If the system is consistent, determine whether it is determinate.
3 If the system is determinate, find its unique solution.
4 If the system is indeterminate, find the set of all solutions.

A.13.2 Equivalent systems of equations and elementary operations

Definition A.17 *The collection of all solutions of a system of equations is called a solution set.*

Note:

1 Inconsistent systems have the empty set as the solution set.
2 Consistent determinate systems have only a single member in the solution set.
3 Indeterminate systems have more than one member in the solution set.

Definition A.18 *Two systems of equations are equivalent if they have the same solution set (i.e., if every solution of one system is also a solution of the other).*

Now note that if we are given a system of linear equations, we can perform a number of operations on them which do not change their solutions. Namely,

1 Interchange any two equations.
2 Multiply any equation by a nonzero constant.
3 Add a multiple of one equation to another.

These are called *elementary operations*. Therefore if we apply a sequence of elementary operations to a system of linear equations we do not change the solution set, so we get an equivalent system of equations.

A.13.3 Gaussian elimination

The Gaussian elimination method (or successive elimination) consists or reducing a system of linear equations to an equivalent system of equations in triangular form. For example, in upper triangular form, the last equation contains only the last unknown; the next to last equation contains only the last two unknowns, and so forth. Thus the equivalent triangular system can be solved for each unknown in turn.

Example A.17

For the system of equations

$$\begin{array}{rcrcrcl} x_1 & + & 2x_2 & + & 3x_3 & = & 7 \\ 3x_1 & + & 2x_2 & + & x_3 & = & 5 \\ x_1 & + & x_2 & + & x_3 & = & 3 \end{array} \quad \text{the augmented matrix is} \quad \left[\begin{array}{ccc|c} 1 & 2 & 3 & 7 \\ 3 & 2 & 1 & 5 \\ 1 & 1 & 1 & 3 \end{array}\right].$$

Step 1 Eliminate x_1 from the second and third equations to get

$$\begin{array}{rcrcrcr} x_1 & + & 2x_2 & + & 3x_3 & = & 7 \\ & - & 4x_2 & - & 8x_3 & = & -16 \\ & - & x_2 & - & 2x_3 & = & -4 \end{array} \quad \text{with augmented matrix} \quad \left[\begin{array}{ccc|c} 1 & 2 & 3 & 7 \\ 0 & -4 & -8 & -16 \\ 0 & -1 & -2 & -4 \end{array}\right].$$

Step 2 Eliminate x_2 from the third equation to get

$$\begin{array}{rcrcrcr} x_1 & + & 2x_2 & + & 3x_3 & = & 7 \\ & & x_2 & + & 2x_3 & = & 4 \end{array} \quad \text{with augmented matrix} \left[\begin{array}{ccc|c} 1 & 2 & 3 & 7 \\ 0 & 1 & 2 & 4 \\ 0 & 0 & 0 & 0 \end{array}\right].$$

These equations, which are in triangular form, are equal to the original set and can be easily solved. In this example we find that the last equation becomes identically zero which means that we did not have three independent equations to start with, so the system is indeterminate. Thus let us choose $x_3 = c$ where c is any constant. The solution is then $x_2 = 4-2c$ from the second equation and $x_1 = 7-2(4-2c)-3c = -1+c$ for any choice of c. In vector form, the solution for any c is

$$\boldsymbol{x} = \begin{bmatrix} -1 \\ 4 \\ 0 \end{bmatrix} + \begin{bmatrix} 1 \\ -2 \\ 1 \end{bmatrix} c$$

which is called the general solution.

Instead of writing out all the equations, we could have applied the elementary operations directly to the rows of the augmented matrix. These are called the *elementary row operations*. In our example we have reduced the augmented matrix to triangular form. That is, the first nonzero entry of the ith row occurs in column i (this is called the pivot for the ith row), and all column entries below the pivot are zero. The form known as row echelon form has all pivots scaled to 1, and all other column entries equal to zero. The general solution is obtained from the row echelon form directly without back substitutions.

In our example starting with the original augmented matrix, we get

$$\left[\begin{array}{ccc|c} 1 & 2 & 3 & 7 \\ 3 & 2 & 1 & 5 \\ 1 & 1 & 1 & 3 \end{array}\right] \rightarrow \left[\begin{array}{ccc|c} 1 & 2 & 3 & 7 \\ 0 & -4 & -8 & -16 \\ 0 & -1 & -2 & -4 \end{array}\right] \rightarrow \left[\begin{array}{ccc|c} 1 & 2 & 3 & 7 \\ 0 & 1 & 2 & 4 \\ 0 & 0 & 0 & 0 \end{array}\right].$$

Note that in Step 1, -3 times the first row was added to the second row and the first row was subtracted from the third row. In Step 2 the pivot for row 2 was scaled to equal 1 by dividing by -4 and the result was added to the third row. To get row echelon form subtract 2 times the second row from the first row of the augmented matrix after Step 2. This gives

$$\left[\begin{array}{ccc|c} 1 & 0 & -1 & -1 \\ 0 & 1 & 2 & 4 \\ 0 & 0 & 0 & 0 \end{array}\right]$$

which corresponds to the equations

$$x_1 - x_3 = -1$$
$$x_2 + 2x_3 = 4.$$

So x_1 and x_2 can be expressed in terms of $x_3 = c$ to give the general solution.

Example A.18

For the equations

$$\begin{array}{rcrcrcrcr} x_1 & + & 2x_2 & & & & & = & 7 \\ 2x_1 & + & 4x_2 & - & x_3 & - & x_4 & = & 3 \\ 3x_1 & + & 6x_2 & - & 2x_3 & - & 2x_4 & = & 29 \end{array}$$ the augmented matrix is

$$\left[\begin{array}{cccc|c} 1 & 2 & 0 & 0 & 5 \\ 2 & 4 & -1 & -1 & 3 \\ 3 & 6 & 2 & 2 & 29 \end{array}\right].$$

Step 1 Subtract 2 times the first row from the second row and 3 times the first row from the third row. This gives the new augmented matrix

$$\left[\begin{array}{cccc|c} 1 & 2 & 0 & 0 & 5 \\ 0 & 0 & -1 & -1 & -7 \\ 0 & 0 & 2 & 2 & 14 \end{array}\right].$$

Step 2 Since the pivot for row 2 is zero we could interchange the second and third columns (this interchanges the positions of the variables x_2 and x_3). We will just use the first nonzero element in row 2 as the new pivot and add 2 times row 2 to row 3 and scale the pivot to 1 by multiplying row 2 by -1. This gives

$$\left[\begin{array}{cccc|c} 1 & 2 & 0 & 0 & 5 \\ 0 & 0 & 1 & 1 & 7 \\ 0 & 0 & 0 & 0 & 0 \end{array}\right].$$

This happens to be in row echelon form and corresponds to the equations

$$x_1 + 2x_2 = 5$$
$$x_3 + x_4 = 7.$$

The general solution is obtained by choosing $x_2 = c_1$ and $x_4 = c_2$ where c_1 and c_2 are arbitrary constants to give

$$\boldsymbol{x} = \begin{bmatrix} 5 \\ 0 \\ 7 \\ 0 \end{bmatrix} + \begin{bmatrix} -2 & 0 \\ 1 & 0 \\ 0 & -1 \\ 0 & 1 \end{bmatrix} \begin{bmatrix} c_1 \\ c_2 \end{bmatrix}.$$

A.13.4 Fundamental theorem for linear equations

Theorem A.11 *The m equations in n unknowns $A\boldsymbol{x} = \boldsymbol{b}$ are consistent if and only if the coefficient matrix A and the augmented matrix $[A \mid \boldsymbol{b}]$ have the same rank.*

Proof: (if) Let rank $A = r$ and suppose $\boldsymbol{a}_1, \boldsymbol{a}_2, \ldots, \boldsymbol{a}_r$ (the first r columns) are the linearly independent columns of A. Since $r = \text{rank } A = \text{rank } [A \mid \boldsymbol{b}]$, this says that $\boldsymbol{a}_1, \boldsymbol{a}_2, \ldots, \boldsymbol{a}_r, \boldsymbol{b}$ are linearly dependent, so we can express $\boldsymbol{b}$ as a linear combination of the vectors $\boldsymbol{a}_1, \boldsymbol{a}_2, \ldots, \boldsymbol{a}_r$

$$\boldsymbol{b} = c_1\boldsymbol{a}_1 + c_2\boldsymbol{a}_2 + \cdots + c_r\boldsymbol{a}_r = [\boldsymbol{a}_1 \mid \boldsymbol{a}_2 \mid \cdots \mid \boldsymbol{a}_r]\begin{bmatrix} c_1 \\ c_2 \\ \vdots \\ c_r \end{bmatrix} = A\begin{bmatrix} c_1 \\ c_2 \\ \vdots \\ c_r \\ 0 \\ \vdots \\ 0 \end{bmatrix}.$$

Therefore, $\begin{bmatrix} \boldsymbol{c} \\ \boldsymbol{0} \end{bmatrix}$ is a solution, so the equations $A\boldsymbol{x} = \boldsymbol{b}$ are consistent.

(only if)

Let $\boldsymbol{x} = \begin{bmatrix} c_1 \\ c_2 \\ \vdots \\ c_n \end{bmatrix}$ be a solution. Then $\Sigma_{i=1}^{n} c_i\boldsymbol{a}_i = \boldsymbol{b}$. Thus $\boldsymbol{b}$ is a linear combination of $\boldsymbol{a}_1, \boldsymbol{a}_2, \cdots, \boldsymbol{a}_n$ so the sets $\boldsymbol{a}_1, \boldsymbol{a}_2, \cdots, \boldsymbol{a}_n$ and $\boldsymbol{a}_1, \boldsymbol{a}_2, \cdots, \boldsymbol{a}_n, \boldsymbol{b}$ have the same number of linearly independent vectors in them. Thus the matrices A and $[A \mid \boldsymbol{b}]$ have the same number of linearly independent columns, so they have the same rank. QED.

Note: A set of homogeneous equations $A\boldsymbol{x} = \boldsymbol{0}$ are always consistent since $\boldsymbol{x} = \boldsymbol{0}$ is a solution. This is called the trivial solution.

We note therefore that to test for consistency we must see whether the rank of $A = \text{rank } [A \mid \boldsymbol{b}]$. The most efficient method for determining the rank of a matrix is to use Gaussian elimination. This will be discussed in detail in a later section.

A.13.5 Properties of the general solution

Suppose we have a consistent system of linear equations $A\boldsymbol{x} = \boldsymbol{b}$ with rank $A = r$. Assume that the equations are arranged so that the $r \times r$ submatrix in the upper left corner has a nonzero determinant. Partition A as shown below:

$$A = \left[\begin{array}{ccc|ccc} a_{11} & \cdots & a_{1r} & a_{1(r+1)} & \cdots & a_{1n} \\ \vdots & \ddots & \vdots & \vdots & \ddots & \vdots \\ a_{r1} & \cdots & a_{rr} & a_{r(r+1)} & \cdots & a_{rn} \\ \hline a_{(r+1)1} & \cdots & a_{(r+1)r} & a_{(r+1)(r+1)} & \cdots & a_{(r+1)n} \\ \vdots & \ddots & \vdots & \vdots & \ddots & \vdots \\ a_{m1} & \cdots & a_{mr} & a_{m(r+1)} & \cdots & a_{mn} \end{array}\right] = \begin{bmatrix} B_1 & B_2 \\ B_3 & B_4 \end{bmatrix}$$

where $\det B_1 \neq 0$ so B_1^{-1} exists. Let

$$\hat{\boldsymbol{x}} = \begin{bmatrix} x_1 \\ \vdots \\ x_r \end{bmatrix} \quad \tilde{\boldsymbol{x}} = \begin{bmatrix} x_{r+1} \\ \vdots \\ x_n \end{bmatrix} \quad \text{and} \quad \hat{\boldsymbol{b}} = \begin{bmatrix} b_1 \\ \vdots \\ b_r \end{bmatrix} \quad \tilde{\boldsymbol{b}} = \begin{bmatrix} b_{r+1} \\ \vdots \\ b_m \end{bmatrix}$$

so

$$B_1\hat{\boldsymbol{x}} + B_2\tilde{\boldsymbol{x}} = \hat{\boldsymbol{b}} \quad \text{and} \quad B_3\hat{\boldsymbol{x}} + B_4\tilde{\boldsymbol{x}} = \hat{\boldsymbol{b}}.$$

Therefore $\hat{\boldsymbol{x}} = B_1^{-1}[\hat{\boldsymbol{b}} - B_2\tilde{\boldsymbol{x}}]$, and for any arbitrary choice of $x_{r+i} = c_i$ for $i = 1, 2, \ldots, n-r$ we get

$$\hat{\boldsymbol{x}} = B_1^{-1}[\hat{\boldsymbol{b}} - B_2\boldsymbol{c}] \tag{A.4}$$

so the solution

$$\boldsymbol{x} = \begin{bmatrix} B_1^{-1}[\hat{\boldsymbol{b}} - B_2\boldsymbol{c}] \\ \boldsymbol{c} \end{bmatrix}$$

satisfies

$$\begin{bmatrix} B_1 & B_2 \end{bmatrix}\boldsymbol{x} = \hat{\boldsymbol{b}} \quad \text{or equivalently} \quad \begin{bmatrix} B_1 & B_2 & \hat{\boldsymbol{b}} \end{bmatrix}\begin{bmatrix} \hat{\boldsymbol{x}} \\ \boldsymbol{c} \\ -1 \end{bmatrix} = \boldsymbol{0} \tag{A.5}$$

for any choice of $\boldsymbol{c}$. We want to show that it also satisfies $[B_3 \quad B_4]\boldsymbol{x} = \tilde{\boldsymbol{b}}$.

Since the system is consistent, we know that

$$\operatorname{rank}\left[\begin{array}{cc|c} B_1 & B_2 & \hat{\boldsymbol{b}} \\ \hline B_3 & B_4 & \tilde{\boldsymbol{b}} \end{array}\right] = r.$$

The first r rows of this matrix are linearly independent since $\det B_1 \neq 0$ (i.e., the rank of $[B_1 \quad B_2 \quad \hat{\boldsymbol{b}}] = r$, so it has r independent rows). Therefore, every row of $[B_3 \quad B_4 \quad \tilde{\boldsymbol{b}}]$ can be written as a linear combination of the rows of $[B_1 \quad B_2 \quad \hat{\boldsymbol{b}}]$. In other words, if we let γ_i^T denote the rows of the augmented matrix $[A\ \boldsymbol{b}]$ then we can write

$$\gamma_i^T = \sum_{j=1}^{r} d_{ji}\gamma_j^T = [d_{1i} \mid \ldots \mid d_{ri}]\begin{bmatrix} \gamma_1^T \\ \vdots \\ \gamma_r^T \end{bmatrix} \quad \text{for } i = r+1, \ldots, m.$$

Writing this out gives

$$\underbrace{\begin{bmatrix} \gamma_{r+1}^T \\ \vdots \\ \gamma_m^T \end{bmatrix} = \begin{bmatrix} d_{1(r+1)} & \cdots & d_{r(r+1)} \\ \vdots & \ddots & \vdots \\ d_{1m} & \cdots & d_{rm} \end{bmatrix}}_{D} \begin{bmatrix} \gamma_1^T \\ \vdots \\ \gamma_r^T \end{bmatrix}$$

which says that

$$\begin{bmatrix} B_3 & B_4 & \tilde{\boldsymbol{b}} \end{bmatrix} = D \begin{bmatrix} B_1 & B_2 & \hat{\boldsymbol{b}} \end{bmatrix}. \tag{A.6}$$

Now, postmultiply (A.6) by the column vector $\begin{bmatrix} \hat{\boldsymbol{c}} \\ \boldsymbol{c} \\ -1 \end{bmatrix}$, and using (A.5) we see that

$$\begin{bmatrix} B_3 & B_4 & \tilde{\boldsymbol{b}} \end{bmatrix} \begin{bmatrix} \hat{\boldsymbol{x}} \\ \boldsymbol{c} \\ -1 \end{bmatrix} = \boldsymbol{0} \quad \text{or equivalently} \quad \begin{bmatrix} B_3 & B_4 \end{bmatrix} \begin{bmatrix} \hat{\boldsymbol{x}} \\ \boldsymbol{c} \end{bmatrix} = \tilde{\boldsymbol{b}}$$

for any choice of $\boldsymbol{c}$. We have thus shown that the vector $\boldsymbol{x} = \begin{bmatrix} \hat{\boldsymbol{x}} \\ \boldsymbol{c} \end{bmatrix}$ (where $\hat{\boldsymbol{x}}$ is given by (A.4)) satisfies the equation $A\boldsymbol{x} = \boldsymbol{b}$ for any arbitrary choice of $\boldsymbol{c}$, so it is a solution.

To verify that any solution has this form, let $\boldsymbol{x}_p$ be some arbitrary solution, so

$$\begin{bmatrix} B_1 & B_2 \\ B_3 & B_4 \end{bmatrix} \begin{bmatrix} \hat{\boldsymbol{x}}_p \\ \tilde{\boldsymbol{x}}_p \end{bmatrix} = \begin{bmatrix} \hat{\boldsymbol{b}} \\ \tilde{\boldsymbol{b}} \end{bmatrix}$$

Therefore $B_1\hat{\boldsymbol{x}}_p + B_2\tilde{\boldsymbol{x}}_p = \hat{\boldsymbol{b}}$, so $\hat{\boldsymbol{x}}_p = B_1^{-1}[\hat{\boldsymbol{b}} - B_2\tilde{\boldsymbol{x}}_p]$. Thus the vector

$$\boldsymbol{x} = \begin{bmatrix} B_1^{-1}[\hat{\boldsymbol{b}} - B_2\boldsymbol{c}] \\ \boldsymbol{c} \end{bmatrix} \tag{A.7}$$

is a general solution of the consistent linear system $A\boldsymbol{x} = \boldsymbol{b}$.

A number of points should be noted.

1 If $r = n$ (i.e., the rank of A equals the number of unknowns), then there is no arbitrary vector $\boldsymbol{c}$, so the solution is unique and the system is determinate. The only solution is $\boldsymbol{x} = B_1^{-1}\hat{\boldsymbol{b}}$ where $B_1 = A$ and $\hat{\boldsymbol{b}} = \boldsymbol{b}$.

2 For a homogeneous system (i.e., $A\boldsymbol{x} = \boldsymbol{0}$) we see that

a When $r = n$, the only solution is $\boldsymbol{x} = \boldsymbol{0}$.

b When $r < n$, nontrivial solutions can exist. They are given by

$$\boldsymbol{x}_h = \begin{bmatrix} -B_1^{-1}B_2\boldsymbol{c} \\ \boldsymbol{c} \end{bmatrix} = \begin{bmatrix} -B_1^{-1}B_2 \\ I_{n-r} \end{bmatrix} \boldsymbol{c}$$

for any arbitrary a.

3 When $r < n$ the system is indeterminate. The general solution (A.7) has the form

$$x = \begin{bmatrix} B_1^{-1}\hat{b} \\ 0 \end{bmatrix} + x_h.$$

We can show that a general solution can also be written as $x = x_p + x_h$, where x_p is any particular solution of the system. This is true because this x satisfies $Ax = b$ and any other solution x_a satisfies $A(x_a - x_p) = 0$, so $x_a - x_p$ is a solution to the homogeneous equation, namely x_h. Thus any solution has the form $x = x_p + x_h$.

A.13.6 Rank using Gaussian elimination

In our earlier discussion of the rank of a matrix, we needed to find the largest square submatrix with nonzero determinant. The rank can actually be determined more easily by using Gaussian elimination.

Theorem A.12 *If a matrix C with rows $c_1^T, c_2^T, \cdots, c_m^T$ is row equivalent to a matrix $\hat{C}$ with rows $\hat{c}_1^T, \hat{c}_2^T, \cdots, \hat{c}_m^T$, then the $\hat{c}_i$ can be expressed as linear combinations of the c_j and vice versa.*

Proof: This follows from the definition of row equivalence, since the matrix $\hat{C}$ can be formed from C by applying only elementary row operations. These row operations give the linear combinations. QED.

Theorem A.13 *If a matrix C is row equivalent to $\hat{C}$, then the rows of C and $\hat{C}$ span the same space.*

Proof: Since each $\hat{c}_i$ can be expressed as a linear combination of the c_j, any vector which can be expressed as a linear combination of the $\hat{c}_j$ can also be expressed as a linear combination of the c_j. That is, since $\hat{c}_i$ is expressible as $\hat{c}_i = \sum_{j=1}^m d_{ij} c_f$, any vector that can be written as $x = \sum_{i=1}^m \alpha_i \hat{c}_i$ can also be written as

$$x = \sum_{i=1}^m \alpha_i \sum_{j=1}^m d_{ij} c_j = \sum_{j=1}^m \left(\sum_{i=1}^m \alpha_i d_{ij} \right) c_j.$$

Exactly the same holds true when c_i is expressed in terms of $\hat{c}_i$. QED.

Therefore to determine the rank of a matrix C (which is the same as the number of linearly independent rows or columns of the matrix), we can get a row equivalent matrix $\hat{C}$ in triangular form. The number of linearly independent rows in this form will be obvious, and it will be the rank of C.

Example A.19
Find the rank of the matrix

$$C = \begin{bmatrix} 1 & 2 & 6 & -2 & -1 \\ -2 & -1 & 0 & -5 & -1 \\ 3 & 1 & -1 & 8 & 1 \\ -1 & 0 & 2 & -4 & -1 \\ -1 & -2 & -7 & 3 & 2 \end{bmatrix}$$

Add appropriate multiples of the first row to the second through fifth rows, in order to put zeros in the first column below the pivot. This gives

$$C = \begin{bmatrix} 1 & 2 & 6 & -2 & -1 \\ 0 & 3 & 12 & -9 & -3 \\ 0 & -5 & -19 & 14 & 4 \\ 0 & 2 & 8 & -6 & -2 \\ 0 & 0 & -1 & 1 & 1 \end{bmatrix}.$$

Scale the pivot of row 2 to have a value of 1 and proceed as before to add multiples of row 2 to the higher numbered rows in order to put zeros in the column below the pivot. Continue in the same way for row 3. At this point we find that the remaining rows are all zero. This results in

$$C = \begin{bmatrix} 1 & 2 & 6 & -2 & -1 \\ 0 & 1 & 4 & -3 & -1 \\ 0 & 0 & 1 & -1 & -1 \\ 0 & 0 & 0 & 0 & 0 \\ 0 & 0 & -1 & 1 & 1 \end{bmatrix} \longrightarrow \begin{bmatrix} 1 & 2 & 6 & -2 & -1 \\ 0 & 1 & 4 & -3 & -1 \\ 0 & 0 & 1 & -1 & -1 \\ 0 & 0 & 0 & 0 & 0 \\ 0 & 0 & 0 & 0 & 0 \end{bmatrix}.$$

We observe that the first 3 rows of the final matrix are linearly independent (as are the first 3 columns); thus rank $C = 3$.

A.14 Eigenvalues and eigenvectors

A.14.1 The eigenvector problem

Consider the following problem:

Given a square matrix F of order n, find the set of scalars λ and of nonzero vectors ξ which satisfy the equation

$$F\xi = \lambda\xi. \tag{A.8}$$

In other words, find the vectors which are transformed by F into a scalar multiple of itself. We allow the scalars and vector components to be complex numbers.

This is called the eigenvector problem. Any scalar for which this problem has a solution is called an eigenvalue or characteristic value of the matrix F. The nonzero vector solution is called the associated eigenvector or characteristic vector.

Note that for (A.8) to have a nonzero solution for $\boldsymbol{\xi}$, the homogeneous equation $[\lambda I - F]\boldsymbol{\xi} = \mathbf{0}$ must have nontrivial solutions. We know this happens only if

$$\det[\lambda I - F] = 0. \tag{A.9}$$

This equation is called the characteristic equation of the matrix F. When $\det[\lambda I - F]$ is expanded out, we obtain an nth-degree polynomial $p\lambda$ in λ called the *characteristic polynomial*

$$p(\lambda) = \det[\lambda I - F] = \lambda^n + \alpha_{n-1}\lambda^{n-1} + \alpha_{n-2}\lambda^{n-2} + \cdots + \alpha_1\lambda + \alpha_0. \tag{A.10}$$

Thus the characteristic equation

$$p(\lambda) = 0 \tag{A.11}$$

is an nth-degree polynomial equation that has n roots (i.e., n solutions for λ). The n roots are the only values of λ that make $\det[\lambda I - F] = 0$ and thus are the only values of λ for which nonzero vectors $\boldsymbol{\xi}$ can be found. These roots of the characteristic equation, denoted as $\lambda_1, \lambda_2, \ldots, \lambda_n$, are the eigenvalues of F.

Example A.20

$$F = \begin{bmatrix} 0 & 1 \\ 1 & 0 \end{bmatrix} \quad \text{so } \det[\lambda I - F] = \det\begin{bmatrix} \lambda & -1 \\ -1 & \lambda \end{bmatrix} = \lambda^2 - 1 = (\lambda+1)(\lambda-1) = 0.$$

Thus $\lambda = \lambda_1 = -1$ and $\lambda = \lambda_2 = 1$ are the two eigenvalues. For $\lambda = -1$, the eigenvector is found by solving the algebraic equations

$$\begin{bmatrix} -1 & -1 \\ -1 & -1 \end{bmatrix} \begin{bmatrix} \xi_1 \\ \xi_2 \end{bmatrix} = \begin{bmatrix} 0 \\ 0 \end{bmatrix} \quad \text{which gives} \quad \xi_1 = -\xi_2.$$

The vector $\boldsymbol{\xi}_1 = k_1 \begin{bmatrix} 1 \\ -1 \end{bmatrix}$ for any $k_1 \neq 0$ is thus an eigenvector associated with $\lambda = -1$. For $\lambda = 1$, we solve

$$\begin{bmatrix} 1 & -1 \\ -1 & 1 \end{bmatrix} \begin{bmatrix} \xi_1 \\ \xi_2 \end{bmatrix} = \begin{bmatrix} 0 \\ 0 \end{bmatrix} \quad \text{which gives} \quad \xi_1 = \xi_2.$$

Thus $\boldsymbol{\xi}_2 = k_2 \begin{bmatrix} -1 \\ 1 \end{bmatrix}$ for any $k_2 \neq 0$ is an eigenvector associated with $\lambda = 1$.

Note: In this example all eigenvectors corresponding to one of the eigenvalues span a one-dimensional space (i.e., all the eigenvectors corresponding to $\lambda_1 = -1$ are colinear with $\begin{bmatrix} 1 \\ -1 \end{bmatrix}$, and all eigenvectors corresponding to λ_2 are colinear with $\begin{bmatrix} 1 \\ 1 \end{bmatrix}$).

Definition A.19 *The space spanned by the eigenvectors corresponding to an eigenvalue λ of a square matrix F is called the eigenspace associated with λ.*

Example A.21
For

$$F = \begin{bmatrix} 1 & 1 \\ 0 & 1 \end{bmatrix} \quad \text{so} \quad \det[\lambda I - F] = \det \begin{bmatrix} \lambda - 1 & -1 \\ 0 & \lambda - 1 \end{bmatrix} = (\lambda - 1)^2 = 0.$$

Thus $\lambda = 1$ is the only eigenvalue, although it is a repeated root of the characteristic equation. Using $\lambda = 1$, the eigenvector is found by solving the algebraic equations

$$\begin{bmatrix} 0 & 1 \\ 0 & 0 \end{bmatrix} \begin{bmatrix} \xi_1 \\ \xi_2 \end{bmatrix} = \begin{bmatrix} 0 \\ 0 \end{bmatrix}$$

which gives ξ_1 = an arbitrary number and $\xi_2 = 0$. The eigenvector associated with $\lambda = 1$ is thus $\boldsymbol{\xi} = k \begin{bmatrix} 1 \\ 0 \end{bmatrix}$ for any $k \neq 0$. Here again the eigenspace associated with $\lambda = 1$ is one-dimensional.

Example A.22
For

$$F = \begin{bmatrix} 2 & 0 \\ 0 & 2 \end{bmatrix} \quad \text{so} \quad \det[\lambda I - F] = \det \begin{bmatrix} \lambda - 2 & 0 \\ 0 & \lambda - 2 \end{bmatrix} = (\lambda - 2)^2 = 0.$$

Therefore $\lambda = 2$ is the only eigenvalue. Using $\lambda = 2$, the eigenvector is found by solving the algebraic equations

$$\begin{bmatrix} 0 & 0 \\ 0 & 0 \end{bmatrix} \begin{bmatrix} \xi_1 \\ \xi_2 \end{bmatrix} = \begin{bmatrix} 0 \\ 0 \end{bmatrix}$$

and we see that any two-dimensional vector $\boldsymbol{\xi}$ will satisfy. Thus we can choose any two linearly independent vectors as the eigenvectors corresponding to $\lambda = 2$. The eigenspace associated with $\lambda = 2$ is two-dimensional.

Definition A.20 *For F a square matrix of order n, with characteristic equation*

$$\det[\lambda I - F] = (\lambda - \lambda_1)^{n_1}(\lambda - \lambda_2)^{n_2} \ldots (\lambda - \lambda_s)^{n_s} = 0$$

with $\lambda_1, \lambda_2, \ldots, \lambda_s$ the distinct eigenvalues and $n_1 + n_2 + \cdots + n_s = n$, we call n_i the algebraic multiplicity of λ_i in the characteristic equation.

In Example A.20, the algebraic multiplicity of both -1 and 1 was one. In Examples A.21 and A.22 the algebraic multiplicity of the single eigenvalue was two.

Definition A.21 *The geometric multiplicity of λ_i is the dimension of the eigenspace associated with λ_i, for a square matrix F of order n, with the distinct eigenvalues $\lambda_1, \lambda_2, \ldots, \lambda_s$. We will denote this as g_i.*

Note: $g_i = n - \text{rank}[F - \lambda_i I]$. In Example A.20, the geometric multiplicity of both -1 and 1 was one. In Examples A.21 the geometric multiplicity of 1 was one. In A.22 the geometric multiplicity of 2 was two. Note that $g_i \leq n_i$ for any i.

Theorem A.14 *Let F be an $n \times n$ matrix and let $\lambda_1, \lambda_2, \ldots, \lambda_s$ be distinct eigenvalues of F. Let $\boldsymbol{\xi}_i$ be any eigenvector corresponding to λ_i. The vectors $\boldsymbol{\xi}_1, \boldsymbol{\xi}_2, \ldots, \boldsymbol{\xi}_s$ are linearly independent.*

Proof: Let j be the largest integer for which $\boldsymbol{\xi}_1, \boldsymbol{\xi}_2, \ldots, \boldsymbol{\xi}_j$ are linearly independent. If $j < s$ then $\boldsymbol{\xi}_1, \boldsymbol{\xi}_2, \ldots, \boldsymbol{\xi}_j, \boldsymbol{\xi}_{j+1}$ are linearly dependent, so we can write $\boldsymbol{\xi}_{j+1}$ as a linear combination of $\boldsymbol{\xi}_1, \boldsymbol{\xi}_2, \ldots, \boldsymbol{\xi}_j$ in the form

$$\boldsymbol{\xi}_{j+1} = d_1\boldsymbol{\xi}_1 + d_2\boldsymbol{\xi}_2 + \cdots + d_j\boldsymbol{\xi}_j \tag{A.12}$$

where not all the $d_i = 0$ since $\boldsymbol{\xi}_{j+1} \neq \mathbf{0}$. Now

$$A\boldsymbol{\xi}_{j+1} = d_1 A\boldsymbol{\xi}_1 + d_2 A\boldsymbol{\xi}_2 + \cdots + d_j A\boldsymbol{\xi}_j$$

so

$$\lambda_{j+1}\boldsymbol{\xi}_{j+1} = d_1\lambda_1\boldsymbol{\xi}_1 + d_2\lambda_2\boldsymbol{\xi}_2 + \cdots + d_j\lambda_j\boldsymbol{\xi}_j. \tag{A.13}$$

Here $\lambda_{j+1} \neq 0$ because if $\lambda_{j+1} = 0$ then $\lambda_i \neq 0$ for $i = 1, 2, \ldots, j$ since the $\lambda_1, \lambda_2, \ldots, \lambda_s$ are distinct, and since not all d_i are zero (A.13) says the $\boldsymbol{\xi}_1, \boldsymbol{\xi}_2, \ldots, \boldsymbol{\xi}_j$ are linearly dependent, which is a contradiction. Therefore we can divide (A.13) by λ_{j+1} to get

$$\boldsymbol{\xi}_{j+1} = \frac{\lambda_1}{\lambda_{j+1}} d_1\boldsymbol{\xi}_1 + \frac{\lambda_2}{\lambda_{j+1}} d_2\boldsymbol{\xi}_2 + \cdots + \frac{\lambda_j}{\lambda_{j+1}} d_j\boldsymbol{\xi}_j. \tag{A.14}$$

Since the representation of $\boldsymbol{\xi}_{j+1}$ as a linear combination of the linearly independent vectors $\boldsymbol{\xi}_1, \boldsymbol{\xi}_2, \ldots, \boldsymbol{\xi}_j$ is unique, comparing (A.12) and (A.14) we see that $d_i(\lambda_i/\lambda_{j+1}) = d_i$ for all $i = 1, 2, \ldots, j$. Since not all d_i are zero there is at least one, say, $d_k \neq 0$, which shows that $\lambda_k = \lambda_{j+1}$. This is a contradiction since the eigenvalues are distinct. Therefore j cannot be less than s. Since it can never be greater than s it must equal s. QED.

Theorem A.15 *Let F be an $n \times n$ matrix and let $\lambda_1, \lambda_2, \ldots, \lambda_s$ be distinct eigenvalues of F. For each λ_i, let $\boldsymbol{\xi}_{i1}, \boldsymbol{\xi}_{i2}, \ldots, \boldsymbol{\xi}_{ig_i}$ be a set of linearly independent eigenvectors corresponding to λ_i. The entire set of vectors.*

$$\boldsymbol{\xi}_{11}, \boldsymbol{\xi}_{12}, \ldots, \boldsymbol{\xi}_{1g1}, \boldsymbol{\xi}_{21}, \boldsymbol{\xi}_{22}, \ldots, \boldsymbol{\xi}_{2g2}, \ldots, \boldsymbol{\xi}_{s1}, \boldsymbol{\xi}_{s2}, \ldots, \boldsymbol{\xi}_{sg_s}$$

are linearly independent.

Proof: Assume they are dependent, then

$$\sum_{i=1}^{s}\sum_{j=1}^{g_i} \alpha_{ij}\boldsymbol{\xi}_{ij} = \mathbf{0} \tag{A.15}$$

where not all the α_{ij} are zero. Let $\boldsymbol{\eta}_i = \sum_{j=1}^{g_i} \alpha_{ij}\boldsymbol{\xi}_{ij}$, so $\boldsymbol{\eta}_i$ is either an eigenvector corresponding to λ_i or it is $\mathbf{0}$. Now if $\boldsymbol{\eta}_i \neq \mathbf{0}$ for some set of values of i, call this set $\mathcal{S}$. We see then that (A.15) says that

$$\sum_{i\in\mathcal{S}} \boldsymbol{\eta}_i = \mathbf{0}$$

which says that the set $\boldsymbol{\eta}_i$ for $i \in \mathcal{S}$ is linearly dependent, but this cannot be true since these are eigenvectors associated with distinct eigenvalues. So $\mathcal{S}$ must be empty and all $\boldsymbol{\eta}_i = \mathbf{0}$. This says that

$$\sum_{j=1}^{g_i} \alpha_{ij}\boldsymbol{\xi}_{ij} = \mathbf{0} \quad \text{for all } i.$$

If any of these $\alpha_{ij} \neq 0$ we get a contradiction because the sets $\boldsymbol{\xi}_{i1}, \boldsymbol{\xi}_{i2}, \ldots, \boldsymbol{\xi}_{ig_i}$ were defined to be linearly independent, so the $\alpha_{ij} = 0$ for all i and j. This says that the original set must be linearly independent. QED.

A.14.2 Diagonalization of matrices

Definition A.22 *Two $n \times n$ matrices A and B are said to be similar if a nonsingular matrix T exists such that $T^{-1}AT = B$. The matrix T is called a similarity transformation.*

Definition A.23 *A square matrix F of order n is called diagonalizable if it is similar to a diagonal matrix. That is, if a nonsingular matrix T exists such that*

$$T^{-1}FT = \Lambda \tag{A.16}$$

where Λ is a diagonal matrix with diagonal elements $\lambda_1, \lambda_2, \ldots, \lambda_n$.

Theorem A.16 *A square matrix F of order n is diagonalizable if and only if it has n linearly independent eigenvectors.*

Proof: (if) If there are n linearly independent eigenvectors $\boldsymbol{\xi}_1, \boldsymbol{\xi}_2, \ldots, \boldsymbol{\xi}_n$, let $T = [\boldsymbol{\xi}_1, \boldsymbol{\xi}_2 \ldots \boldsymbol{\xi}_n]$.

Then

$$FT = [F\boldsymbol{\xi}_1\ F\boldsymbol{\xi}_2\ \dots\ F\boldsymbol{\xi}_n] = [\lambda_1\boldsymbol{\xi}_1\ \lambda_2\boldsymbol{\xi}_2\ \dots\ \lambda_n\boldsymbol{\xi}_n]$$

$$= [\boldsymbol{\xi}_1\ \boldsymbol{\xi}_2\ \dots\ \boldsymbol{\xi}_n] \begin{bmatrix} \lambda_1 & 0 & \dots & 0 \\ 0 & \lambda_2 & \dots & 0 \\ \vdots & \vdots & \ddots & \vdots \\ 0 & 0 & \dots & \lambda_n \end{bmatrix} = T\Lambda$$

so $T^{-1}FT = \Lambda$. Thus the diagonalizing matrix T is the matrix having the n linearly independent eigenvectors as its columns. The diagonal matrix Λ which is similar to F has the eigenvalues of F as its diagonal elements.

(only if) Suppose F is diagonalizable, then there exists a T such that $T^{-1}FT = \Lambda$ so $FT = T\Lambda$ which says that $F\boldsymbol{t}_i = \lambda_i\boldsymbol{t}_i$ for all $i = 1, 2, \dots, n$ where $\boldsymbol{t}_1, \boldsymbol{t}_2, \dots, \boldsymbol{t}_n$ denote the columns of T. Thus we see λ_i is an eigenvalue of F and $\boldsymbol{t}_i$ is the corresponding eigenvector. Since $\boldsymbol{t}_1, \boldsymbol{t}_2, \dots, \boldsymbol{t}_n$ are the columns of a nonsingular matrix, they are linearly independent. QED.

Theorem A.17 *A square matrix F of order n is diagonalizable if and only if $g_i = n_i$ for all i (i.e., the geometric and algebraic multiplicity of each eigenvalue are the same).*

Proof: (if) We know that for all the distinct eigenvalues $\lambda_1, \lambda_2, \dots, \lambda_s$ of F, we can obtain $g_1 + g_2 + \cdots + g_s$ linearly independent eigenvectors

$$\boldsymbol{\xi}_{11}, \boldsymbol{\xi}_{12}, \dots, \boldsymbol{\xi}_{1g_1}, \boldsymbol{\xi}_{21}, \boldsymbol{\xi}_{22}, \dots, \boldsymbol{\xi}_{2g_2}, \dots, \boldsymbol{\xi}_{s1}, \boldsymbol{\xi}_{s2}, \dots, \boldsymbol{\xi}_{sg_s}.$$

Now $n_1 + n_2 + \cdots + n_s = n$, so when $g_i = n_i$ for all i this says that $g_1 + g_2 + \cdots + g_s = n$, so there are n linearly independent eigenvectors, and thus F is diagonalizable.

(only if) When F is diagonalizable, Theorem A.16 says that F has a set of n linearly independent eigenvectors. Call these

$$\boldsymbol{\xi}_{11}, \boldsymbol{\xi}_{12}, \dots, \boldsymbol{\xi}_{1p_1}, \boldsymbol{\xi}_{21}, \boldsymbol{\xi}_{22}, \dots, \boldsymbol{\xi}_{2p_2}, \dots, \boldsymbol{\xi}_{s1}, \boldsymbol{\xi}_{s2}, \dots, \boldsymbol{\xi}_{sp_s}$$

where $\boldsymbol{\xi}_{i1}, \boldsymbol{\xi}_{i2}, \dots, \boldsymbol{\xi}_{ip_i}$ are eigenvectors corresponding to the ith distinct eigenvalue and $p_1 + p_2 + \cdots + p_s = n$. Now $p_i \le g_i$ for each i, since g_i is the dimension of the eigenspace associated with λ_i. Thus

$$n = \sum_{i=1}^{s} p_i \le \sum_{i=1}^{s} g_i \le \sum_{i=1}^{s} n_i = n$$

since $g_i \le n_i$ for each i. Thus we see that $g_i = n_i$ for each i. QED.

Corollary to Theorem A.17 *A square matrix F of order n which has n distinct eigenvalues is diagonalizable.*

This follows directly from Theorem A.16: since all n eigenvalues are distinct, the n corresponding eigenvectors are linearly independent, so F is diagonalizable.

Example A.23
For

$$F = \begin{bmatrix} 1 & 0 & 0 \\ 0 & 2 & 1 \\ 0 & 0 & 1 \end{bmatrix} \quad \text{then} \quad \det(\lambda I - F) = (\lambda - 1)^2(\lambda - 2)$$

so $\lambda = 1$ has algebraic multiplicity $n_1 = 2$, and $\lambda = 2$ has algebraic multiplicity $n_2 = 1$. For $\lambda = 2$ solve

$$[F - 2I]\boldsymbol{\xi} = \begin{bmatrix} -1 & 0 & 0 \\ 0 & 0 & 1 \\ 0 & 0 & -1 \end{bmatrix} \begin{bmatrix} \xi_1 \\ \xi_2 \\ \xi_3 \end{bmatrix} = \begin{bmatrix} 0 \\ 0 \\ 0 \end{bmatrix}$$

so $\xi_1 = 0$, $\xi_3 = 0$, and ξ_2 is arbitrary. Thus $\boldsymbol{\xi}_1 = k_1 \begin{bmatrix} 0 \\ 1 \\ 0 \end{bmatrix}$ is an eigenvector corresponding to $\lambda = 2$. For $\lambda = 1$ solve

$$[F - I]\boldsymbol{\xi} = \begin{bmatrix} 0 & 0 & 0 \\ 0 & 1 & 1 \\ 0 & 0 & 0 \end{bmatrix} \begin{bmatrix} \xi_1 \\ \xi_2 \\ \xi_3 \end{bmatrix} = \begin{bmatrix} 0 \\ 0 \\ 0 \end{bmatrix}$$

so $\xi_2 + \xi_3 = 0$ and $\boldsymbol{\xi}_1$ is arbitrary. The general solution has the form $\boldsymbol{\xi} = c_1 \begin{bmatrix} 1 \\ 0 \\ 0 \end{bmatrix} + c_2 \begin{bmatrix} 0 \\ 1 \\ -1 \end{bmatrix}$ so $\boldsymbol{\xi}_2 = k_2 \begin{bmatrix} 1 \\ 0 \\ 0 \end{bmatrix}$ and $\boldsymbol{\xi}_3 = k_3 \begin{bmatrix} 0 \\ 1 \\ -1 \end{bmatrix}$ are two independent eigenvectors corresponding to $\lambda = 1$. We see that $g_2 = 2$. F can be diagonalized using the transformation

$$T = \begin{bmatrix} 0 & 1 & 0 \\ 1 & 0 & 1 \\ 0 & 0 & -1 \end{bmatrix} \quad \text{so} \quad T^{-1} = \begin{bmatrix} 0 & 1 & 1 \\ 1 & 0 & 0 \\ 0 & 0 & -1 \end{bmatrix} \quad \text{and} \quad T^{-1}FT = \begin{bmatrix} 2 & 0 & 0 \\ 0 & 1 & 0 \\ 0 & 0 & 1 \end{bmatrix}.$$

A.15 Generalized eigenvectors

We showed that a square matrix of order n can be diagonalized if and only if $g_i = n_i$ for all i. When this is not true, we cannot find n linearly independent eigenvectors.

Example A.24

$$A = \begin{bmatrix} 1 & 1 & 2 \\ 0 & 1 & 3 \\ 0 & 0 & 2 \end{bmatrix} \quad \text{so } \det[\lambda I - A] = (\lambda - 1)^2(\lambda - 2)$$

and we see that $\lambda_1 = 1$ has $n_1 = 2$ and $\lambda_2 = 2$ has $n_2 = 1$. Next we find for $\lambda = \lambda_1 = 1$ that

$$[A - I]\boldsymbol{\xi} = \begin{bmatrix} 0 & 1 & 2 \\ 0 & 0 & 3 \\ 0 & 0 & 1 \end{bmatrix} \begin{bmatrix} \xi_1 \\ \xi_2 \\ \xi_3 \end{bmatrix} = \begin{bmatrix} 0 \\ 0 \\ 0 \end{bmatrix}$$

so $\xi_2 = -2\xi_3$ and $\xi_3 = 0$ so $\xi_2 = 0$. We find therefore that $\boldsymbol{\xi} = k_1 \begin{bmatrix} 1 \\ 0 \\ 0 \end{bmatrix}$, so there is only one linearly independent eigenvector corresponding to $\lambda_1 = 1$, so $g_1 = 1 \neq n_1$. For $\lambda = \lambda_2 = 2$ that

$$[A - 2I]\boldsymbol{\xi} = \begin{bmatrix} -1 & 1 & 2 \\ 0 & -1 & 3 \\ 0 & 0 & 0 \end{bmatrix} \begin{bmatrix} \xi_1 \\ \xi_2 \\ \xi_3 \end{bmatrix} = \begin{bmatrix} 0 \\ 0 \\ 0 \end{bmatrix}$$

so $\xi_2 = 3\xi_3$ and $\xi_1 = \xi_2 + 2\xi_3 = 5\xi_3$. A general solution of these equations has the form $\boldsymbol{\xi} = k_2 \begin{bmatrix} 5 \\ 3 \\ 1 \end{bmatrix}$. We see that there is one degree of freedom (i.e., one arbitrary constant), so $g_2 = 1$.

Since A is third order and has only two linearly independent eigenvectors, it cannot be diagonalized.

Definition A.24 *A vector $\boldsymbol{\xi}_i^{(k)}$ is called a generalized eigenvector of rank k corresponding to eigenvalue λ_i if*

$$[A - \lambda_i I]^{(k-1)}\boldsymbol{\xi}_i^{(k)} \neq \mathbf{0} \quad \textit{but} \quad [A - \lambda_i I]^k \boldsymbol{\xi}_i^{(k)} = \mathbf{0}.$$

Note: The eigenvectors we have considered until now are of rank 1. Consider the set of vectors generated as follows:

$$\begin{aligned} [A - \lambda_i I]\boldsymbol{\xi}_i^{(1)} &= \mathbf{0} \leftarrow \text{the usual eigenvector} \\ [A - \lambda_i I]\boldsymbol{\xi}_i^{(2)} &= \boldsymbol{\xi}_i^{(1)} \\ [A - \lambda_i I]\boldsymbol{\xi}_i^{(3)} &= \boldsymbol{\xi}_i^{(2)}. \\ \vdots \quad & \quad \vdots \\ [A - \lambda_i I]\boldsymbol{\xi}_i^{(k)} &= \boldsymbol{\xi}_i^{(k-1)}. \end{aligned}$$

Now note that for any $k \geq 1$,

$$[A - \lambda_i I]^{(k-1)}\boldsymbol{\xi}_i^{(k)} = [A - \lambda_i I]^{(k-2)}\boldsymbol{\xi}_i^{(k-1)} = \cdots [A - \lambda_i I]\boldsymbol{\xi}_i^{(2)} = \boldsymbol{\xi}_i^{1} \neq \mathbf{0}. \quad \text{(A.17)}$$

Thus multiplying (A.17) by $[A - \lambda_i I]$ gives

$$[A - \lambda_i I]^k \boldsymbol{\xi}_i^{(k)} = [A - \lambda_i I]\boldsymbol{\xi}_i^{(1)} = \mathbf{0}$$

so we see that each vector $\boldsymbol{\xi}_i^{(k)}$ generated in this way is a generalized eigenvector of rank k corresponding to eigenvalue λ_i.

Theorem A.18 *Eigenvectors of different rank corresponding to λ_i are linearly independent.*

Proof: Suppose

$$c_1\boldsymbol{\xi}_i^{(1)} + c_2\boldsymbol{\xi}_i^{(2)} + \cdots + c_k\boldsymbol{\xi}_i^{(k)} = \mathbf{0}.$$

Then premultiply both sides by $[\lambda_i I - A]^{(k-1)}$ to give

$$c_k[\lambda_i I - A]^{(k-1)}\boldsymbol{\xi}_i^{(k)} = \mathbf{0}$$

and since $[\lambda_i I - A]^{(k-1)}\boldsymbol{\xi}_i^{(k)} \neq \mathbf{0}$, this means that $c_k = 0$. Continue in this way to show successively that $c_{k-1} = 0, c_{k-2} = 0, \ldots, c_1 = 0$, so the vectors are linearly independent. QED.

A.16 Jordan canonical form

For square matrices A which cannot be diagonalized, a canonical form more general than a diagonal matrix can be obtained by using as the similarity transformation a matrix T, obtained from the linearly independent eigenvectors and generalized eigenvectors obtained from them. The resulting matrix $J = T^{-1}AT$ is called the *Jordan canonical matrix*. It has the block diagonal form

$$J = \begin{bmatrix} J_1 & 0 & \cdots & 0 \\ 0 & J_2 & \cdots & 0 \\ \vdots & \vdots & \ddots & \vdots \\ 0 & 0 & \cdots & J_{s'} \end{bmatrix} \quad \text{(A.18)}$$

where each block J_k is a square $(\ell_n \times \ell_k)$ matrix with the form

$$J_k = \begin{bmatrix} \lambda_k & 1 & 0 & \cdots & 0 \\ 0 & \lambda_k & 1 & \cdots & 0 \\ \vdots & \ddots & \ddots & \ddots & \vdots \\ 0 & \cdots & \cdots & \lambda_k & 1 \\ 0 & \cdots & \cdots & \cdots & \lambda_k \end{bmatrix}. \quad \text{(A.19)}$$

Note that all elements on the main diagonal equal λ_k and all elements on the first superdiagonal (i.e., where the column index is one greater than the row index) equal 1. The order ℓ_k of J_k is equal to the highest rank generalized eigenvector which can be generated from $\boldsymbol{\xi}_k^{(1)}$. There are as many blocks J_k as there are linearly independent ordinary eigenvectors (i.e., of rank 1), so by this definition $s' \geq s =$ the number of distinct eigenvalues. Note that here the λ_k are not all distinct. The index k changes for each linearly independent eigenvector of rank 1. Each distinct eigenvalue λ_i for $i = 1, \ldots, s$ has g_i blocks associated with it, where g_i is the geometric multiplicity of λ_i. When each distinct eigenvalue has only one independent eigenvector of rank 1 (i.e., $g_i = 1$ for all λ_i), then $s' = s$, $\ell_i = n_i$ the algebraic multiplicity of λ_i where $\Sigma_{i=1}^{s} n_i = n$.

For diagonalizible matrices, since there exist n linearly independent eigenvectors, each Jordan block J_i is of order 1, so the matrix J above is diagonal. In other words, a diagonal matrix is a special case of the more general Jordan canonical matrix.

We will now show that when we form the similarity transformation T as

$$T = [\boldsymbol{\xi}_i^{(1)}, \ldots, \boldsymbol{\xi}_i^{(\ell_1)}, \boldsymbol{\xi}_2^{(1)}, \ldots, \boldsymbol{\xi}_2^{(\ell_2)}, \ldots, \boldsymbol{\xi}_{s'}^{1}, \ldots, \boldsymbol{\xi}_{s'}^{(\ell_{s'})}], \tag{A.20}$$

then $T^{-1}AT$ is in the Jordan canonical form above. Now since $[A - \lambda_k I]\boldsymbol{\xi}_k^{(i)} = \boldsymbol{\xi}_k^{(i-1)}$ this means that $A\boldsymbol{\xi}_k^{(i)} = \lambda_k I \boldsymbol{\xi}_k^{(i)} + \boldsymbol{\xi}_k^{(i-1)}$, so

$$\begin{aligned}
AT =& [\lambda_1 \boldsymbol{\xi}_1^{(1)}, \lambda_1 \boldsymbol{\xi}_1^{(2)}, \boldsymbol{\xi}_1^{(1)}, \ldots, \lambda_1 \boldsymbol{\xi}_1^{(\ell_1)} \\
& + \boldsymbol{\xi}_i^{(\ell_1 - 1)}, \lambda_2 \boldsymbol{\xi}_2^{(1)}, \lambda_2 \boldsymbol{\xi}_2^{(2)} + \boldsymbol{\xi}_2^{(1)}, \ldots, \lambda_{s'} \boldsymbol{\xi}_{s'}^{(\ell^{s'})}, \boldsymbol{\xi}_{s'}^{(\ell^{s'} - 1)} \\
=& T \left[\begin{array}{ccccc|ccccc|c|ccccc}
\lambda_1 & 1 & 0 & \cdots & 0 & \cdot & \cdot & \cdots & \cdot & & \cdots & \cdot & \cdot & \cdots & & 0 \\
0 & \lambda_1 & 1 & \cdots & 0 & \cdot & \cdot & \cdots & \cdot & & \cdots & \cdot & \cdot & \cdots & & 0 \\
\vdots & \vdots & \ddots & & \vdots & \vdots & \vdots & \ddots & \vdots & & \ddots & \vdots & \vdots & \ddots & & \vdots \\
0 & 0 & \cdots & \lambda_1 & 1 & \cdot & \cdot & \cdots & \cdot & & \cdots & \cdot & \cdot & \cdots & & 0 \\
0 & 0 & \cdot & \cdots & \lambda_1 & \cdot & \cdot & \cdots & \cdot & & \cdots & \cdot & \cdot & \cdots & & 0 \\
\hline
0 & 0 & \cdot & \cdots & \cdot & \lambda_2 & 1 & 0 & \cdots & 0 & \cdots & \cdot & \cdot & \cdots & & 0 \\
0 & 0 & \cdot & \cdots & \cdot & 0 & \lambda_2 & 1 & \cdots & 0 & \cdots & \cdot & \cdot & \cdots & & 0 \\
\vdots & \vdots & \vdots & \ddots & \vdots & \vdots & \vdots & \ddots & \ddots & \vdots & \ddots & \vdots & \vdots & \ddots & & \vdots \\
0 & \cdot & \cdot & \cdots & \cdot & 0 & 0 & \cdots & \lambda_2 & 1 & \cdots & \cdot & \cdot & \cdots & & 0 \\
0 & \cdot & \cdot & \cdots & \cdot & 0 & 0 & \cdot & \cdots & \lambda_2 & \cdots & \cdot & \cdot & \cdots & & 0 \\
\hline
\vdots & \vdots & \vdots & \ddots & \vdots & \vdots & \vdots & \vdots & \ddots & \vdots & \ddots & \vdots & \vdots & \vdots & \ddots & \vdots \\
\hline
0 & \cdot & \cdot & \cdots & \cdot & \cdot & \cdot & \cdots & & \cdot & \cdots & \lambda_{s'} & 1 & 0 & \cdots & 0 \\
0 & \cdot & \cdot & \cdots & \cdot & \cdot & \cdot & \cdots & & \cdot & \cdots & 0 & \lambda_{s'} & 1 & \cdots & 0 \\
\vdots & \vdots & \vdots & \ddots & \vdots & \vdots & \vdots & \ddots & \ddots & \vdots & \ddots & \vdots & \vdots & \vdots & \ddots & \vdots \\
0 & \cdot & \cdot & \cdots & \cdot & \cdot & \cdot & \cdots & & \cdot & \cdots & \cdot & \cdot & \cdot & \lambda_{s'} & 1 \\
0 & \cdot & \cdot & \cdots & \cdot & \cdot & \cdot & \cdots & & \cdot & \cdots & \cdot & \cdot & \cdot & \cdots & \lambda_{s'}
\end{array}\right] \\
=& TJ
\end{aligned}$$

so $J = T^{-1}AT$.

Example A.25

Consider the matrix $A = \begin{bmatrix} 3 & -1 & 1 \\ 0 & 2 & 0 \\ 1 & -1 & 1 \end{bmatrix}$. From

$$\det[\lambda I - A] = \det \begin{bmatrix} \lambda - 3 & 1 & -1 \\ 0 & \lambda - 2 & 0 \\ -1 & 1 & \lambda - 1 \end{bmatrix} = (\lambda-2)(\lambda^2-4\lambda+4) = (\lambda-2)^3 = 0$$

we see that $\lambda - 2$ is the only eigenvalue with algebraic multiplicity $\alpha(2) = 3$. Now using $[A - 2I]\boldsymbol{\xi} = \mathbf{0}$ (which is the same as $[2I - A]\boldsymbol{\xi} = \mathbf{0}$ we get

$$\begin{bmatrix} 1 & -1 & -1 \\ 0 & 0 & 0 \\ 1 & -1 & -1 \end{bmatrix} \begin{bmatrix} \xi_1 \\ \xi_2 \\ \xi_3 \end{bmatrix} = \begin{bmatrix} 0 \\ 0 \\ 0 \end{bmatrix}.$$

There is only one independent equation $\xi_1 - \xi_2 - \xi_3 = 0$. Since one equation in three unknowns gives two degrees of freedom, we can choose any two linearly independent solutions. Let us choose $\xi_1 = \xi_2 = 1, \xi_3 = 0$ as one solution to give the eigenvector $\xi_1^{(1)} = \begin{bmatrix} 1 \\ 1 \\ 0 \end{bmatrix}$ and choose $\xi_1 = \xi_3 = 1, \xi_2 = 0$ as the other solution to give a second independent eigenvector $\boldsymbol{\xi}_2^{(1)} = \begin{bmatrix} 1 \\ 0 \\ 1 \end{bmatrix}$. We see that the geometric multiplicity is $g(2) = 2$.

To get a generalized eigenvector, let us try to generate one from $[A - 2I]\boldsymbol{\xi}_1^{(2)} = \boldsymbol{\xi}_1^{(1)}$. Thus

$$\begin{bmatrix} 1 & -1 & -1 \\ 0 & 0 & 0 \\ 1 & -1 & -1 \end{bmatrix} \begin{bmatrix} \xi_1 \\ \xi_2 \\ \xi_3 \end{bmatrix} = \begin{bmatrix} 1 \\ 1 \\ 0 \end{bmatrix}.$$

We can see that this has no solution, so no higher rank generalized eigenvector can be generated from $\boldsymbol{\xi}_1^{(1)}$. Using $[A - 2I]\boldsymbol{\xi}_2^{(2)} = \boldsymbol{\xi}_2^{(1)}$, we find

$$\begin{bmatrix} 1 & -1 & -1 \\ 0 & 0 & 0 \\ 1 & -1 & -1 \end{bmatrix} \begin{bmatrix} \xi_1 \\ \xi_2 \\ \xi_3 \end{bmatrix} = \begin{bmatrix} 1 \\ 0 \\ 1 \end{bmatrix}$$

which gives one independent equation, $\xi_1 - \xi_2 - \xi_3 = 1$. Any solution of this will give us a generalized eigenvector of rank two. Let us choose $\xi_1 = 1, \xi_2 = \xi_3 = 0$. Note that any other solution will be linearly dependent on the eigenvectors and generalized eigenvector already chosen. The canonical transformation we have found is

$$T = [\boldsymbol{\xi}_1^{(1)}, \boldsymbol{\xi}_2^{(1)}, \boldsymbol{\xi}_2^{(2)}] = \begin{bmatrix} 1 & 1 & 1 \\ 1 & 0 & 0 \\ 0 & 1 & 0 \end{bmatrix}$$

and

$$T^{-1}AT = \begin{bmatrix} 0 & 1 & 0 \\ 0 & 0 & 1 \\ 1 & -1 & -1 \end{bmatrix} \begin{bmatrix} 3 & -1 & -1 \\ 0 & 2 & 0 \\ 1 & -1 & 1 \end{bmatrix} \begin{bmatrix} 1 & 1 & 1 \\ 1 & 0 & 0 \\ 0 & 1 & 0 \end{bmatrix} = \begin{bmatrix} 2 & 0 & 0 \\ 0 & 2 & 1 \\ 0 & 0 & 2 \end{bmatrix} = J$$

which is in Jordan canonical form

Theorem A.19 *If A is a Hermitian matrix (recall that a real symmetric matrix is Hermitian) then*

1 *All eigenvalues of A are real and*
2 *The eigenvectors $\boldsymbol{\xi}_j$ and $\boldsymbol{\xi}_k$ associated with distinct eigenvalues are orthogonal (perpendicular), that is, $\boldsymbol{\xi}_j^T\boldsymbol{\xi}_k = 0$.*

Proof:

1 Let λ_i be an eigenvalue and $\boldsymbol{\xi}_i$ its associated eigenvector. Then $A\boldsymbol{\xi}_i = \lambda_1\boldsymbol{\xi}_i$, and using $\bar{\boldsymbol{\xi}}_i$ to denote the complex conjugate of $\boldsymbol{\xi}_i$ we see that $\bar{\boldsymbol{\xi}}_i^T A\boldsymbol{\xi}_i = \lambda_i\bar{\boldsymbol{\xi}}_i^T\boldsymbol{\xi}_i$. Taking thc complcx conjugate of this gives $\boldsymbol{\xi}_i^T\bar{A}\bar{\boldsymbol{\xi}}_i^T = \bar{\lambda}_1\boldsymbol{\xi}_i^T\bar{\boldsymbol{\xi}}_i$, and taking the transpose gives $\bar{\boldsymbol{\xi}}_i^T\boldsymbol{\xi}_i = \bar{\lambda}_i\bar{\boldsymbol{\xi}}_i^T\boldsymbol{\xi}_i$. But since A is Hermitian, $\bar{A}^T = A$, so we see that $\bar{\lambda}_i = \lambda_i$ which says that λ_i is real.
2 Using $A\boldsymbol{\xi}_j = \lambda_j\boldsymbol{\xi}_j$ and $A\boldsymbol{\xi}_k = \lambda_k\boldsymbol{\xi}_k$ where $\lambda_j \neq \lambda_k$ we see that

$$\bar{\boldsymbol{\xi}}_k^T A\boldsymbol{\xi}_j = \lambda_j\bar{\boldsymbol{\xi}}^T{}_k\boldsymbol{\xi}_j^T \quad \text{and} \quad \bar{\boldsymbol{\xi}}_j^T A\boldsymbol{\xi}_k = \lambda_k\bar{\boldsymbol{\xi}}_k^T\boldsymbol{\xi}_k.$$

Now take the complex conjugate, transpose the second equation, and use $\bar{A}^T = A$ to get

$$\lambda_j\bar{\boldsymbol{\xi}}_k^T\boldsymbol{\xi}_j^T = \bar{\boldsymbol{\xi}}_k^T A\boldsymbol{\xi}_j = \lambda_j\bar{\boldsymbol{\xi}}_k^T\boldsymbol{\xi}_j^T.$$

Now if $\bar{\boldsymbol{\xi}}_k^T\boldsymbol{\xi}_j \neq 0$ the above equation says that $\lambda_j = \lambda_k$, but since λ_j and λ_k are distinct, this is not true, so we see that $\bar{\boldsymbol{\xi}}_k^T\boldsymbol{\xi}_j = 0$. QED.

Definition A.25 *P is called an orthogonal matrix if $P^{-1} = P^T$.*

Theorem A.20 *If A is real and symmetric then an orthogonal matrix P exists such that P^TAP is diagonal with diagonal elements equal to the eigenvalues of A.*

Note in the proof that follows we will assume that the eigenvalues of A are distinct. This is not needed in a more complicated proof of the theorem.

Proof: We know that using $T = [\boldsymbol{\xi}_1, \boldsymbol{\xi}_2, \cdots, \boldsymbol{\xi}_n]$ will give $T^{-1}AT = \Lambda$. If we just define the columns of P to be $\boldsymbol{p}_i = \boldsymbol{\xi}_i/||\boldsymbol{\xi}_i||$ (i.e., the $\boldsymbol{p}_i$ are unit-length eigenvectors) then $P^{-1}AP = \Lambda$. But since $\boldsymbol{p}_i^T\boldsymbol{p}_i = 1$ and $\boldsymbol{p}_j^T\boldsymbol{p}_i = 0$ for $j \neq i$, we see that $P^TP = I$, so $P^T = P^{-1}$. QED.

A.17 The Cayley-Hamilton theorem

This theorem states a very useful property of square matrices.

Theorem A.21 *Every square matrix A satisfies its own characteristic equation. That is, for $p(\lambda) = \det(\lambda I - A) = \lambda^n + a_{n-1}\lambda^{n-1} + \cdots + a_1\lambda + a_0$, the matrix $p(A)$ defined by*

$$p(A) = A^n + a_{n-1}A^{n-1} + \cdots + a_1 A + a_0 I$$

equals the zero matrix Θ.

Proof: Let us denote the adjoint matrix $\mathrm{Adj}[\lambda I - A]$ by $\Gamma(\lambda)$ and note that since each element $\Gamma_{ij}(\lambda)$ of $\Gamma(\lambda)$ is obtained from the determinant of an $(n-1) \times (n-1)$ matrix, each $\Gamma_{ij}(\lambda)$ is a polynomial in λ of degree no greater than λ^{n-1}. Thus we can write for $\Gamma(\lambda)$

$$\Gamma(\lambda) = K_{n-1}\lambda^{n-1} + K_{n-2}\lambda^{n-2} + \cdots + K_1\lambda + K_0 \tag{A.21}$$

where K_i are $(n \times n)$ constant matrices. From

$$\frac{\Gamma(\lambda)}{p(\lambda)} = [\lambda I - A] = I \tag{A.22}$$

we see that

$$\begin{aligned}(K_{n-1}\lambda^{n-1} + K_{n-2}\lambda^{n-2} + \cdots + K_1\lambda + K_0)[\lambda I - A]\\ = (\lambda^n + a_{n-1}\lambda^{n-1} + \cdots + a_1\lambda + a_0)I.\end{aligned} \tag{A.23}$$

In order for (A.23) to hold for all λ the coefficients of each power of λ must be equal. Thus we get

$$\begin{aligned} K_{n-1} &= I \\ K_{n-2} - K_{n-1}A &= a_{n-1}I \\ K_{n-3} - K_{n-2}A &= a_{n-2}I \\ &\vdots \\ K_0 - K_1 A &= a_1 I \\ -K_0 A &= a_0 I. \end{aligned} \tag{A.24}$$

Now postmultiply the first equation of (A.24) by A_n, the second by A_{n-1}, the third by A_{n-2}, and so on, and add all the resulting equations to get

$$A^n + a_{n-1}A^{n-1} + \cdots + a_1 A + a_0 1 = \Theta. \tag{A.25}$$

This is the desired result. QED.

When all the eigenvectors of A have a rank of one, there is a shorter proof since

$$p(A)\xi_i = A^n\xi_i + a_{n-1}A^{n-1}\xi_i + \cdots + a_1 A\xi_i + a_0\xi_i = p(\lambda_i)\xi_i = \mathbf{0}.$$

Therefore $p(A)[\xi_1, \xi_2, \ldots, \xi_n] = \Theta$, and since the matrix $T = [\xi_1, \xi_2, \ldots, \xi_n]$ is nonsingular, it follows that $p(A) = \Theta$.

The fact that the nth degree polynomial function $p(A) = \Theta$ means that not all powers of A are linearly independent. In fact, it says that the nth power of A can be expressed as a unique linear combination of the powers A^i for $i = 0, 1, \ldots, n-1$. Namely,

$$A^n = -\sum_{i=0}^{n-1} a_i A^i \tag{A.26}$$

thus

$$\begin{aligned} A^{n+1} &= -\sum_{i=0}^{n-1} a_i A^{i+1} = -a_{n-1}A^n - \sum_{i=0}^{n-1} a_{j-i} A^j \\ &= \sum_{i=0}^{n-1} a_{n-1}a_i A^i - \sum_{i=0}^{n-1} a_{j-i}A^j \triangleq \sum_{i=0}^{n-1} \alpha_{n+1,i} A^i. \end{aligned} \tag{A.27}$$

This shows that $A^{(n+1)}$ can also be expressed as a unique linear combination of the powers A^i for $i = 0, \ldots, n-1$. Proceeding in the same way, we can thus show (by induction) that for any $N \geq n$, we can express A^N as a unique linear combination of the powers A^i for $i = 0, \ldots, n-1$. Therefore.

$$A^N = \sum_{i=0}^{n-1} \alpha_{N,i} A^i \quad \text{for any } N \geq n. \tag{A.28}$$

We can also use the theorem to obtain an alternate way to calculate the inverse of A. Just multiply (A.25) by A^{-1} and isolate the resulting A^{-1} term to give

$$A^{-1} = -\frac{1}{a_0}\sum_{i=0}^{n-1} a_{i+1} A^i$$

where $a_n = 1$. Note that the inverse exists if and only if $a_0 \neq 0$. This is to be expected since a_0 equals the product of the eigenvalues which is nonzero if and only if all the eigenvalues are nonzero.

Example A.26

For $A \begin{bmatrix} 0 & 1 \\ -2 & -3 \end{bmatrix}$ we find $p(\lambda) = \lambda^2 + 3\lambda + 2 = 0$, so we get $A^2 + 3A + 2I = \Theta$.

Therefore

$$A^{-1} = -\frac{1}{2}A - \frac{3}{2}I = \begin{bmatrix} -\frac{3}{2} & -\frac{1}{2} \\ 1 & 0 \end{bmatrix}.$$

A.18 Leverrier's algorithm

As a by-product of the general proof of the Cayley-Hamilton theorem, we have an algorithm for getting Adj $[\lambda I - A]$, since the equations (A.24) correspond to the recursion equation

$$K_{j-1} = K_j A + a_j I \quad \text{for } j = n-1, n-2, \ldots, 1 \quad \text{with } K_{n-1} = I. \tag{A.29}$$

Thus we have a procedure for iteratively generating the matrices K_j when the coefficients a_j are known. *Note:* The equation for $j = 0$ can be considered as a check on the roundoff error in computation since K_{-1} should be the zero matrix.

It turns out that we do not even have to evaluate $\det[\lambda I - A]$ to get the coefficients a_j since these can also be calculated iteratively. We can show that

$$\frac{dp(\lambda)}{d\lambda} = \text{tr}\Gamma(\lambda) \tag{A.30}$$

where tr denotes the trace, defined as the sum of the diagonal elements of a square matrix. To do this we consider a matrix F(λ) that has elements which depend on a scalar parameter λ, so that

$$F(\lambda) = \begin{bmatrix} f_{11}(\lambda) & f_{12}(\lambda) & \cdots & f_{1n}(\lambda) \\ f_{21}(\lambda) & f_{22}(\lambda) & \cdots & f_{2n}(\lambda) \\ \vdots & \vdots & \ddots & \vdots \\ f_{n1}(\lambda) & f_{n2}(\lambda) & \cdots & f_{nn}(\lambda) \end{bmatrix}$$

or in terms of its column $\boldsymbol{f}_i(\lambda)$,

$$F(\lambda) = [\boldsymbol{f}_1(\lambda) \mid \boldsymbol{f}_2(\lambda) \mid \cdots \mid \boldsymbol{f}_n(\lambda)].$$

Then by the rule for differentiation of determinants,

$$\frac{d\det F(\lambda)}{d\lambda} = \sum_{i=1}^{n} \det[\boldsymbol{f}_1(\lambda) \mid \boldsymbol{f}_2(\lambda) \mid \cdots \mid \boldsymbol{f}_{i-1}(\lambda) \mid \boldsymbol{f}_i'(\lambda) \mid \boldsymbol{f}_{i+1}(\lambda) \mid \cdots \mid \boldsymbol{f}_n(\lambda)] \tag{A.31}$$

where $\boldsymbol{f}_i'(\lambda)$ denotes the derivative of $\boldsymbol{f}_i(\lambda)$ with respect to λ. Now when $F(\lambda) = [\lambda I - A]$, we see that, since λ appears only in the diagonal elements, $\boldsymbol{f}_i'(\lambda) = \boldsymbol{e}_i =$ a vector with 1 in its ith position and zeros everywhere else. Therefore when the ith term in (A.31) is expanded along the ith column we get

$$\frac{d\det[\lambda I - A]}{d\lambda} = \sum_{i=1}^{n} C_{ii}(\lambda) \tag{A.32}$$

where $C_{ij}(\lambda)$ is the cofactor of the ij element of $[\lambda I - A]$. Note that (A.32) involves only the cofactors of the diagonal elements of $[\lambda I - A]$, and recalling that

$$\Gamma(\lambda) = \text{Adj}[\lambda I - A] = \begin{bmatrix} C_{11}(\lambda) & C_{12}(\lambda) & \cdots & C_{n1}(\lambda) \\ C_{12}(\lambda) & C_{22}(\lambda) & \cdots & C_{n2}(\lambda) \\ \vdots & \vdots & \ddots & \vdots \\ C_{1n}(\lambda) & C_{2n}(\lambda) & \cdots & C_{nn}(\lambda) \end{bmatrix}$$

we see that the sum of the cofactors of the diagonal elements in (A.32) is just the trace of $\Gamma(\lambda)$. This is the result stated in (A.30).

From (A.21) and (A.30) we get

$$\begin{aligned} &\text{tr}(K_{n-1}\lambda^{n-1} + K_{n-2}\lambda^{n-2} + \cdots + K_1\lambda + K_0) \\ &\quad = n\lambda^{n-1} + (n-1)a_{n-1}\lambda^{n-2} + \cdots + 2a_2\lambda + a_1. \end{aligned} \tag{A.33}$$

Again, in order for (A.33) to be valid for all λ, the coefficients of each power of λ must be equal. Thus

$$\begin{aligned} \text{tr } K_{n-1} &= n \\ \text{tr } K_{n-2} &= (n-1)a_{n-1} \\ &\vdots \\ \text{tr } K_0 &= a_1 \end{aligned}$$

which for $a_n = 1$ corresponds to the recursion equation

$$\text{tr }(K_{j-1}) = ja_j \quad \text{for } j = n, n-1, \ldots, 1. \tag{A.34}$$

Taking the trace of equations (A.29) gives

$$\text{tr } K_{j-1} = \text{tr}(K_jA) + na_j.$$

Subtract (A.34) from this to give

$$0 = \text{tr}(K_jA) + na_j - ja_j$$

or

$$a_j = -\frac{1}{n-j}\text{tr}(K_jA) \quad \text{for } j = n-1, \ldots, 0. \tag{A.35}$$

This iterative procedure, known as Leverrier's algorithm, can be summarized as follows:

1 Start with $K_{n-1} = I$ (we know that $a_n = 1$). Then for $j = n - 1$
2 Calculate $a_j = -\frac{1}{n-j}\text{tr }(K_jA)$,
3 Calculate $K_{j-1} = K_jA + a_jI$,

4 Decrease j by 1 and return to (2) if $j \geq 0$. (This also gives the check on $K_{-1} = \Theta$.)

Example A.27

For the matrix $A = \begin{bmatrix} 0 & 1 \\ -2 & -3 \end{bmatrix}$ we find directly that $\det[\lambda I - A] = p(\lambda) = \lambda^2 + 3\lambda + 2$. We also find directly that

$$\operatorname{Adj}\begin{bmatrix} \lambda & -1 \\ 2 & \lambda+3 \end{bmatrix} = \begin{bmatrix} \lambda+3 & 1 \\ -2 & \lambda \end{bmatrix} = \begin{bmatrix} 1 & 0 \\ 0 & 1 \end{bmatrix}\lambda + \begin{bmatrix} 3 & 1 \\ -2 & 0 \end{bmatrix} = K_1\lambda + K_0.$$

Using the algorithm we find

$$K_1 = \begin{bmatrix} 1 & 0 \\ 0 & 1 \end{bmatrix}, \quad K_1 A = \begin{bmatrix} 0 & 1 \\ -2 & -3 \end{bmatrix}, \quad a_1 = -\operatorname{tr}(K_1 A) = 3$$

then

$$K_0 = K_1 A + a_1 I = \begin{bmatrix} 3 & 1 \\ -2 & 0 \end{bmatrix}, \quad K_0 A = \begin{bmatrix} -2 & 0 \\ 0 & -2 \end{bmatrix}, \quad a_0 = -\frac{1}{2}\operatorname{tr}(K_0 A) = 2$$

which are indeed the values found directly.

Example A.28

$$A = \begin{bmatrix} 3 & 0 & 3 \\ 2 & 4 & 1 \\ 1 & 1 & 2 \end{bmatrix}$$

so $n = 3$, $K_2 = I$, and $a_3 = 1$. We get

$$a_2 = -\operatorname{tr}A = -9$$

$$K_1 = A - 9I = \begin{bmatrix} -6 & 0 & 3 \\ 2 & -5 & 1 \\ 1 & 1 & -7 \end{bmatrix}$$

$$a_1 = -\frac{1}{2}\operatorname{tr}(K_1 A) = -\frac{1}{2}\operatorname{tr}\begin{bmatrix} -15 & 3 & -12 \\ -3 & -19 & 3 \\ -2 & -3 & -10 \end{bmatrix} = 22$$

$$K_0 = K_1 A + a_1 I = \begin{bmatrix} 7 & 3 & -12 \\ -3 & 3 & 3 \\ -2 & -3 & 12 \end{bmatrix}$$

$$a_0 = = \frac{1}{3}\operatorname{tr}(K_0 A) = -\frac{1}{3}[15 + 15 + 15] = -15.$$

We check the result by finding

$$K_0 A + a_0 I = \begin{bmatrix} 15 & 0 & 0 \\ 0 & 15 & 0 \\ 0 & 0 & 15 \end{bmatrix} - 15I = \Theta$$

which is correct. Using this result gives the inverse matrix

$$[\lambda I - A]^{-1} = \frac{1}{\lambda^3 - 9\lambda^2 + 22\lambda - 15}\begin{bmatrix} \lambda^2 - 6\lambda + 7 & 3 & 3\lambda - 12 \\ 2\lambda - 3 & \lambda^2 - 5\lambda + 3 & \lambda + 3 \\ \lambda - 2 & \lambda - 3 & \lambda^2 - 7\lambda + 12 \end{bmatrix}.$$

A.19 The minimum polynomial

We have seen that the Cayley-Hamilton theorem states that $p(A) = \Theta$. Thus there exists a polynomial function $p(\lambda)$ of degree n, which gives the zero matrix Θ when evaluated at A. It should be apparent that there are other polynomial functions of the square matrix A which also equal Θ. In fact, for any polynomial function $q(A)$ the function $p(A)q(A) = \Theta$ since $p(A) = \Theta$. Thus there are many polynomials which equal Θ when evaluated at A. These are called annihilating polynomials for A.

Definition A.26 *The minimum polynomial $p_m(\lambda)$ of a square matrix A is the polynomial of lowest degree m such that $p_m(A) = \Theta$.*

Theorem A.22 *If λ_j is an eigenvalue of A, then it is a zero of $p_m(\lambda)$ (i.e., $p_m(\lambda_j) = 0$).*

Proof: Let

$$p_m(\lambda) = \sum_{i=0}^{m} b_i \lambda^i \quad \text{then } p_m(A) = \sum_{i=0}^{m} b_i A^i = \Theta$$

where m is the degree of $p_m(\lambda)$. Therefore, for $\boldsymbol{\xi}_j$ an eigenvector of A,

$$p_m(A)\boldsymbol{\xi}_j = \sum_{i=0}^{m} b_i A^i \boldsymbol{\xi}_j \sum_{i=0}^{m} b_i \lambda_j^i \boldsymbol{\xi}_j = p_m(\lambda_j)\boldsymbol{\xi}_j = \mathbf{0}.$$

Since $\boldsymbol{\xi}_j \neq \mathbf{0}$ this tells us that $p_m(\lambda_j) = 0$. QED.

Note: If A has n distinct roots, then $p_m(\lambda)$ must have n roots which implies that $p_m(\lambda) = p(\lambda)$. Only when A has repeated roots can m be smaller than n.

Example A.29

For $A = \begin{bmatrix} 2 & 0 & 0 & 0 \\ 0 & 2 & 0 & 0 \\ 0 & 0 & 2 & 0 \\ 0 & 0 & 0 & 2 \end{bmatrix}$ we find that $p(\lambda) = (\lambda - 2)^4$. Since $p_m(\lambda)$ must have a zero at $\lambda = 2$, we try $p_m(\lambda) = (\lambda - 2)$. We can check that $p_m(A) = A - 2I = \Theta$, so $p_m(\lambda) = (\lambda - 2)$ is correct.

Example A.30

For $A = \begin{bmatrix} 2 & 1 & 0 & 0 \\ 0 & 2 & 0 & 0 \\ 0 & 0 & 2 & 1 \\ 0 & 0 & 0 & 2 \end{bmatrix}$ we find that $p(\lambda) = (\lambda - 2)^4$ as before. Again, to see whether $p_m(\lambda) = (\lambda - 2)$, we check $A - 2I$ and find

$$A - 2I = \begin{bmatrix} 0 & 1 & 0 & 0 \\ 0 & 0 & 0 & 0 \\ 0 & 0 & 0 & 1 \\ 0 & 0 & 0 & 0 \end{bmatrix} \neq \Theta$$

so this is not correct. Next let us try $p_m(\lambda) = (\lambda - 2)^2$. We find that

$$p_m(A) = [A - 2I]^2 = \Theta$$

so $p_m(\lambda) = (\lambda - 2)^2$ is the minimum polynomial.

A.20 Functions of a matrix

We have seen that polynomial functions of a square matrix can be defined directly. In Chapter 4 (Section 4.3.3) we also showed how arbitrary functions which can be expressed as a power series can be defined on matrix arguments. We derived two methods (Methods 2 and 3) for diagonalizable matrices. In this section we will derive the generalization of those results, which were stated in Chapter 4 (without proof), for matrices whose Jordan canonical form is not diagonal.

First note that since the Jordan matrix J in (A.18) is block diagonal we can easily see that for any power i

$$J^i = \begin{bmatrix} J_1^i & 1 & \cdots & 0 \\ 0 & J_2^i & \cdots & 0 \\ \vdots & \vdots & \ddots & \vdots \\ 0 & 0 & \cdots & J_{s'}^i \end{bmatrix} \tag{A.36}$$

where the order ℓ_k of each block J_k^i equals the rank of the highest rank eigenvector that can be generated from $\xi_k^{(1)}$. Therefore in the same way as for diagonalizable matrices, any function which can be expanded in a power series such as $f(\lambda) = \sum_{i=0}^{\infty} a_i \lambda^i$ can be expressed in block diagonal form as

$$f(J) = \sum_{i=0}^{\infty} a_i J^i = \begin{bmatrix} f(J_1) & 1 & \cdots & 0 \\ 0 & f(J_2) & \cdots & 0 \\ \vdots & \vdots & \ddots & \vdots \\ 0 & 0 & \cdots & f(J_{s'}) \end{bmatrix} \tag{A.37}$$

where each block $f(J_k) \triangleq \sum_{i=0}^{\infty} a_i J_k^i$ is $\ell_k \times \ell_k$. If we look at a typical block J_k, we see that

$$J_k^2 = \begin{bmatrix} \lambda_k & 1 & 0 & \cdots & 0 \\ 0 & \lambda_k & 1 & \cdots & 0 \\ \vdots & \ddots & \ddots & \ddots & \vdots \\ 0 & \cdots & \cdots & \lambda_k & 1 \\ 0 & \cdots & \cdots & \cdots & \lambda_k \end{bmatrix} \begin{bmatrix} \lambda_k & 1 & 0 & \cdots & 0 \\ 0 & \lambda_k & 1 & \cdots & 0 \\ \vdots & \ddots & \ddots & \ddots & \vdots \\ 0 & \cdots & \cdots & \lambda_k & 1 \\ 0 & \cdots & \cdots & \cdots & \lambda_k \end{bmatrix}$$

$$= \begin{bmatrix} \lambda_k^2 & 2\lambda_k & 1 & 0 & 0 & \cdots & 0 \\ 0 & \lambda_k^2 & 2\lambda_k & 1 & 0 & \cdots & 0 \\ 0 & 0 & \lambda_k^2 & 2\lambda_k & 1 & \cdots & 0 \\ \vdots & \vdots & \ddots & \ddots & \ddots & \ddots & \vdots \\ 0 & \cdot & \cdots & \cdots & \lambda_k^2 & 2\lambda_k & 1 \\ 0 & \cdot & \cdots & \cdots & \cdot & \lambda_k^2 & 2\lambda_k \\ 0 & \cdot & \cdots & \cdots & \cdot & \cdot & \lambda_k^2 \end{bmatrix}. \quad \text{(A.38)}$$

Note that in (A.38) the first two superdiagonals above the main diagonal are nonzero. (If r = the row index of an element and c = the column index then the jth superdiagonal consists of all elements for which $c = r + j$.) Multiply J_k^2 by J_k to get

$$J_k^3 = \begin{bmatrix} \lambda_k^2 & 2\lambda_k & 1 & 0 & 0 & \cdots & 0 \\ 0 & \lambda_k^2 & 2\lambda_k & 1 & 0 & \cdots & 0 \\ 0 & 0 & \lambda_k^2 & 2\lambda_k & 1 & \cdots & 0 \\ \vdots & \vdots & \ddots & \ddots & \ddots & \ddots & \vdots \\ 0 & \cdot & \cdots & \cdots & \lambda_k^2 & 2\lambda_k & 1 \\ 0 & \cdot & \cdots & \cdots & \cdot & \lambda_k^2 & 2\lambda_k \\ 0 & \cdot & \cdots & \cdots & \cdot & \cdot & \lambda_k^2 \end{bmatrix} \begin{bmatrix} \lambda_k & 1 & 0 & \cdots & 0 \\ 0 & \lambda_k & 1 & \cdots & 0 \\ \vdots & \ddots & \ddots & \ddots & \vdots \\ 0 & \cdots & \cdots & \lambda_k & 1 \\ 0 & \cdots & \cdots & \cdots & \lambda_k \end{bmatrix}$$

$$= \begin{bmatrix} \lambda_k^3 & 3\lambda_k^2 & 3\lambda_k & 1 & 0 & 0 & \cdots & 0 \\ 0 & \lambda_k^3 & 3\lambda_k^2 & 3\lambda_k & 1 & 0 & \cdots & 0 \\ 0 & 0 & \lambda_k^3 & 3\lambda_k^2 & 3\lambda_k & 1 & \cdots & 0 \\ \vdots & \vdots & \ddots & \ddots & \ddots & \ddots & \ddots & \vdots \\ 0 & \cdot & \cdots & \cdots & \lambda_k^3 & 3\lambda_k^2 & 3\lambda_k & 1 \\ 0 & \cdot & \cdots & \cdots & \cdot & \lambda_k^3 & 3\lambda_k^2 & 3\lambda_k \\ 0 & \cdot & \cdots & \cdots & \cdot & \cdot & \lambda_k^3 & 3\lambda_k^2 \\ 0 & \cdot & \cdots & \cdots & \cdot & \cdot & \cdot & \lambda_k^3 \end{bmatrix}. \quad \text{(A.39)}$$

This procedure can be continued. We note that for any power i, each term on the main diagonal is λ_k^i, each term on the first superdiagonal is $d\lambda_k^i/d\lambda_k$, each term on the second superdiagonal is $(1/2!)(d^2\lambda_k^i/d\lambda_k^2)$, and each term on the third superdiagonal is $(1/3!)(d^3\lambda_k^i/d\lambda_k^3)$. In general, we see that all terms on the jth superdiagonal are $(1/j!)(d^j\lambda_k^i/d\lambda_k^j)$. Therefore when we form $f(J_k) = \sum_{i=0}^{\infty} a_i J_k^i$, we find that all terms on the main diagonal are

$$\sum_{i=0}^{\infty} a_i \lambda_k^i = f(\lambda_k)$$

all terms on the first superdiagonal are

$$\sum_{i=0}^{\infty} a_i \frac{d\lambda_k^i}{d\lambda_k} = \frac{d}{d\lambda_k}\left[\sum_{i=0}^{\infty} a_i \lambda_k^i\right] = \frac{df(\lambda_k)}{d\lambda_k}$$

all terms on the second superdiagonal are

$$\sum_{i=0}^{\infty} a_i \frac{1}{2!}\frac{d^2\lambda_k^i}{d\lambda_k^2} = \frac{1}{2!}\frac{d^2}{d\lambda_k^2}\left[\sum_{i=0}^{\infty} a_i \lambda_k^i\right] = \frac{1}{2!}\frac{d^2 f(\lambda_k)}{d\lambda_k^2}$$

and, in general all terms on the jth superdiagonal are

$$\sum_{i=0}^{\infty} a_i \frac{1}{j!}\frac{d^j\lambda_k^i}{d\lambda_k^j} = \frac{1}{j!}\frac{d^j}{d\lambda_k^j}\left[\sum_{i=0}^{\infty} a_i \lambda_k^i\right] = \frac{1}{j!}\frac{d^j f(\lambda_k)}{d\lambda_k^j}.$$

Thus we find that

$$f(J_k) = \sum_{i=0}^{\infty} a_i J_k^i$$
$$= \begin{bmatrix} f(\lambda_k) & f'(\lambda_k) & \frac{1}{2!}f''(\lambda_k) & \frac{1}{3!}f^{(3)}(\lambda_k) & \cdots & \frac{1}{(\ell_k-1)!}f^{(\ell_k-1)}(\lambda_k) \\ 0 & f(\lambda_k) & f'(\lambda_k) & \frac{1}{2!}f''(\lambda_k) & \cdots & \frac{1}{(\ell_k-2)!}f^{(\ell_k-2)}(\lambda_k) \\ \vdots & \ddots & \ddots & \ddots & \ddots & \vdots \\ 0 & \cdots & \cdots & f(\lambda_k) & f'(\lambda_k) & \frac{1}{2!}f''(\lambda_k) \\ 0 & \cdots & \cdots & \cdot & f(\lambda_k) & f'(\lambda_k) \\ 0 & \cdots & \cdots & \cdot & \cdot & f(\lambda_k) \end{bmatrix}$$

where $f^{(\nu)}(.)$ indicates the νth derivative. Thus $f(J_k)$ is an $\ell_k \times \ell_k$ matrix which can be written as the finite sum

$$f(J_k) = \sum_{j=0}^{\ell_k-1} \frac{d^j f(\lambda_k)}{d\lambda_k^j}\hat{E}_{kj} \tag{A.40}$$

where $\hat{E}_{kj}$ is an $\ell_k \times \ell_k$ matrix with 1 all along its jth superdiagonal and zeros everywhere else (note that $\hat{E}_{k0}$ is the identity matrix since the 0th superdiagonal is the main diagonal). Let us rewrite (A.37) as the finite sum

$$f(J) = \sum_{j=0}^{s'} \hat{I}_k^T f(J_k)\hat{I}_k \tag{A.41}$$

where $\hat{I}_k$ is an $\ell_k \times n$ matrix with an $\ell_k \times \ell_k$ identity matrix in the kth block position and zeros everywhere else. Inserting (A.40) into (A.41) we get

$$f(J) = \sum_{k=1}^{s'} \sum_{j=0}^{\ell_k - 1} \frac{d^j f(\lambda_k)}{d\lambda_k^j} \hat{I}_k^T \hat{E}_{kj} \hat{I}_k = \sum_{k=1}^{s'} \sum_{j=0}^{\ell_k - 1} \frac{d^j f(\lambda_k)}{d\lambda_k^j} E'_{kj} \tag{A.42}$$

where $E'_{kj} = (\hat{I}_k)^T \hat{E}_{kj} \hat{I}_k$.

When we use the canonical transformation T of (A.20), we know that $J = T^{-1}AT$, so substituting for J in the summation in (A.37) we get

$$f(J) = \sum_{i=0}^{\infty} a_i (T^{-1}AT)^i = T^{-1} \left[\sum_{i=0}^{\infty} a_i A^i \right] T. \tag{A.43}$$

We define the function $f(A)$ for the matrix A as

$$f(A) \triangleq \sum_{i=0}^{\infty} a_i A^i \tag{A.44}$$

so we see from (A.43) and (A.44) that $f(A) = Tf(J)T^{-1}$. When we substitute the $f(J)$ from (A.42) into this we get

$$f(A) = \sum_{k=1}^{s'} \sum_{j=0}^{\ell_k - 1} \frac{d^j f(\lambda_k)}{d\lambda_k^j} T E'_{kj} T^{-1} = \sum_{k=1}^{s'} \sum_{j=0}^{\ell_k - 1} \frac{d^j f(\lambda_k)}{d\lambda_k^j} Z'_{kj}. \tag{A.45}$$

This is not quite the equation used in Chapter 4 for Method 2, when A cannot be diagonalized, because here s' is not the number of distinct eigenvalues.

To get that equation we must combine all the terms corresponding to the same eigenvalue. Thus let $\lambda_1, \lambda_2, \ldots, \lambda_s$ denote the distinct eigenvalues, and observe that we can define $\hat{E}_{kj}$ in (A.40) to be the zero matrix when $j \geq \ell_k$ (i.e., since an $\ell_k \times \ell_k$ matrix has no superdiagonal greater than $\ell_k - 1$, so $\hat{E}_{kj} = \Theta$ for $j \geq \ell_k$). Thus for each k corresponding to the same eigenvalue λ_i we rewrite (A.40) as

$$f(J_k) = \sum_{j=0}^{m_i - 1} \frac{d^j f(\lambda_i)}{d\lambda_i^j} \hat{E}_{ij} \tag{A.46}$$

where $m_i = \max_{k \text{ corres } \lambda_i} (\ell_k)$ is the size of the largest ℓ_k corresponding to λ_i (*Note:* m_i is also the multiplicity of λ_i in the minimum polynomial of A). Then take all the terms k corresponding to λ_i to form an $n_i \times n_i$ block diagonal matrix where n_i is the algebraic multiplicity of λ_i:

$$f(J_i) = \sum_{j=0}^{m_i - 1} \frac{d^j f(\lambda_i)}{d\lambda_i^j} \tilde{E}_{ij} \tag{A.47}$$

where $\tilde{E}_{ij}$ is the $n_i \times n_i$ block diagonal matrix formed from the $\hat{E}_{kj}$ for k corresponding to λ_i. Then following exactly the same steps which lead from (A.40) to

(A.45) we get

$$f(A) = \sum_{i=1}^{s} \sum_{j=0}^{m_i-1} \frac{d^j f(\lambda_i)}{d\lambda_i^j} T E_{ij} T^{-1} = \sum_{i=1}^{s} \sum_{j=0}^{m_i-1} \frac{d^j f(\lambda_i)}{d\lambda_i^j} Z_{ij} \tag{A.48}$$

where $E_{ij} = (\tilde{I}_i)^T \tilde{E}_{ij} \tilde{I}_i$; $\tilde{I}_i$ is an $n_i \times n$ matrix with an $n_i \times n_i$ identity matrix in the ith block position and zeros everywhere else. This is the equation used in Chapter 4 for Method 2, when A cannot be diagonalized.

For Method 3 we saw that any function $f(\lambda)$ which is expressible as a convergent power series, gives a matrix function $f(A)$ which can be represented in terms of an $n-1$ degree polynomial function $g(A)$

$$f(A) = \gamma_{n-1} A^{n-1} + \gamma_{n-2} A^{n-2} + \cdots + \gamma_1 A + \gamma_0 I \triangleq g(A). \tag{A.49}$$

For this representation to be useful we must be able to find the coefficients γ_ℓ. When A cannot be diagonalized, we postmultiply (A.49) by the canonical transformation T of (A.20) and premultiply by T^{-1} to get

$$f(J) = \gamma_{n-1} J^{n-1} + \gamma_{n-2} J^{n-2} + \cdots + \gamma_1 J + \gamma_0 I \triangleq g(J). \tag{A.50}$$

Since J is block diagonal we find that

$$f(J_k) = \gamma_{n-1} J_k^{n-1} + \gamma_{n-2} J_k^{n-2} + \cdots + \gamma_1 J_k + \gamma_0 I = g(J_k). \tag{A.51}$$

When the geometric multiplicity $g_i = 1$ for each distinct eigenvalue λ_i, then $s' = s$ and $\ell_i = n_i$ equals the algebraic multiplicity of λ_i, and equations (A.40) and (A.47) become the same, namely,

$$f(J_i) = \sum_{j=0}^{n_i-1} \frac{d^j f(\lambda_i)}{d\lambda_i^j} \hat{E}_{ij}. \tag{A.52}$$

This equation can be applied to any other function $g(\lambda)$ to give

$$g(J_i) = \sum_{j=0}^{n_i-1} \frac{d^j g(\lambda_i)}{d\lambda_i^j} \hat{E}_{ij} \tag{A.53}$$

so we see that if

$$\frac{d^j f(\lambda_i)}{d\lambda_i^j} = \frac{d^j g(\lambda_i)}{d\lambda_i^j} \quad \text{for } i = 1, \ldots, s;\ j = 0, \ldots, n_i \tag{A.54}$$

then $f(J_i) = g(J_i)$ for all i, so $f(J) = g(J)$. In particular, when $g(\lambda) = \Sigma_{j=0}^{n-1} \gamma_j \lambda^j$, the n coefficients $\gamma_0, \ldots, \gamma_{n-1}$ can be found using the $\Sigma_{i=1}^{s} n_i = n$ equations given by (A.54). This is the equation used in Chapter 4 for Method 3, when A cannot be diagonalized.

When the geometric multiplicity g_i is not equal to 1 for some λ_i, then we need to find fewer coefficients, namely, m the degree of the minimum polynomial. Then

m can be used instead of n in (A.49) and m_i instead of n_i in (A.54). It is usually not worth the extra effort to find m and the m_i, so it is easier to use (A.54) in this case also.

A.21 Quadratic forms

A polynomial such as

$$f(\boldsymbol{x}) = 2x_1^2 + 6x_1x_2 - 7x_2^2 \tag{A.55}$$

with real coefficients and each term of degree 2 in x_i is called a real quadratic form in x_1 and x_2. In general for n variables $x_1, \ldots, x_n$, a real quadratic form can be written as

$$f(\boldsymbol{x}) = \boldsymbol{x}^T A \boldsymbol{x} = \sum_{i=1}^{n}\sum_{j=1}^{n} x_i a_{ij} x_j \tag{A.56}$$

where A is a real $n \times n$ matrix. Here we will be concerned only with real quadratic forms, so we will usually omit the explicit modifier "real."
Note: When a quadratic form is expressed as in (A.56), many matrices A could be used to represent the quadratic form. For example, the $f(\boldsymbol{x}) = \boldsymbol{x}^T A \boldsymbol{x}$ in (A.55) could be represented by any of the matrices

$$A_1 = \begin{bmatrix} 2 & 4 \\ 2 & -7 \end{bmatrix} \quad A_2 = \begin{bmatrix} 2 & 9 \\ -3 & -7 \end{bmatrix} \quad A_3 = \begin{bmatrix} 2 & 3 \\ 3 & -7 \end{bmatrix}.$$

We will *always* take A to be symmetric. That is, we always make $a_{ij} = a_{ji} = 1/2$ the coefficient of the $x_i x_j$ term in the quadratic form, where a_{ij} is the entry in the ith row and jth column of A.

Consider a change of variables from $\boldsymbol{x}$ to $\boldsymbol{y}$ where $\boldsymbol{x} = P\boldsymbol{y}$ and P is nonsingular, so $\boldsymbol{y} = P^{-1}\boldsymbol{x}$. Then $\boldsymbol{x}^T A \boldsymbol{x} = \boldsymbol{y}^T P^T A P \boldsymbol{y} = \boldsymbol{y}^T \hat{A} \boldsymbol{y}$ where $\hat{A}$ is the matrix of the quadratic form for the new variables $\boldsymbol{y}$.

Note: When P is chosen as the orthogonal matrix which makes $P^T A P = \Lambda$ a diagonal matrix, then the quadratic form becomes $\boldsymbol{y}^T \Lambda \boldsymbol{y} = \Sigma_{i=1}^{n} \lambda_i y_i^2$ where the λ_i are the eigenvalues of A (this can always be done for a real symmetric A; see Theorem A.20).

A.21.1 Definite quadratic forms

Definition A.27 *A quadratic form is called positive definite if it is positive except when all x_i are zero. Thus*

$\boldsymbol{x}^T A \boldsymbol{x}$ *is positive definite* $\leftrightarrow \boldsymbol{x}^T A \boldsymbol{x} > 0 \quad$ *for* $\boldsymbol{x} \neq \boldsymbol{0}$.

It is called positive semidefinite or nonnegative definite if $\mathbf{x}^T A\mathbf{x} \geq 0$ for $\mathbf{x} \neq \mathbf{0}$ (i.e., it is never negative and can equal 0 for a nonzero $\mathbf{x}$).

$x_1^2 + x_2^2$ *is positive definite.*

$$x_1^2 - x_1x_2 + \frac{1}{4}x_2^2 = \left(x_1 - \frac{1}{2}x_2\right)^2 \text{ is positive semidefinite.}$$

Definition A.28 *A real symmetric matrix A is called positive definite if the quadratic form $\mathbf{x}^T A\mathbf{x}$ is positive definite (similarly A is positive semidefinite if its quadratic form is positive semidefinite).*

Theorem A.23 *A is a positive definite matrix if and only if all the eigenvalues of A are positive. (It is positive semidefinite if all eigenvalues are nonnegative.)*

Proof:

$$\mathbf{x}^T A\mathbf{x} = \mathbf{y}^T P^T AP\mathbf{y} = \mathbf{y}^T \Lambda \mathbf{y} = \sum_{i=1}^{n} \lambda_i y_i^2$$

for some orthogonal matrix P. Since $\mathbf{x} = P\mathbf{y}$ and P is nonsingular, $\mathbf{x} \neq \mathbf{0}$ if and only if $\mathbf{y} \neq 0$. For A positive definite the left-hand side is greater than zero whenever $\mathbf{x} \neq 0$, and therefore the right-hand side is also greater than zero whenever $\mathbf{y} \neq 0$, which says that all $\lambda_i > 0$. The same reasoning in the reverse direction shows that all $\lambda_i > 0$ implies that A is positive definite. (For semidefinite matrices, the proof is the same except that "greater than or equal to" ($\geq$) is used instead of "greater than" ($>$).) QED.

Theorem A.24 *If A is positive definite then det A > 0 (if semidefinite then det A ≥ 0).*

Proof: For A positive definite, an orthogonal P exists such that $P^T AP = \Lambda$ where Λ is diagonal with all $\lambda_i > 0$. Now $\det \Lambda = \lambda_1\lambda_2 \cdots \lambda_n > 0$ and $\det P = \pm 1$ (since $(\det P)^2 = 1$). Therefore

$$\det(P^T AP) = (\det P)^2 \det A = \det A \quad \text{so} \quad \det A = \det \Lambda > 0.$$

(For semidefinite replace all $>$ with $\geq$.) QED.

Definition A.29 *The kth-order principal submatrix of A, denoted by A_k, is the $k \times k$ matrix formed by deleting the last $n - k$ columns and the last $n - k$ rows of A.*

Example A.31
For the matrix

$$\begin{bmatrix} a_{11} & a_{12} & a_{13} & a_{14} \\ a_{21} & a_{22} & a_{23} & a_{24} \\ a_{31} & a_{32} & a_{33} & a_{34} \\ a_{41} & a_{42} & a_{43} & a_{44} \end{bmatrix}$$

the principal submatrices are

$$A_1 = a_{11} \quad A_2 = \begin{bmatrix} a_{11} & a_{12} \\ a_{21} & a_{22} \end{bmatrix} \qquad A_3 = \begin{bmatrix} a_{11} & a_{12} & a_{13} \\ a_{21} & a_{22} & a_{23} \\ a_{31} & a_{32} & a_{33} \end{bmatrix}$$

and of course $A_4 = A$. Note that the diagonal elements of a principal submatrix are diagonal elements of A, and the principal submatrices of any A_k are all the lower order principal submatrices.

Theorem A.25 *A real symmetric matrix A is positive definite if and only if every principal submatrix has a positive determinant.*

Note: The determinant of a principal submatrix is called a principal minor, so the theorem says that A is positive definite if and only if every principal minor is positive.

Proof: Necessity (only if): when A is positive definite, then

$$\boldsymbol{x}^T A\boldsymbol{x} = \sum_{i=1}^{n}\sum_{j=1}^{n} a_{ij}x_i x_j > 0$$

for any $\boldsymbol{x} \neq \boldsymbol{0}$. For any choice of k satisfying $0 < k \leq n$, if we make $x_{k+1} = x_{k+2} = \cdots = x_n = 0$ then

$$\boldsymbol{x}^T A\boldsymbol{x} = \sum_{i=1}^{k}\sum_{j=1}^{k} a_{ij}x_i x_j > 0$$

for $x_1, x_2, \ldots, x_k$ not all zero. Therefore the quadratic form

$$\begin{bmatrix} x_1 & \cdots & x_k \end{bmatrix} \begin{bmatrix} a_{11} & \cdots & a_{1k} \\ \vdots & \ddots & \vdots \\ a_{k1} & \cdots & a_{kk} \end{bmatrix} \begin{bmatrix} x_1 \\ \vdots \\ x_k \end{bmatrix} > 0$$

so the matrix of this form A_k, which is a principal submatrix of A, is positive definite and $\det A_k > 0$.

Sufficiency (if): we will prove this part by induction. We see that for $n = 1$ the quadratic form $a_{11}x_1^2 > 0$ for $x_1 \neq 0$ if $a_{11} > 0$; therefore A_1 is positive definite. We could go on to show that the theorem is also true for $n = 2$. Instead we will assume that we have shown it to be true for $(n-1) \times (n-1)$ matrices, and prove that it is then true for the $n \times n$ matrix A.

Form the matrix $B = [\boldsymbol{a}_1 \mid \boldsymbol{b}_2 \mid \boldsymbol{b}_3 \mid \cdots \mid \boldsymbol{b}_n]$ from the columns $\boldsymbol{a}_k$ of A by using

$$\boldsymbol{b}_k = \boldsymbol{a}_k - \tfrac{a_{1k}}{a_{11}}\boldsymbol{a}_1 \quad \text{for } k = 2, \ldots, n \tag{A.57}$$

Note:

1 We can divide by a_{11} since it is greater than zero.

2 B is formed from A by subtracting a multiple of the first column of A from each of the other columns of A. Therefore each principal submatrix B_k of B has been formed by subtracting a multiple of the first column of A_k from each of the other columns of A_k. Since the determinant of a matrix does not change when such column operations are performed, we conclude that *all the principal minors of A and B are the same.*

3 The $(n-1) \times (n-1)$ matrix

$$C = \begin{bmatrix} b_{22} & \cdots & b_{2n} \\ \vdots & \ddots & \vdots \\ b_{n2} & \cdots & b_{nn} \end{bmatrix} \tag{A.58}$$

is symmetric since (A.57) says that the elements b_{ik} are formed by

$$b_{ik} = a_{ik} - \frac{a_{1k}}{a_{11}} a_{i1} \quad \text{for } i = 1, \ldots, n, k = 2, \ldots, n. \tag{A.59}$$

Since A is symmetric (i.e., $a_{ik} = a_{ki}$) we see that $b_{ik} = b_{ki}$ for $i, k = 2, \ldots, n$, so C is symmetric. Equation (A.59) also shows that $b_{1k} = 0$ for $k = 2, \ldots, n$.

4 The operations (A.57) are equivalent to postmultiplying A by the matrix

$$Q = \begin{bmatrix} 1 & -\frac{a_{12}}{a_{11}} & -\frac{a_{13}}{a_{11}} & \cdots & -\frac{a_{1n}}{a_{11}} \\ 0 & 1 & 0 & \cdots & 0 \\ \vdots & \vdots & \ddots & \ddots & \vdots \\ 0 & 0 & \cdots & 1 & 0 \\ 0 & 0 & 0 & \cdots & 1 \end{bmatrix} \tag{A.60}$$

so $B = AQ$.

The resulting matrix looks like

$$B = \begin{bmatrix} a_{11} & 0 & \cdots & 0 \\ a_{21} & b_{22} & \cdots & b_{2n} \\ \vdots & \vdots & \ddots & \vdots \\ a_{n1} & b_{n2} & \cdots & b_{nn} \end{bmatrix}.$$

Next we form another matrix $\hat{B}$ from B by subtracting multiples of the first row of B ($\boldsymbol{b}_1^T$) from the other rows of B ($\boldsymbol{b}_i^T$) in order to put zeros into the first column below a_{11}.

This is done using

$$\hat{\boldsymbol{b}}_i^T = \boldsymbol{b}_i^T - \frac{a_{i1}}{a_{11}}\boldsymbol{b}_1^T \quad \text{for } i = 2, \ldots, n. \tag{A.61}$$

where $\hat{\boldsymbol{b}}_i^T$ is the ith row of the matrix $\hat{B}$. This is equivalent to $\hat{B} = Q^T B = Q^T A Q$ for Q given by (A.60) since $a_{1i} = a_{i1}$. We see that since Q is nonsingular, A is positive definite if and only if $\hat{B}$ is positive definite. We will show that $\hat{B}$ is positive definite.

The resulting matrix looks like.

$$\hat{B} = \begin{bmatrix} a_{11} & 0 & \cdots & 0 \\ 0 & b_{22} & \cdots & b_{2n} \\ \vdots & \vdots & \ddots & \vdots \\ 0 & b_{n2} & \cdots & b_{nn} \end{bmatrix}.$$

Here also, B and $\hat{B}$ have the same principal minors, so A and $\hat{B}$ have the same principal minors which are all positive.

We see that we can express the principal minors of $\hat{B}$ as

$$\det \hat{B}_k = a_{11} \det C_{(k-1)} \quad \text{for } k = 2, \ldots, n. \tag{A.62}$$

where C_i are the principal submatrices of C in (A.58). Equation (A.62) shows that all the principal minors of C are positive, because all the principal minors of $\hat{B}$ are positive and $a_{11} > 0$. Therefore by the induction assumption, the $(n-1) \times (n-1)$ matrix C is positive definite. We can now see that for any n-vector $\boldsymbol{x} = \begin{bmatrix} x_1 \\ \hat{\boldsymbol{x}} \end{bmatrix}$

$$\boldsymbol{x}^T \hat{B} \boldsymbol{x} = a_{11} x_1^2 + \hat{\boldsymbol{x}}^T C \hat{\boldsymbol{x}} > 0$$

for any $\boldsymbol{x} \neq 0$ because $a_{11} > 0$ and C is positive definite. Thus $\hat{B}$ is positive definite and so is A. QED.

Note: Theorem A.25 does *not* hold for positive semidefinite matrices when all principal minors are nonnegative. The following matrix is not positive semidefinite even though the principal minors are nonnegative:

$$A = \begin{bmatrix} 2 & 0 & 0 \\ 0 & 0 & 0 \\ 0 & 0 & -1 \end{bmatrix}.$$

Definition A.30 *A quadratic form $f(\boldsymbol{x})$ is called negative definite if $-f(\boldsymbol{x})$ is positive definite ($f(\boldsymbol{x})$ is negative semidefinite if $-f(\boldsymbol{x})$ is positive semidefinite).*

Note: A is negative definite if and only if $(-A)$ is positive definite.

Theorem A.26 *A real symmetric matrix A is positive definite if and only if a nonsingular real matrix T exists such that $A = T^T T$.*

Proof: (if) When T is a nonsingular matrix, then for $\boldsymbol{y} = T\boldsymbol{x}$ we see that

$$\boldsymbol{x}^T T^T T\boldsymbol{x} = \boldsymbol{y}^T\boldsymbol{y} > 0 \quad \text{for } \boldsymbol{y} \neq \boldsymbol{0}$$

but since T is nonsingular $\boldsymbol{y} \neq \boldsymbol{0} \rightarrow \boldsymbol{x} \neq \boldsymbol{0}$. Therefore $\boldsymbol{x}^T(T^T T)\boldsymbol{x} > 0$ for $\boldsymbol{x} \neq \boldsymbol{0}$, so $T^T T$ is positive definite.
(only if) When A is positive definite then an orthogonal P exists such that $P^T A P = \Lambda$ is diagonal where all $\lambda_i > 0$. Let

$$D = \begin{bmatrix} \frac{1}{\sqrt{\lambda_1}} & 0 & \cdots & 0 \\ 0 & \frac{1}{\sqrt{\lambda_2}} & \cdots & 0 \\ \vdots & \vdots & \ddots & \vdots \\ 0 & 0 & \cdots & \frac{1}{\sqrt{\lambda n}} \end{bmatrix}$$

so $D^T \Lambda D = I$. Thus $D^T P^T A P D = I$, so $A = [D^T P^T]^{-1}[PD]^{-1} = ([PD]^{-1})^T[PD]^{-1}$, and we see that $A = T^T T$ where $T = [PD]^{-1}$ is nonsingular. QED.

REFERENCES

Agashe, S. D. "A New General Routh-Like Algorithm to Determine the Number of RHP Roots of a Real or Complex Polynomial." *IEEE Trans. Aut. Control*, vol. AC-30, pp. 406–9, April 1985.

Cohn, A. "Über die Anzahl der Wurzeln einer algebraischen Gleichung in einem Kreise." *Mat. Zeitschrift*, vol. 14, pp. 110–18, 1922.

Faddeev D. K., and Sominskii, I. S. *Problems in Higher Algebra*. Gostekhizdat, Moscow, 1949.

Frame, J. S. "A Simple Recursion Formula for Inverting a Matrix." *Bull. Amer. Math. Soc.*, vol. 55, pp. 1045, 1949.

Frazer, R. A., Duncan, W. J., and Collar, A. R. *Elementary Matrices and Some Applications to Dynamics and Differential Equations*. Cambridge University Press, Cambridge, 1938.

Gantmacher, F. R. *The Theory of Matrices*, vols. 1–2. Chelsea, New York, 1959.

Hahn, W. *Theory and Application of Liapunov's Direct Method*. Prentice-Hall, Englewood Cliffs, NJ, 1963.

Householder, A. S. *The Theory of Matrices in Numerical Analysis*. Blaisdell Publishing, New York, 1964.

Hurwitz, A. "Über die Bedingungen unter welchen eine Gleichung nur Wurzeln mit negativen reelen Teilen besitzt." *Mat. Ann.*, vol. 46, pp. 273–84, 1895.

Jury, E. I., "A Stability Test for Linear Discrete Systems Using Simple Division." *Proc. IRE*, vol. 49, no. 12, pp. 1947–8, Dec. 1961.

Jury, E. I., and Blanchard, J. "A Stability Test for Linear Discrete Systems in Table Form." *Proc. IRE*, vol. 49, no. 12, pp. 1947–8, Dec. 1961.

Kalman, R. E., "Contributions to the Theory of Optimal Control." *Bol. Soc. Mat. Mexicana*, vol. 5, pp. 102–19, 1960.

Kalman, R. E., "Mathematical Description of Linear Dynamical Systems., *SIAM J. Control*, vol. 1, pp. 152–92, 1963.

Kantorovich, L. V., and Akilov. G. P. *Functional Analysis in Normed Spaces*. Pergamon Press Oxford, 1964.

Kreindler, E., and Sarachik, P. E. "On the Concepts of Controllability and Observability of Linear Systems." *IEEE Trans. Aut. Control*, vol. 9, pp. 129–36, 1964.

Leverrier, U. J. J. "Sur les variations séculaires des éléments elliptiques des sept planétes principales." *J. Math. Pures Appl.*, vol. 5, pp. 220–54, 1840.

Liénard, A., and Chipart, M. H. "Sur le signe de la partie réelle des racines d'une équation algébrique." *J. Math Pures Appl.*, vol. 10, pp. 291–346, 1914.

Liusternik, L. A., and Sobolev, V. J. *Elements of Functional Analysis*. Ungar Publishing, New York, 1961.

Lyapunov, A. M. "Probléme général de la stabilité du mouvement" (reprint of a French translation of the 1892 Russian paper from *Comm. Soc. Math. Kharkov*). *Annals of Math. Studies*, no. 17, Princeton University Press, Princeton, 1947.

Parks, P. C. "A New Proof of the Routh-Hurwitz Stability Criterion Using the 'Second Method' of Lyapunov." *Proc. Cambridge Phil. Soc.*, vol. 58, pt. 4, pp. 694–720, 1962.

Routh, E. J. *A Treatise on the Stability of a Given State of Motion Particularly Steady Motion*. Macmillan, New York, 1877.

Schwarz, R. J., and Friedland, B. *Linear Systems*. McGraw-Hill, New York, 1965.

Silverman, L. M., and Meadows, H. E. "Controllability and Observability in Time-Variable Linear Systems." *SIAM Journal on Control*, vol. 5, pp. 64–73, 1967.

Souriau, J. M. "Une méthode pour la décomposition spectrale et l'inversion des matrices." *Compt. Rend. Acad. Sci. Paris*, vol. 227, pp. 1010–11, 1948.

Zadeh, L. A., and Desoer, C. A. "Linear System Theory: The State Space Approach." McGraw-Hill, New York, 1963.

INDEX

absolute convergence of $\mathcal{Z}$-transform, 112
adder, 26
additivity, 7–9
adjoint matrix, 230
adjoint system equations, 161
advance operator, 30
Agashe theorem, 208–11
analog function, 9
analog system, 9
  components for simulation diagram, 26–7
  with impulse modulated inputs, 139–43
  with sampled inputs, 90–1
  state equations for, 27–30, 60
anticipative and nonanticipative system, 3
asymptotic stability, 185
  Lyapunov function for, 193
A to D converter, 132

basis of a linear space, 233
BIBO stability, 188–92
bounded equilibrium state, 184
bounded input bounded output stability (BIBO), 188–92
bounded time function, 61, 111

canonical forms
  controllability, 167
  Jordan, 255–7
  Kalman's, 174–5
  observability, 171
capacitor, 16
causal and noncausal system, 3
Cayley-Hamilton theorem, 75–6, 259–60
characteristic equation, 68, 248
characteristic polynomial (*see also* minimum polynomial), 68, 248
characteristic roots or values (*see also* eigenvalues)
  of A matrix, 68
  determining stability, 187

circle of convergence, 112
cofactor of a matrix, 230
complex exponential, 98
composition
  in analog system, 45
  in discrete system, 51
consistent and inconsistent linear equations, 239, 243
continued fraction expansion, 199
continued fraction test for determining stability, 199–202
continuous-time function, 9
continuous-time system, *see* analog system
controllability, 151–8
  criteria for, 152–60
  decomposition, 167
  matrix, 156
controllable pair, 156
controllable subspace, 165–9
controllable system, 151
convergence theorem for $\mathcal{Z}$-transform, 111
convolution
  graphical interpretation of, 47–8
  integral, 47
  property for $\mathcal{Z}$-transform, 118
  summation, 52

damping, 22
decomposition property, 8
delay element, 26
delta response, 52
delta response matrix, 54
determinate and indeterminate linear equations, 239
diagonalizable matrix, 251
diagonalization of matrices, 251–3
difference equations, 30, 32, 129–31
  method for obtaining simulation diagram of, 30

difference equations (*cont.*)
in normal form, 15, 85, 126
$\mathcal{Z}$-transform of, 127–30
differential equations, 28–30, 34–7, 103–6
Laplace transform of, 101–6
method for obtaining simulation diagram of, 28–30
in normal form, 15, 60
dimension of a linear space, 233
discrete delta function, 51–3
discrete step function, 52
discrete system, 9
composition in, 51
delta response of, 51–3
simulation of, 26, 30–2
stability of, 187, 192, 214–18
state equations for, 85–7
discrete-time function, 9
discrete transfer function, 125
discrete transfer function matrix, 126–7, 130
discrete variable, 9
D to A converter, 132
dual system equations, 162
dynamic system, 4

eigenspace, 249
eigenvalues, 68, 247–51
controllable and uncontrollable, 167
observable and unobservable, 171
eigenvectors, 68, 247–51
elementary functions, 10, 43, 98, 125
equilibrium state, 6, 184
bounded, 184
stable and unstable, 185
equivalent systems of equations, 240

Faddeev's algorithm, *see* Leverrier's algorithm
final-value theorem of $\mathcal{Z}$-transform, 120
finite memory system, 4
fixed discrete system (*see also* discrete system)
fundamental property of, 125
superposition summation for, 52
fixed system, 6
fundamental property of, 98
superposition integral for, 47
forced systems
solution of state equations for, 62–3
stability of, 188–92
Fourier series of impulse train, 136
Frame's algorithm, *see* Leverrier's algorithm
frequency domain analysis
fixed analog systems, 98–106
fixed discrete systems, 125–31
sampled data systems, 131–45
function
discrete delta, 51
elementary, of frequency domain analysis, 11, 98, 125
elementary, of time domain analysis, 11, 44, 51
impulse, 44
of a matrix, 265–70
step, 43
transfer, 99, 100, 125–6
fundamental matrix, *see* transition matrix

Gaussian elimination, 20, 240–2
generalized eigenvector, 253–5
global stability, 185–6

hold unit, 132
homogeneity, 7
zero-input, 9
zero-state, 8
Hurwitz matrix, 213
Hurwitz polynomial, 198
Hurwitz test for stability, 213

impulse function, 43, 44
area under, 44, 134
properties of, 44
resolution into, 44
impulse modulation and modulator, 135
impulse response, 44
of causal systems, 46
of fixed system, 46
matrix, 53
relation to step response, 45
impulse train, 135
Fourier series of, 136
inductor, 16
initial-value theorem of $\mathcal{Z}$-transform, 120
input, 1
input-output relation, 2–3
for analog system in terms of impulse response, 44
for analog system in terms of step response, 45
for discrete system, 51
input-output state relation, 5
input vector, 2
instantaneous system, 4
integrator, 26
invariant subspace, 166
inverse $\mathcal{Z}$-transform, 114–18
inversion formula for $\mathcal{Z}$-transform, 115–18

Jordan canonical form, 255–7
Jury stability theorem, 216

Kalman decomposition, 175
Kronecker delta (*see also* discrete delta function), 51

Laurent series expansion for $\mathcal{Z}$-transform, 114, 116
Leverrier's algorithm, 106–8, 261–3
Liapunov, *see* Lyapunov, *entries*
Lienard-Chipart theorem, 214
linear dependence and independence, 159, 232–6
linearity, 7, 9
  zero-input, 9
  zero-state, 8
linearity of $\mathcal{Z}$-transform, 118
linear space, 232
linear system, 7–9
  conditions for stability, *see* stability
  equivalence of boundedness and stability in, 186
linear system analysis, techniques of, 10–12
long division method of inverting $\mathcal{Z}$-transform, 114–5
lumped system, 15
Lyapunov equation, 196
Lyapunov function, 193, 196–8
Lyapunov's second method, 192–8
Lyapunov stability theorem, 193

mass, 22
matricant, *see* transition matrix
matrix, 221
  addition, 222
  adjoint, 230
  augmented, 239
  cofactor, 230
  diagonal, 226
  diagonalizable, 251
  of discrete delta responses, 54
  equality, 222
  Hermitian, 226, 258
  Hurwitz, 213
  of impulse responses, 53
  inverse, 229–31
  Jordan canonical form of, 255–7
  multiplication, 223–5
  norm of, 186
  orthogonal, 258
  partitioning, 227–9
  principal minor of, 272
  principal submatrix of, 271–2
  rank of, 236–9
  scalar multiple of, 222
  similarity of, 251
  singular and nonsingular, 231
  superdiagonal of, 256, 266–7
  symmetric and skew symmetric, 226
  of transfer functions, *see* transfer function matrix
  transition, *see* transition matrix
  transpose, 225
matrizant, *see* transition matrix
memory, 4
minimum polynomial, 264–5
modified $\mathcal{Z}$-transform, 144
multiple input multiple output system, 53–4
multiplicity
  algebraic, 249
  of eigenvalues, 73
  geometric, 250

negative definite function, 193, 275
nonanticipative system, *see* causal and noncausal system
noncausal system, 3
norm
  of matrix, 186
  of vector, 184
normal form of state equations, 15
  procedure for determining, 16–25, 26–37

observability, 158
  decomposition, 171
  matrix, 160
  relation to controllability, 160–2
  tests for, 158–60
observable pair, 160
observable subspace, 169–71
one-sided $\mathcal{Z}$-transform, *see* $\mathcal{Z}$-transform
order of system, 15
orthogonal matrix, 258
output, 1
output vector, 2

periodicity of $F^*(s)$, 137
physically realizable system, 3
piecewise constant function, 91
  in sampled-data system, 133
positive definite function, 193, 270–1
positive definite matrix, 154, 271–2
positive semidefinite function, 193, 271

quadratic form, 270–1

radius of absolute convergence of $\mathcal{Z}$-transform, 112
rank
  of generalized eigenvectors, 254
  of a matrix, 236–9

region of convergence of $\mathcal{Z}$-transform, 112
residue formula for inverse $\mathcal{Z}$-transform, 116
resistor, 16
resolution
  into damped exponentials for analog functions, 11, 99
  into discrete deltas for discrete functions, 51
  into impulses for analog functions, 11, 44
  into unit step functions for analog functions, 45
response
  impulse, 44
  step, 44
Routh array, 203
Routh stability test, 202–7
  extension of Agashe, 208–11

sampled data system, 90–1, 131–43
sampled function, 90–1
sampler, 133
  replacement of, by impulse modulator, 134 5
scale change theorem of $\mathcal{Z}$-transform, 121
second method of Lyapunov, 192–8
sifting property of impulse, 44
similarity transformation, 251
similar matrices, 251
simulation diagram, 26–8
  elements for, 26–7
simultaneous difference equations, 129–31
simultaneous differential equations, 34–7, 103–6
singularity functions, 43
Souriau's algorithm, *see* Leverrier's algorithm
spanning a linear space, 233
spectrum,
  amplitude, 137
  of impulse-modulated function, 137
  of a matrix, 74, 84
spring, 22
stability, 184–218
  asymptotic (AS), 185
  bounded input bounded output (BIBO), 188–92
  determined by eigenvalue locations, 187
  of discrete-time system, 187, 192, 214–18
  of equilibrium state, 185
  of fixed linear system, 190–2
  global, 185–6
  in Lyapunov sense, 184
  of unforced linear system, 185–7
  of unforced system, 185
  uniform, 185
  zero-input, 185–7
  zero-state, *see* BIBO stability
stability determination
  by Agashe's method, 207–13
  by continued fraction test, 199–202
  by Hurwitz test, 213–14
  by Jury's methods for discrete systems, 215–18
  by Routh's test, 202–7
state, 5
  equations, 6, 15
  equilibrium, *see* equilibrium state
  transition equation, 5
  zero, 6
state variables, 16
  choice of, 16–17, 22, 25, 27
state vector, 5, 15
step function, 43
  resolution into, 44
step response, 45, 52
  of causal system, 46
  of fixed system, 46
  relation to impulse response, 45
stored energy
  in an electrical circuit, 17
  in a mechanical system, 22
subspace
  controllable and uncontrollable, 165, 167
  observable and unobservable, 170
summing element, 26
superdiagonal of a matrix, 256, 266–7
superposition integral, 44–7, 53
  for multiple inputs and outputs, 53
  in terms of impulse response, 44
  in terms of step response, 44
superposition summation, 51–2, 54
Sylvester's formula, 72
system, 1
  analog, *see* analog system
  causal, 3
  continuous-time, *see* analog system
  controllable, 151
  discrete-time, *see* discrete system
  dynamic, 4
  finite memory, 4
  fixed, 6
  instantaneous, 4
  linear, 7–9
  lumped, 15
  memory of, 4
  with multiple inputs and outputs, 53–4
  observable, 158
  order of, 15
  physically realizable, 3
  stable, *see* stability

time-invariant, 6
time-varying, 6
zero-memory, 4

time-domain analysis, 11, 43–54
time-invariant system, 6
time translation property of $\mathcal{Z}$-transforms, 119
time-varying system, 6
trace of a matrix, 261
transfer function, 99
discrete, 125
relation to delta response, 126
relation to impulse response, 100
transfer function matrix
for analog system, 101–2
for discrete system, 126–7
transition matrix, 61–85
calculation for analog system with constant A, 66–79
calculation for analog system with time varying A, 64–6, 80–5
calculation for discrete system, 86–8
Laplace transform of, for analog system, 102, 106
physical significance of, 63–4
properties of, 61–2, 88
relation to delta response, 87
relation to impulse response, 63
$\mathcal{Z}$-transform of, for discrete system, 128

uniform boundedness, 184
uniform stability, 185

unit impulse, *see* impulse function
unit-step function, 43
unstable equilibrium state, 185
unstable system, 185

vector, 2, 221
input, 2
norm of, 184
output, 2
state, 15

zero-input response, 6
zero-memory system, 4
zero-order hold, 133
zero-state, 6
zero-state response, 6
$\mathcal{Z}$-transform, 111–22
absolute convergence of, 111–12
of convolution, 118
definition of, 111
of delta response, 126
final value property of, 120
initial value property of, 120
inversion of, 114–18
inversion formula for, 116
linearity of, 118
one-sided, 111
radius of absolute convergence for, 112
region of convergence of, 112
scale change property of, 121
of summation, 122
time multiplication property of, 122
time translation property of, 119

Printed in the United States
By Bookmasters